T0092590

Kasthurirangan Gopalakrishnan, Siddhartha Kumar Khaitan,
and Soteris Kalogirou (Eds.)

Soft Computing in Green and Renewable Energy Systems

Studies in Fuzziness and Soft Computing, Volume 269

Editor-in-Chief

Prof. Janusz Kacprzyk
Systems Research Institute
Polish Academy of Sciences
ul. Newelska 6
01-447 Warsaw
Poland
E-mail: kacprzyk@ibspan.waw.pl

Further volumes of this series can be found on our homepage: springer.com

Vol. 256. Robert Jeansoulin, Odile Papini, Henri Prade, and Steven Schockaert (Eds.)
Methods for Handling Imperfect Spatial Information, 2010
ISBN 978-3-642-14754-8

Vol. 257. Salvatore Greco, Ricardo Alberto Marques Pereira, Massimo Squillante, Ronald R. Yager, and Janusz Kacprzyk (Eds.)
Preferences and Decisions,2010
ISBN 978-3-642-15975-6

Vol. 258. Jorge Casillas and Francisco José Martínez López
Marketing Intelligent Systems Using Soft Computing, 2010
ISBN 978-3-642-15605-2

Vol. 259. Alexander Gegov
Fuzzy Networks for Complex Systems, 2010
ISBN 978-3-642-15599-4

Vol. 260. Jordi Recasens
Indistinguishability Operators, 2010
ISBN 978-3-642-16221-3

Vol. 261. Chris Cornelis, Glad Deschrijver, Mike Nachtegael, Steven Schockaert, and Yun Shi (Eds.)
35 Years of Fuzzy Set Theory, 2010
ISBN 978-3-642-16628-0

Vol. 262. Zsófia Lendek, Thierry Marie Guerra, Robert Babuška, and Bart De Schutter
Stability Analysis and Nonlinear Observer Design Using Takagi-Sugeno Fuzzy Models, 2010
ISBN 978-3-642-16775-1

Vol. 263. Jiuping Xu and Xiaoyang Zhou
Fuzzy-Like Multiple Objective Decision Making, 2010
ISBN 978-3-642-16894-9

Vol. 264. Hak-Keung Lam and Frank Hung-Fat Leung
Stability Analysis of Fuzzy-Model-Based Control Systems, 2011
ISBN 978-3-642-17843-6

Vol. 265. Ronald R. Yager, Janusz Kacprzyk, and Prof. Gleb Beliakov (eds.)
Recent Developments in the Ordered Weighted Averaging Operators: Theory and Practice, 2011
ISBN 978-3-642-17909-9

Vol. 266. Edwin Lughofer
Evolving Fuzzy Systems – Methodologies, Advanced Concepts and Applications, 2011
ISBN 978-3-642-18086-6

Vol. 267. Enrique Herrera-Viedma, José Luis García-Lapresta, Janusz Kacprzyk, Mario Fedrizzi, Hannu Nurmi, and Sławomir Zadrożny
Consensual Processes, 2011
ISBN 978-3-642-20532-3

Vol. 268. Olga Poleshchuk and Evgeniy Komarov
Expert Fuzzy Information Processing, 2011
ISBN 978-3-642-20124-0

Vol. 269. Kasthurirangan Gopalakrishnan, Siddhartha Kumar Khaitan, and Soteris Kalogirou (Eds.)
Soft Computing in Green and Renewable Energy Systems, 2011
ISBN 978-3-642-22175-0

Kasthurirangan Gopalakrishnan,
Siddhartha Kumar Khaitan, and
Soteris Kalogirou (Eds.)

Soft Computing in Green and Renewable Energy Systems

Editors

Dr. Kasthurirangan Gopalakrishnan
Research Assistant Professor of
Civil Engineering
Associate, Ames Lab,
US Department of Energy
Research Affiliate,
Iowa Bioeconomy Institute
354 Town Engineering Building
Iowa State University
Ames, IA 50011, USA
E-mail: rangan@iastate.edu

Dr. Siddhartha Kumar Khaitan
Post-Doctoral Research Associate
Department of Electrical and
Computer Engineering
1113 Coover Hall
Iowa State University
Ames, IA 50011, USA
E-mail: skhaitan@iastate.edu

Dr. Soteris Kalogirou
Cyprus University of Technology
Department of Mechanical Engineering
and Materials Sciences and Engineering
P.O. Box 50329
3603 Limassol
Cyprus
E-mail: Soteris.kalogirou@cut.ac.cy

ISBN 978-3-642-22175-0 e-ISBN 978-3-642-22176-7

DOI 10.1007/978-3-642-22176-7

Studies in Fuzziness and Soft Computing ISSN 1434-9922

Library of Congress Control Number: 2011932574

Typeset & Cover Design: Scientific Publishing Services Pvt. Ltd., Chennai, India.

Printed on acid-free paper

9 8 7 6 5 4 3 2 1

springer.com

Preface

Renewable energy and energy efficient technologies have been attracting much attention in recent years due to the soaring energy crisis and environmental problems associated with the depletion of natural resources. Electricity generation from burning of fossil fuels is a major source of greenhouse gas (GHG) emissions leading to global warming. Renewable energy resources such as solar, wind, biomass, hydrogen, geothermal, ocean and hydropower not only help conserve fossil resources for future generations, but are also considered clean sources of energy that are constantly replenished. The growth of renewable energy sources and their integration into the grid necessitate proper characterization of these systems and components for optimal performance under economic, environmental, and operational constraints.

However, the highly variable and site-specific nature of renewable energy sources has also increased the level of uncertainty in the operation of power systems and the unpredictability of load situations. Soft computing (SC) techniques offer an effective solution for studying and modeling the stochastic behavior of renewable energy generation, operation of grid-connected renewable energy systems, and sustainable decision-making among alternatives. The tolerance of SC techniques to imprecision, uncertainty, partial truth and approximation make them useful alternatives to conventional techniques.

This carefully edited book covers the application of SC in diverse area of renewable energy studies. Application areas include characterization of photovoltaic (PV) systems and grid-connected PV plants, study of operational characteristics of various renewable sources in multi-criteria decision-making, study of thermal energy systems and absorption cooling systems, probabilistic load flow problems, diagnosis and prediction of desert dust transport episodes for improved operation of renewable energy systems utilizing solar radiation, short-term wind forecasting based on time series analysis, and renewable energy hydrogen hybrid systems. A brief description of each chapter follows.

The chapter entitled “Soft Computing Applications in Thermal Energy Systems” presents a comprehensive review of applications of NNs, genetic algorithms (GAs), fuzzy logic (FL), and cluster analysis (CA) in thermal energy systems. The usefulness of such SC applications is demonstrated for modeling, prediction, and control of a range of energy systems which may be difficult or even impossible to do by conventional techniques.

The chapter entitled “Use of Soft Computing Techniques in Renewable Energy Hydrogen Hybrid Systems” reviews the application of soft computing techniques to renewable energy hybrid systems that consists of different technologies (photovoltaic and wind, electrolyzers, fuel cells, hydrogen storage, piping, thermal and electrical/electronic control systems) capable as a whole of converting solar energy, storing it as chemical energy (in the form of hydrogen) and turning it back into electrical and thermal energy. Single or mixed implementation of a range of SC applications, including FL decision-making methodologies, NNs, GAs, and particle swarm optimization (PSO), are discussed.

The chapter entitled "Soft Computing in Absorption Cooling Systems" presents a wide overview of SC techniques in system modeling, control, optimization and determination of working fluids properties of absorption cooling systems which uses thermal energy to operate its compressor in place of a conventional system's compressor, which uses electricity.

The chapter entitled "A Comprehensive Overview of Short Term Wind Forecasting Models based on Time Series Analysis" presents several different approaches to short term wind forecasting and re-examines them with an eye towards setting automated procedures to clarify "grey" areas in their application. In addition, some recent applications of localized linear models and clustering algorithms coupled with linear and nonlinear models and the development of a customized regime model which captures the impact of changing synoptic weather characteristics are presented.

The chapter entitled "Load Flow with Uncertain Loading and Generation in Future Smart Grids" covers a variety of approaches to solve stochastic load flow problems, ranging from currently deployed state-of-the-art procedures to the newest advances in probabilistic load flow calculation and determination. The robustness and real-time issues of the proposed algorithms to deal with highly dynamic Smart Grid scenarios resulting from power feed-in from renewable sources are discussed.

The chapter entitled "Evaluation of Green and Renewable Energy System Alternatives Using a Multiple Attribute Utility Model: The Case of Turkey" discusses the use of multi-attribute utility theory (MUAT) to determine the most appropriate renewable energy alternative among solar, wind, hydropower, biomass, and geothermal. Based on utilities of criteria, the proposed MUAT methodology determines the most appropriate renewable energy alternative for Turkey.

The chapter entitled "A Novel Fuzzy-based Methodology for Biogas Fuelled Hybrid Energy Systems Decision Making" discusses the use of fuzzy multi-rules and fuzzy multi-sets to evaluate the main operational characteristics of five types of renewable sources fuelled by biogas. Using several criteria, including, costs, efficiency, cogeneration, life-cycle, technical maturity, power application range, and environmental impacts, the chapter illustrates the use of fuzzy-based methodology for biogas fuelled hybrid energy systems sustainable decision making.

The chapter entitled "Two New Applications of Artificial Neural Networks: Estimation of Instantaneous Performance Ratio and of the Energy Produced by PV Generators" discusses the application of NNs for estimating the instantaneous performance ratio, a fundamental parameter in the characterization of PV systems; and compare the results of conventional as well as NN-based methods for estimating the annual energy produced by a PV generator with different setting and types of modules.

The chapter entitled "Optimization of Fuzzy Logic Controller Design for Maximum Power Point Tracking in Photovoltaic Systems" presents the design and optimization of a FL controller (FLC) with a minimum rule base for maximum power point tracking in PV systems. The use of GAs is proposed for automated design and optimization of the FLC.

The chapter entitled "Application of Artificial Neural Networks for the Prediction of a 20-kWp Grid-connected Photovoltaic Plant Power Output" describes a simplified NN configuration used for estimating the power produced by a 20-kWp grid-connected PV (GCPV) plants. The development of four multilayer-perceptron (MLP) NN models using a database of experimentally measured climate (irradiance and air temperature) and electrical data (power delivered to the grid) for nine months are discussed.

The chapter entitled "Artificial Neural Networks for the Diagnosis and Prediction of Desert Dust Transport Episodes" discusses the practical applications of NNs in the study of atmospheric pollution by particulate matter due to desert dust transport episodes which profoundly affect the use of renewable energy systems utilizing solar radiation.

Researchers, educators, practitioners and students interested in the study of renewable energy systems will find this book very useful. This book will also serve as an excellent state-of-the-art reference material for graduate and postgraduate students with an interest in soft computing in green and renewable energy systems.

Kasthurirangan (Rangan) Gopalakrishnan
Siddhartha Kumar Khaitan
Soteris Kalogirou

Contents

About the Editors

Kasthurirangan Gopalakrishnan, Ph.D.

Prof. Kasthurirangan Gopalakrishnan is a Research Assistant Professor in the Department of Civil, Construction and Environmental Engineering at Iowa State University. He received his Ph.D. in Civil Engineering from the University of Illinois at Urbana-Champaign in 2004. His research interests include sustainable infrastructure, green engineering technology, bio-inspired computing, and smart pavements. Dr. Gopalakrishnan is the author of a recent e-book, *Sustainable Highways, Pavements and Materials: An Introduction* and is also the lead editor of Springer's *Intelligent and Soft Computing in Infrastructure Systems Engineering: Recent Advances, Sustainable and Resilient Critical Infrastructure Systems: Simulation, Modeling, and Intelligent Engineering*, and *Nanotechnology in Civil Infrastructure: A Paradigm Shift.*

Siddhartha Khaitan, Ph.D.

Dr. Siddhartha Kumar Khaitan is a Research Associate in the Department of Electrical and Computer Engineering at Iowa State University. He received his Ph.D. in Electrical Engineering at Iowa State University in 2008. He was awarded the *ISU Research Excellence Award* for significant contribution to the power systems community through his Ph.D. research. His research interests include power system dynamic simulation, cascading, green and renewable energy systems, soft computing and optimization, linear algebra, energy storage and parallel computing. Dr. Khaitan is also the lead editor of a forthcoming book on high-performance computing in energy systems to be published by Springer.

Soteris Kalogirou, Ph.D.

Dr. Soteris Kalogirou is a Lecturer in the Department of Mechanical Engineering and Materials Sciences and Engineering at the Cyprus University of Technology. He received his Ph.D. in Mechanical Engineering from the University of Glamorgan in 1995. For more than 25 years, he has been actively involved in research in the area of solar energy. His publication record includes 26 books and book contributions and more than 200 peer-reviewed papers. He is the Executive Editor of *Energy*, Associate Editor of *Renewable Energy* and an editorial board member of 12 other journals. He is the editor of the book *Artificial Intelligence in Energy and Renewable Energy Systems*, published by Nova Science Inc. and author of the book *Solar Energy Engineering: Processes and Systems*, published by Academic Press of Elsevier.

List of Contributors

Almonacid, Florencia
Research Group "IDEA",
Department of Electronics
Engineering, Polytechnics
School of Jaén, University of Jaén,
Jaén, Spain

Barin, Alexandre
Federal University of Santa
Maria/CEEMA, Brazil

Canha, Luciane N.
Federal University of Santa
Maria/CEEMA, Brazil

Hamam, Yskandar
Tshwane University of Technology,
Pretoria, South Africa; ESIEE-Paris,
Paris-Est University, LISV, UVSQ,
France

Hontoria, Leocadio
Research Group "IDEA",
Department of Electronics
Engineering, Polytechnics
School of Jaén, University of Jaén,
Jaén, Spain

Kahraman, Cengiz
Department of Industrial
Engineering, Istanbul Technical
University, Maçka, Istanbul,
Turkey

Kalogirou, Soteris A.
Department of Mechanical
Engineering and Materials Science
and Engineering, Cyprus University
of Technology, Limassol, Cyprus

Kaya, İhsan
Department of Industrial
Engineering, Yıldız Technical
University, Yıldız, Istanbul, Turkey

Krause, Olav
School of Information Technology
and Electrical Engineering,
The University of Queensland,
Brisbane, Queensland, Australia

Lehnhoff, Sebastian
Department of Computing Science,
Carl von Ossietzky University,
Oldenburg, Germany

Letting, Lawrence K.
Tshwane University of Technology,
Pretoria, South Africa

Magnago, Karine M.
Federal University of Santa
Maria/CEEMA, Brazil

Matos, Manuel A.
Institute for Systems and Computer
Engineering of Porto, Portugal

Mellit, Adel
Department of Electronics,
Faculty of Sciences and
Technology, Jijel University,
Ouled-aissa, Jijel, Algeria

Michaelides, Silas
Meteorological Service, Nicosia,
Cyprus

Munda, Josiah L.
Tshwane University of Technology, Pretoria, South Africa

Pacheco-Vega, Arturo
Department of Mechanical Engineering, California State University, Los Angeles, CA, USA

Paronis, Dimitris
Institute for Space Applications & Remote Sensing, National Observatory of Athens, Greece

Pavan, Alessandro Massi
Department of Materials and Natural Resources, University of Trieste Via A. Valerio, Trieste, Italy

Pedrazzi, Simone
University of Modena e Reggio Emilia, Via Vignolese, Modena, Italy

Pérez-Higueras
Pedro. Research Group "IDEA", Department of Electronics Engineering, Polytechnics School of Jaén, University of Jaén, Jaén, Spain

Retalis, Adrianos
Institute for Environmental Research & Sustainable Development, National Observatory of Athens, Greece

Rus, Catalina
Research Group "IDEA", Department of Electronics Engineering, Polytechnics School of Jaén, University of Jaén, Jaén, Spain

Şencan, Arzu Şahin
Department of Mechanical Education, Technical Education Faculty, Süleyman Demirel University, Isparta, Turkey

Sfetsos, Athanasios
Environmental Research Laboratory, Institute of Nuclear Technology and Radiation Protection, National Centre for Scientific Research Demokritos, Ag. Paraskevi, Greece

Tartarini, Paolo
University of Modena e Reggio Emilia, Via Vignolese, Modena, Italy

Tymvios, Filippos
Meteorological Service, Nicosia, Cyprus

Wottrich, Breno
Federal University of Santa Maria/CEEMA, Brazil

Zini, Gabriele
University of Modena e Reggio Emilia, Via Vignolese, Modena, Italy

Soft Computing Applications in Thermal Energy Systems

Arturo Pacheco-Vega

California State University, Los Angeles
Los Angeles, CA 90032, USA
apacheco@calstatela.edu

Abstract. Soft computing methodologies, of which artificial neural networks (ANNs), genetic algorithms (GAs), fuzzy logic (FL), and cluster analysis (CA) are elements, have gained much attention in recent years as practical tools to analyze complex problems in real-world applications. This chapter presents a review of SC applications in energy systems that belong to the field of thermal engineering. Special attention is devoted to the analysis, design and control of heat exchangers. For each methodology considered, the principles of operation are briefly described and discussed. Various applications to other energy systems are also mentioned.

1 Introduction

Energy systems are engineered systems that deal with the conversion of energy from one form (e.g. chemical, nuclear, mechanical and electromagnetic), to another (e.g. thermal or electrical, etc). The use of these systems is essential to a wide variety of applications where the purpose of the system is to achieve specific conditions in energy conversion, for human comfort, to fulfill energy demand, to avoid damage of sensitive equipment, or to preserve the quality of valuable products. Examples of energy systems include fuel and photovoltaic cells, wind turbines and heat exchangers, among others. For these systems it may be desirable to predict the performance under specific conditions of operation, or even to control it to achieve a specific objective. This task is frequently difficult due to the complexity of either the occurring physical phenomena, like turbulent fluid flow, or the geometry of the system which make the resulting mathematical model impossible to solve in real time.

Soft computing (SC) comprises a specific set of techniques within the framework of artificial intelligence (AI) that have received much attention as feasible methods for dealing with practical problems. Soft computing includes neural networks (ANNs), genetic algorithms (GAs), fuzzy logic (FL) and cluster analysis

K. Gopalakrishnan et al. (Eds.): Soft Comput. in Green & Renew. Ener. Sys., STUDFUZZ 269, pp. 1–35.
springerlink.com

(CA). Their ability to handle imprecise information has been a key factor for their increasing demand. These technologies have been successfully applied to a variety of disciplines like biology, marketing, medicine, manufacturing, science and engineering, where the central point in all of them has been the difficulty of modeling the system from first principles or finding accurate solutions in real time.

This chapter is written to provide an overview of the application range of soft computing technologies in the area of thermal engineering and energy systems. Much of the content is based on applications by the author to thermal systems, particularly heat exchangers, using a subset of SC, i.e., artificial neural networks, genetic algorithms and programming, fuzzy logic and cluster analysis. In the sections below, each technique will be first described in outline, and later applied to illustrate its usefulness in addressing the complexity of the thermal system for modeling, prediction or control. At the end of each section, a set of interesting applications to other energy systems are also included.

The reader unfamiliar with these methodologies is referred to the monographs by Sen and Yang [1] and Sen and Goodwine [2] which present an account of ANNs, GAs and Fuzzy Logic, and to Jain et al. [3] for the topic of cluster analysis, all of which are fundamentally tutorial in nature. Excellent books on the different areas of SC are also available. Schalkoff [4] and Haykin [5] cover artificial neural networks; Goldberg [6] and Koza [7] present, respectively, an exposition on genetic algorithms and genetic programming; Chen and Pham [8] provide a broad account of fuzzy logic; Everitt et al. [9] cover the topic of cluster analysis; Jang et al. [10] introduce the neuro-fuzzy hybrid technique. The texts by Tettamanzi [11], and Karray and De Silva [12] provide a good introduction to the broad field of soft computing.

2 Artificial Neural Networks

The artificial neural network (ANN) is perhaps the most celebrated technique comprising SC methodologies. It is rooted in the biological network of the brain in an attempt to mimic its operation, and has been successfully applied to a variety of disciplines, such as: philosophy, psychology, economics, science and engineering, among others, where the common factor is complexity. Information about the subject is available in introductory texts, among which those of Schalkoff [4] and Haykin [5] cover its history, its mathematical background and implementation procedures. A brief description of the technique is given next.

2.1 Description

Several types of ANNs exist in the literature, but the feedforward fully-connected multilayer architecture is by far the most popular in engineering applications

[1, 13]. A fully-connected ANN consists of a large number of interconnected processing elements, also known as neurons or nodes, that are organized in layers. The structure of a feedforward ANN is comprised of an input layer, one or several hidden layers and an output layer. In this type of architecture, all the nodes of each layer are usually connected to all the nodes of the adjacent layer by means of synaptic weights. The typical structure of a feedforward ANN is illustrated in Figs. 1(b) and 2(b).

The configurations shown have one input layer, two hidden layers and one output layer. During the feedforward stage, a set of input data is supplied to the input nodes and the information is transferred forward through the network to the nodes in the output layer. The nodes perform nonlinear input-output transformations by means of an activation function. Though the sigmoidal function is very common, several other activation functions have also been studied [4, 5]. The training process is carried out by comparing the output of the network to the given data. The weights and biases are changed in order to minimize the error between the output values and the data using the well-known backpropagation algorithm [14]. Feedforward followed by backpropagation of all the data comprises a training cycle.

The configuration of the ANN is set by selecting the number of hidden layers and the number of nodes in each hidden layer, since the number of nodes in the input and output layers are determined from physical variables. All variables are usually normalized in the [0.15, 0.85] range. A clear account on the implementation issues of the methodology is provided in the monograph by Sen and Yang [1]. The following section illustrates the usefulness of ANNs to model the behavior of complex thermal systems.

2.2 *Application to Compact Heat Exchangers*

The specific problem to be addressed by the method is the steady state performance of heat exchangers, which are essential components of energy systems. This is an example of a system that is complex both from the perspective of the physics involved and its geometry. Factors like turbulence, property dependence on temperatures, change of phase, and the number of parameters, make the system difficult to compute. The analysis then relies on compression of experimental information in terms of dimensionless power-law correlations of the two transfer coefficients, from which the heat rate is determined. Though useful in practice, correlations have their own problems and many times do not effectively compute the desired output. It will be shown that an ANN can be used to make accurate estimations of the system performance.

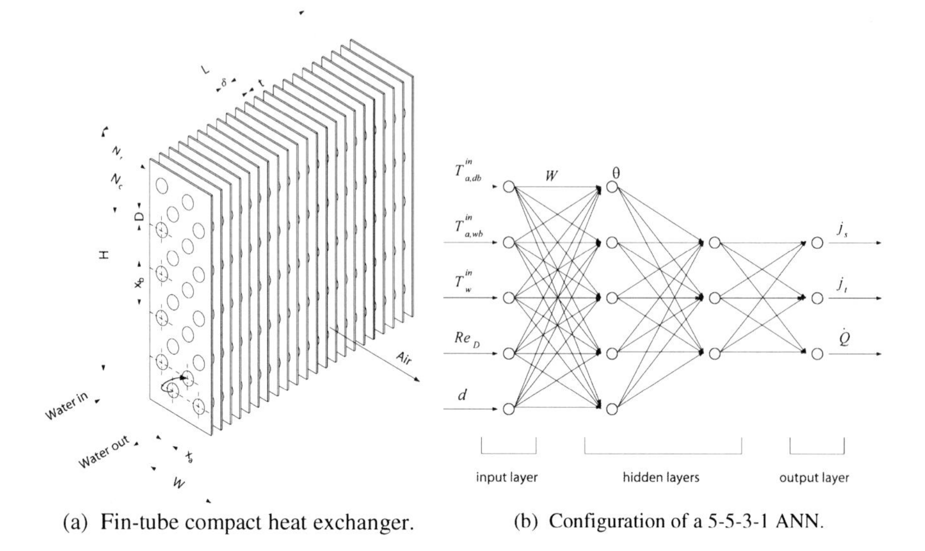

(a) Fin-tube compact heat exchanger. (b) Configuration of a 5-5-3-1 ANN.

Fig. 1 Schematic of a heat exchanger and its ANN representation.

The first example deals with the multirow multicolumn plate-fin type heat exchanger shown in Fig. 1(a). Using chilled water flowing inside the tubes, and warm air as the external fluid, the system was extensively studied by experimental measurements and correlations of the Colburn j-factors by McQuiston [15, 16, 17]. The data collected included the inlet water temperature T_w^{in}, the chilled-water mass flow rate in the form of Reynolds numbers Re_D, the dry-bulb and wet-bulb inlet air temperatures $T_{a,db}^{in}$ and $T_{a,wb}^{in}$, the fin spacing δ, and the heat rate $\dot{Q}$, for operating conditions in which condensation on the fins could occur. The reported data sets conform to three surface conditions: dry, dropwise condensation, and film condensation. The heat rate is a function $\dot{Q} = (T_w^{in}, Re_D, T_{a,db}^{in}, T_{a,wb}^{in}, \delta)$.

After several trials, the ANN chosen for the analysis was the fully-connected 5-5-3-3 configuration shown in Fig. 1(b), consisting of five inputs, i.e., the set of independent variables $\{T_w^{in}, Re_D, T_{a,db}^{in}, T_{a,wb}^{in}, \delta\}$, and three outputs, the sensible and total j-factors j_s and j_t, and the heat transfer rate $\dot{Q}$, all independent of each other. Three neural networks were first individually trained with data corresponding to each of the surface conditions. Later an additional ANN was trained with the combined data sets to assess its robustness in handling the different physics involved. Details of this analysis, including a study of the data-separation issue for training and testing, and the development of global-based correlating equations, can be found in Pacheco-Vega et al. [18].

Root-mean square (rms) values of the percentage deviations between the ANN results and the experimental data are given in Table 1. Also included are the predictions by correlations developed from the same data [17, 19]. The table shows that for all

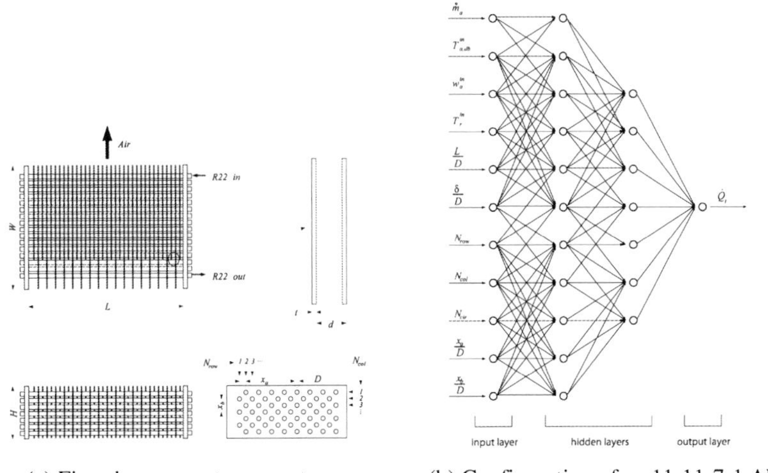

(a) Fin-tube compact evaporator. (b) Configuration of an 11-11-7-1 ANN.

Fig. 2 Schematic of an evaporator and its ANN representation.

Table 1 Percentage errors in j_s, j_t and $\dot{Q}$ predictions by correlations and ANNs.

Surface	*Prediction method*	j_s	j_t	$\dot{Q}$
Dry surface	McQuiston [17]	14.57	14.57	6.07
	Gray and Webb [19]	11.62	11.62	4.95
	ANN	1.002	1.002	0.928
Dropwise condensation	McQuiston [17]	8.50	7.55	8.10
	ANN	3.32	3.87	1.446
Film condensation	McQuiston [17]	9.01	14.98	10.25
	ANN	2.58	3.15	1.960
Combined	ANN	4.58	5.05	2.69

three surfaces, the ANN predictions are much better than any of the correlations. Estimations of the heat rate are especially good, indicating that the ANN was able to correctly recognize the input-output relationship of $\dot{Q}$ with the other physical variables. It should be noted that since the physical phenomena associated with condensation are more complex, the ANN predictions for wet surfaces have larger errors than those for dry cases. However, even in the case of the ANN trained with the complete set of data, the error in the heat rate estimations is very small.

The capability of the ANN to model complex phenomena is now illustrated by its application to a fin-tube evaporator, this time when only very few data sets are available, a common situation in industry. The heat exchanger geometry is illustrated in Fig. 2(a). A total of 38 experimental measurements were performed, and the data collected, under limited number of operating conditions and large range of geometrical parameters. The fluids used correspond to refrigerant R-22 flowing inside the tubes and air flowing through the fin passages. Much of the information in this section is in Pacheco-Vega et al. [20].

The complete set of experimental runs were taken to train a fully-connected 11-11-7-1 ANN shown in Fig. 2(b), where seven of the 11 input parameters correspond to the geometrical quantities $\{L, \delta, x_a, x_b, N_{row}, N_{col}, N_{cir}\}$, scaled by tube diameter D, four comprise the set of operating variables $\{\dot{m}_a, T_{a,db}^{in}, w_a^{in}, T_r^{in}\}$, and the total heat rate $\dot{Q}_t$ is the output node. For prediction purposes, the resulting function

$$\dot{Q}_t = \dot{Q}_t(\dot{m}_a, T_{a,db}^{in}, w_a^{in}, T_r^{in}, {}^{L}/_{D}, {}^{\delta}/_{D}, N_{row}, N_{col}, N_{cir}, {}^{x_a}/_{D}, {}^{x_b}/_{D})$$

is a manifold in a twelve-dimensional parameter space. The training process for the ANN was achieved with 400,000 cycles. The prediction of $\dot{Q}_t$ is plotted against the available experimental data in Fig. 3(a). The ANN results are almost perfect, with rms errors of ±1.5% that are within the experimental uncertainties.

Though the ANN results are remarkable, it should be noted that, due to the high dimensionality of the parameter space and the limited number of experiments used for building the ANN model, the reliability of its predictive (generalization) capability is arguable. In fact, it is known that errors in ANN estimations would increase as the number of training data sets decreases, and that ANNs would perform poorly if applied beyond the domain of the data available to support the predictions. This problem is fundamental if a neural network is to be used as a reliable tool for analysis and design of energy systems.

This problem was addressed by an ANN-based error-estimate procedure developed based on a variant of cross-validation [21]. The methodology (see details in [20]) computes the relative importance of each data point in the limited data set, and hence the validity of the ANN predictions. The results are shown in Fig. 3(b) as a bar plot of the hyper-surface S_{cv} versus each data-set ordered according to its Euclidean distance R in 11-dimensional space from the centroid of all the data. It is observed that the error values are in the [0, 60%] range. Small errors evidence the existence of enough data supporting the ANN predictions, whereas large errors indicate both the need for additional experiments, and where these should be done, to improve the reliability of the ANN prediction capability. The methodology actually provides an upper bound of the expected error.

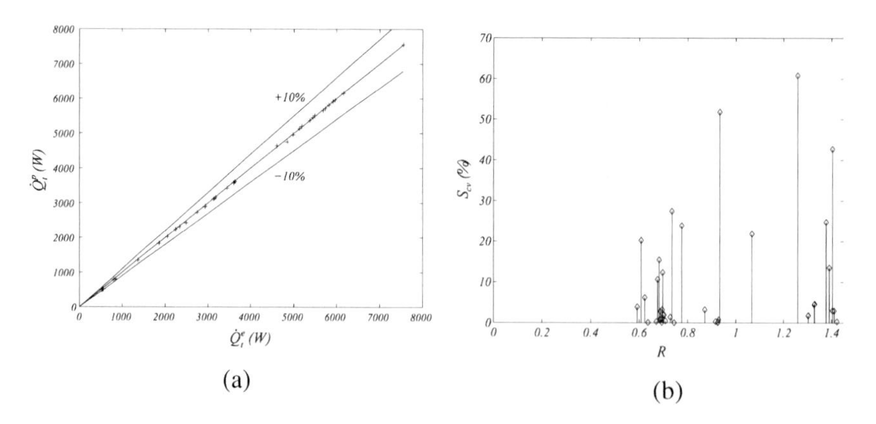

Fig. 3 (a) ANN estimation of $\dot{Q}_t$under limited data. (b)Bar diagram of error estimation.

2.3 Other Applications in Energy Systems

The application of neural networks for the prediction and/or control of different types of energy systems is vast, as seen by the substantial number of literature reviews that have appeared in recent years. The monographs by Sen and Yang [1] and Yang [22], cover applications on a broad range of thermal engineering systems, whereas more system-specific reviews include those of Kalogirou [23, 24] on the ANN modeling of combustion, photovoltaic and green energy systems, and Sen and Yang [25] for applications to multiphase systems. More recently, the use of ANNs has expanded to correlate in-tube heat transfer data in the transition region [26], predict the performance of cooling towers [27], evaluate alternative fuels in engines [28], estimate thermohydraulic behavior of nuclear system components [29], and to develop models for thermodynamic properties of new refrigerants [30], among several energy-related applications.

3 Evolutionary Algorithms

Evolutionary algorithms (EAs), of which the genetic algorithm (GA) and programming (GP) are typical examples, are adaptive stochastic computational techniques inspired in the Darwinian evolutionary principle of natural selection wherein the fittest members of a species survive and are favored to produce offspring. EAs are commonly used for the purpose of optimization, a process fundamental to the design of engineering systems. The GA encodes a set of candidate solutions as binary strings to search for the best; i.e., the global optimum, to a specific problem. GP, on the other hand, is a symbolic extension that works with a set of possible functions to find the best fit to a given set of data. These methodologies have been used in a variety of applications, including finance, electronic design, signal processing and system identification. GAs are discussed in detail by Holland [31], Goldberg [6] and Michalewicz [32], whereas the book of Koza [7] and the monograph by Sette and Boullart [33] are excellent sources for GP.

3.1 Description

Though many variants of the GA technique exist, the brief outline provided here is based on the binary representation. In GAs, the members of a species are regarded as candidate solutions to an optimization problem. The vocabulary used within the technique is borrowed from that of natural genetics. A solution is encoded as binary strings, the collection of possible solutions is a *population*, the objective function is called *fitness*, and a *generation* is an iteration of the algorithm. The idea behind GAs is that after creating an initial population of possible solutions, the fittest members are favored to combine amongst themselves to form the next generation of solutions which, in average, give better results. The evolution is achieved by the so-called *crossover*, where parts of binary strings are switched between parents, and *mutation* which works by randomly changing a digit from a selected string.

The procedure to find the global optimum of a function $f(x)$ in a domain $a \leq x \leq b$, shown schematically in Fig. 4, is summarized as follows.

- An initial population of M members $x_1, \cdots, x_M \in [a, b]$ is randomly generated.
- For each member, the value of the fitness, i.e., the objective function $f(x)$ is computed.
- The parents are selected based on their fitness values.
- Crossover is applied to the parents based on a preselected probability p_c.
- Mutation is performed on a bit by bit basis with a preselected probability p_m.

Once crossover and mutation have been applied to the complete population, a new solution set that keeps the fittest member of the previous generation is created, and the process is continued until some criterion based on convergence or maximum number of generations G_{max}, is achieved. The index j, in the figure, refers to a member in the population of size M (an even number), and G indicates the current iteration.

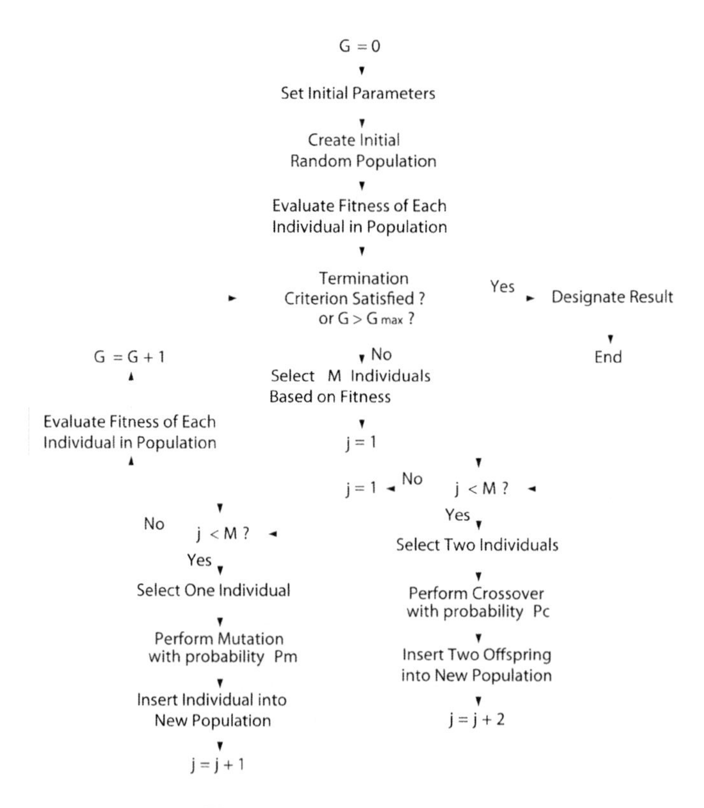

Fig. 4 Flow chart of GA and GP.

Compared to the GA, in GP functions take the place of numbers in an attempt to find the best solution to a particular problem by genetically recombining a population of individuals that portray candidate solutions. This is achieved by using tree-structured representations of functions; an example of the function $5x\cos(5x+1)$ is shown in Fig. 5(a). Branch nodes may be operators with one or two arguments (such as sin, cos, exp, log, +, -, *, /, ^), or may be Boolean (such as AND, OR, NOT) or conditional (IF-THEN-ELSE, etc.) operators. Leaf or terminal nodes $x_j = 1, \cdots, N_v$, on the other hand, are the variables in a particular problem, or constants to be determined. It is to be noted that, the representation of functional forms conserving a correct syntax depends on the programming language being used for its coding. As an example, Fig. 5(b) shows the function $5x\cos(5x+1)$ coded as an array for a MATLAB-based GP program. Details are in Cai et al. [34].

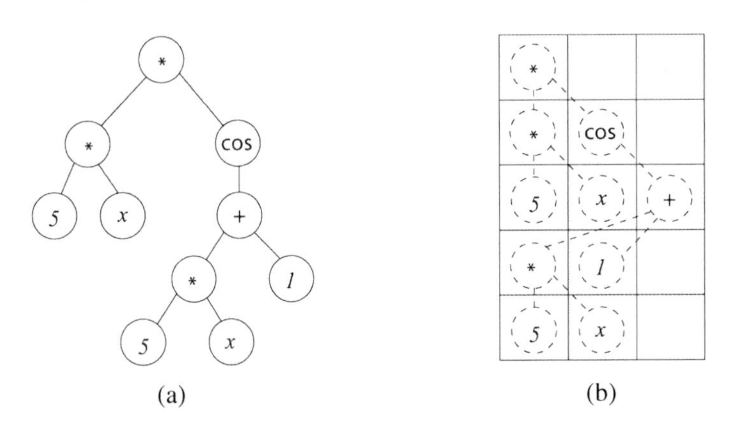

Fig. 5 Representation of $5x\cos(5x+1)$ as (a) parse tree, and (b) array.

The procedure to search for the best solution in the functional space using GP is essentially the same outlined above for the GA. The main difference is that in addition to the numerical parameters, in GP each population consists of a set of functional forms.

3.2 Application to Compact Heat Exchangers

The heat exchanger problem considered before is used again here as a platform to illustrate the application of evolutionary algorithms. In the traditional approach, correlation equations built from experimental measurements in terms of dimensionless groups like the Nusselt Nu, Reynolds Re and Prandtl Pr numbers (sometimes also geometrical parameters are included), are the typical models used to estimate the system performance. From an assumed functional form, the objective is to find the parameters in the correlation that best fits the data performing regression analysis.

From the experimental measurements reported by McQuiston [16], only the dry-surface data are considered for this analysis; the procedures for the other surface conditions are in [18]. The correlation proposed by McQuiston [17] for the Colburn j-factor is

$$j_s = n_1 + n_2 Re^{-n_3} A_r{}^{-n_4}, \tag{1}$$

where A_r is a geometrical parameter representing an area ratio, Re the Reynolds number and $(n_1, n_2, n_3, n_4) = (0.0014, 0.2618, 0.4, 0.15)$ are the values of the constants. A slightly different correlation was reported by Gray and Webb [19] using data from [16] and other sources. Additional information is in Pacheco-Vega et al. [18, 35].

To find the correlation constants from an assumed form, an optimization technique attempts to minimize the objective function, defined as the variance of the error between predictions $j^p_{s,i}$, and measurements $j^e_{s,i}$, which is given as

$$S_{j_s} = \frac{1}{N} \sum_1^N (j^e_{s,i} - j^p_{s,i})^2, \tag{2}$$

where $S_{j,s}$ is a manifold in a five-dimensional parameter space holding multiple local minima, two of which are shown in Fig. 6. This multiplicity of solutions, either arising from the assumed mathematical form of the correlation or from the experimental procedure to decouple the transfer coefficients, is a main reason for their lack of accuracy in predictions.

With a population of $M = 40$, $p_c = 1$ and $p_m = 0.3$, a GA code [36] was used to find the set of constants in the domain $(-0.6, 0.6)$ for all the unknowns. The values found are: $(n_1, n_2, n_3, n_4) = (-0.0218, 0.0606, 0.0778, 0.0187)$, which conform to the global minimum in $S_{j,s}$, labeled **A** in the figure. The corresponding results are shown qualitatively in Fig. 7, and quantitatively in Table 2. As expected, it is clearly seen that the global-regression-based correlation provides smaller rms percentage errors than the correlations developed by McQuiston [17] and Gray and Webb [19].

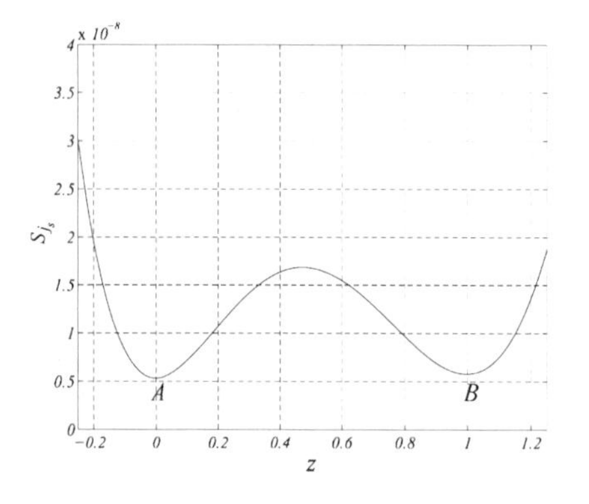

Fig. 6 Section of $S_{j,s}(n_1, n_2, n_3, n_4)$; **A** is the global minimum; **B** is a local minimum.

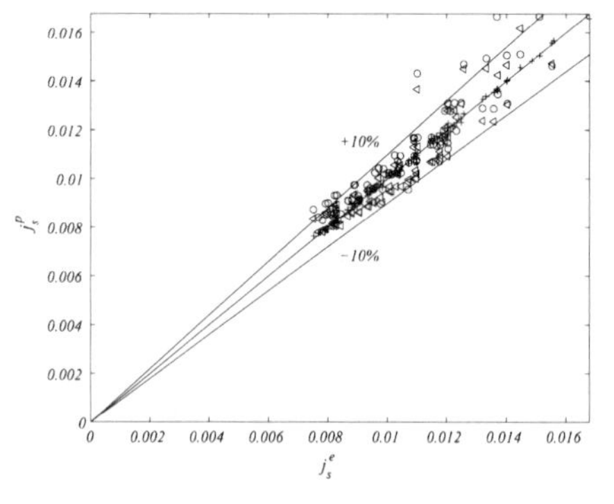

Fig. 7 Predicted j_s for a heat exchanger; Δ McQuiston [17]; ○ Gray and Webb [19]; + ANN.

Table 2 RMS errors in j_s from correlations.

Prediction method	*Error (%)*
McQuiston [17]	14.57
Gray and Webb [19]	11.62
Pacheco-Vega et al. [18]	6.21
Cai et al. Eq. (3) [34]	6.18

The natural extension of the correlation procedure illustrated above is to find the functional form that provides the best fit to the data. It will now be shown that GP is suitable for the purpose of symbolic regression from experimental data. As before, the objective function (fitness) is defined by Eq. (2), where $j^p_{s,i}$ are now the predicted values from each of the M correlations in the population. The method seeks a correlation function and the corresponding constants that minimize S_{j_s}. The chosen parameters are: $M = 100$, $G_{max} = 800$, $p_c = 0.8$, $p_m = 0.2$. The terminal sets include the variables $x_1 = Re$, and $x_2 = A_r$. Additional details are in Cai et al. [34]. The correlation resulting from the symbolic regression procedure is given as

$$j_s = \frac{2205.32}{1.39\times10^5 + 24.16Re + A_r Re}. \tag{3}$$

As observed, its mathematical form is different from that of Eq. (1), but provides better predictions, as shown in Table 2. It can actually be noticed that though the global-regression-based correlation discussed before is the best possible that can be obtained from the assumed functional form, Eq. (3) is seen to give a slightly smaller rms error, and hence, the best overall.

3.3 Other Applications in Energy Systems

Applications of evolutionary algorithms to different types of energy systems have increased substantially in the last fifteen years. GAs have been used for the purpose of optimization in many different areas, among them, thermal engineering. The monographs by Sen and Yang [1] and Sen and Goodwine [2] review applications in a diversity of energy systems, whereas the review by Gosselin et al. [37] covers the utilization of GAs in the areas of inverse heat transfer, design and modeling of thermal systems. Some recent applications, in which the process of optimization is carried out, include power-generation [38], HVAC systems in buildings [39, 40], desalination system design [41], control-based efficiency-enhancement of energy systems [42], heat exchanger networks [43], optimum performance solar heaters [44] and the design of photovoltaic systems [45]. Applications of genetic programming to energy systems are scarce. The development of semi-empirical models for chemical-process systems [46], critical heat flux in round pipes [47], heat exchangers [34], helically-finned tubes [48], fuel cells [49] and energy forecasting [50] are among the very few investigations.

4 Fuzzy Logic

Fuzzy logic (FL) is a methodology rooted upon the theory of fuzzy sets [51], that allows the description of complex systems and their performance by means linguistic variables and inference rules. Developed by Lofti Zadeh in the early 1960s as a way to model the uncertainty of natural language [52, 53], the technique has been successfully applied to several areas of science and technology [54, 55], particularly to system control [2]. The main advantage offered by FL is its ability to handle *imprecise* or *noisy* data in order to find definitive solutions of a particular problem. This is done by formulating mappings between given system inputs and its output from which decisions about the system behavior can be inferred. The mathematical foundations of fuzzy sets and fuzzy logic can be found in several introductory texts, including those of Mordeson and Nair [56], Klir and Yuan [57], and Chen and Pham [8]. A brief description of the technique towards the development of a controller is provided next within the context of thermal control. Much of the information reported here is in Pacheco-Vega et al. [58].

4.1 Description

The FL methodology involves three mechanisms: (1) *fuzzification* in which the input variables, defined as linguistic, are maped into fuzzy sets according to a specific degree or membership, (2) *inference* where the fuzzy sets are processed according to a library of expert-based *if-then* rules, and (3) *defuzzification* in which the fuzzy outputs are mapped back to their crisp values. For the specific problem of thermal control in heat exchangers, a fuzzy-based controller aims to regulate the outlet temperature of one of the fluids, e.g., that of the cold water T_c^{out}, as a function of the mass flow rate of only one fluid, e.g., the cold fluid $\dot{m}_c$. This constitutes a single-input single-output (SISO) system, where $\dot{m}_c$ is the manipulated (control) variable and T_c^{out} the controlled (output) one. The linguistic variables describing the system are the temperature difference between the set-point T_{set} and T_c^{out}, i.e., $E_{\Delta T} = T_{set} - T_c^{out}$, and the percentage of opening in the control valve $V_{\Delta T}$, which modifies $\dot{m}_c$. For both $E_{\Delta T}$ and $V_{\Delta T}$, five membership functions, shown in Figures 8(a) and 8(b), were selected. The fuzzy sets were defined as $\mu_{E_{\Delta T}} = (\mu_{NL}, \mu_{NS}, \mu_Z, \mu_{PS}, \mu_{PL})$ and $\mu_{V_{\Delta T}} = (\mu_{VL}, \mu_L, \mu_M, \mu_H, \mu_{VH})$. The corresponding ranges were chosen as: $[-1, 1]$ °C for $E_{\Delta T}$, and $[0, 3.5]$ Volts for $V_{\Delta T}$. The temperature-control decision table, constructed from expert-based *if-then rules*, and the common Mamdani inference system [59], is illustrated in Table 3. The "response line" that defines $V_{\Delta T}$ as a function of $E_{\Delta T}$, shown in Fig. 9, is calculated using the well-known center-of-gravity technique, though other defuzzification models exist in the literature [8, 57].

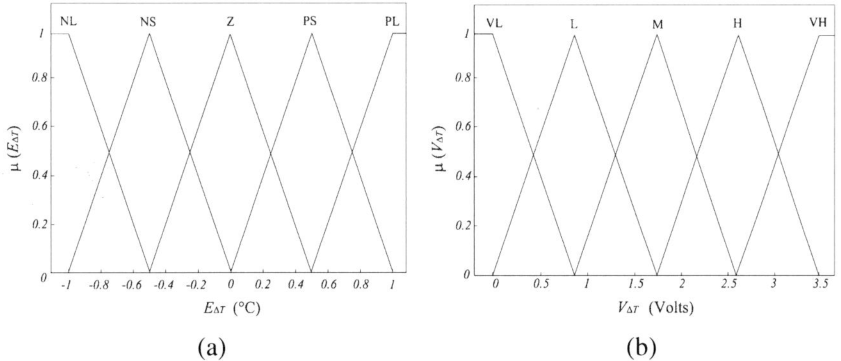

Fig. 8 Fuzzy sets for controller. (a) Temperature error; (b) Voltage in control valve.

The control actions of the fuzzy controller are outlined below.

1. The input variables for the controller, i.e., $E_{\Delta T}$ (and $dE_{\Delta T}/dt$ for the extended-input controller –not included in this chapter), are first measured by the appropriate sensor (i.e., a thermocouple), and then fuzzified by computing their degree of membership in each fuzzy set.

2. The if-then rules in the rule base are evaluated in parallel. The output from each corresponds to the associated fuzzy set of $V_{\Delta T}$ (Mamdani inference engine [59]). Each of the output fuzzy sets is "cut-off" according to the associated membership obtained from logical operations of the states and "added" to conform the aggregated output fuzzy-set for $V_{\Delta T}$.

3. The crisp value of $V_{\Delta T}$ (defuzzification) is computed. Here, the crisp output value is the centroid of the aggregated fuzzy-output-set.

The structure of the fuzzy controller, and the closed-loop SISO system for the control of T_c^{out} using $\dot{m}_c$, are respectively shown in Figs. 10(a) and 10(b).

Table 3 Decision table.

$E_{\Delta T}$	$V_{\Delta T}$
Negative large (NL)	Very low (VL)
Negative Small (NS)	Low (L)
Zero (Z)	Medium (M)
Positive small (P)	High (H)
Positive large (PL)	Very high (VH)

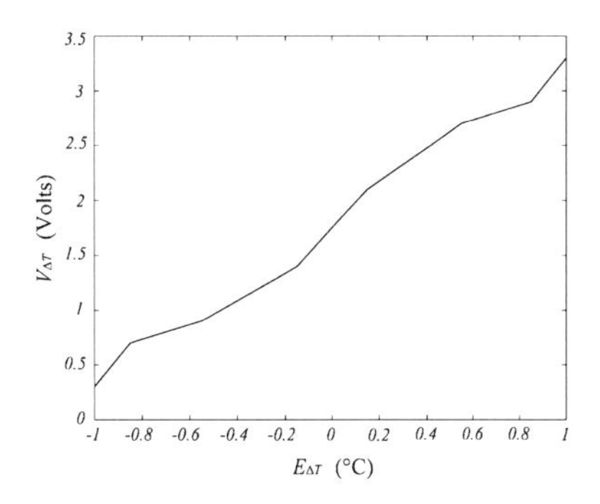

Fig. 9 Response line in control valve.

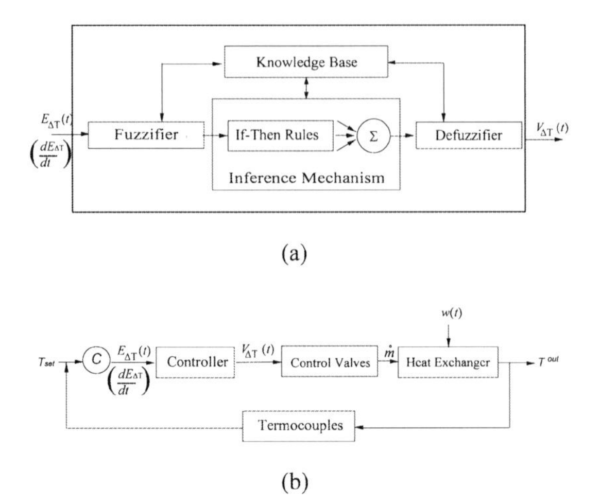

Fig. 10 Close-loop fuzzy control. (a) Fuzzy controller; (b) Closed-loop configuration.

4.2 Application to Temperature Control of a Heat Exchanger

To illustrate the robustness of the fuzzy controller, the tests were done in an concentric-tubes heat exchanger experimental facility [60], a schematic of which is shown in Fig. 11. Hot water flows inside the inner-tube and cold water at room temperature flows in the annulus. Measurements of hot and cold water mass flow rates $\dot{m}_h$ and $\dot{m}_c$, inlet and outlet hot temperatures T_h^{in} and T_h^{out}, and inlet and outlet cold temperatures T_c^{in} and T_c^{out}, were taken and recorded. Measurements and control actions between system and controller were interfaced via LabVIEW to a personal computer (PC).

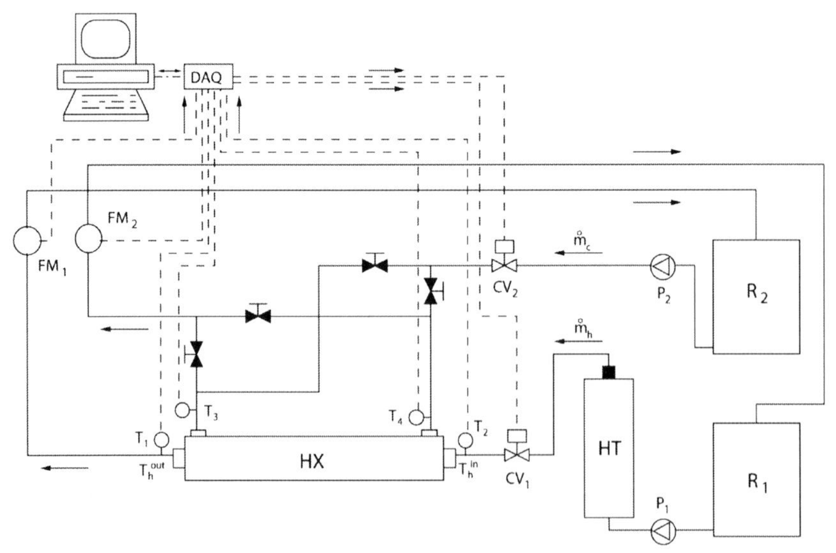

Fig. 11 Schematic of experimental facility.

In the first test, the system was initially subjected to a sudden change in the set-point, from $T_c^{out} = 29°C$ to 32°C, and later to a gradual increase in the inlet hot-water-circuit temperature, from $T_h^{out} = 41°C$ to 49°C. The results, illustrated in Figs. 12(a) and 12(b), show that the controller is able to maintain T_c^{out} within 0.2°C of the setpoint, with some oscillations arising from the controller response to fluctuations in T_h^{in}. It can be also seen that even when T_h^{in} is substantially increased, the fuzzy controller never loses control of the system. The offset between the controller and the setpoint is somehow expected since only information about the error in temperature is provided to it. Further tests (not shown here) have demonstrated that, as more information about the system is given to the controller, its ability to achieve the control objective improves.

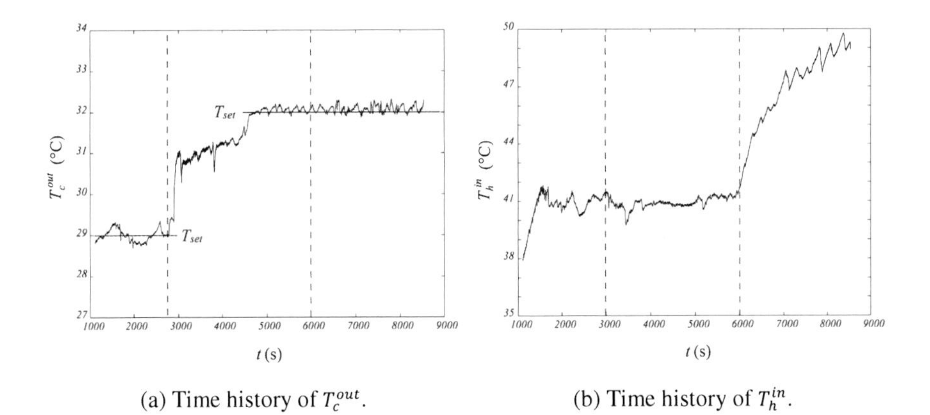

(a) Time history of T_c^{out}.

(b) Time history of T_h^{in}.

Fig. 12 Cold-water temperature control with fuzzy controller.

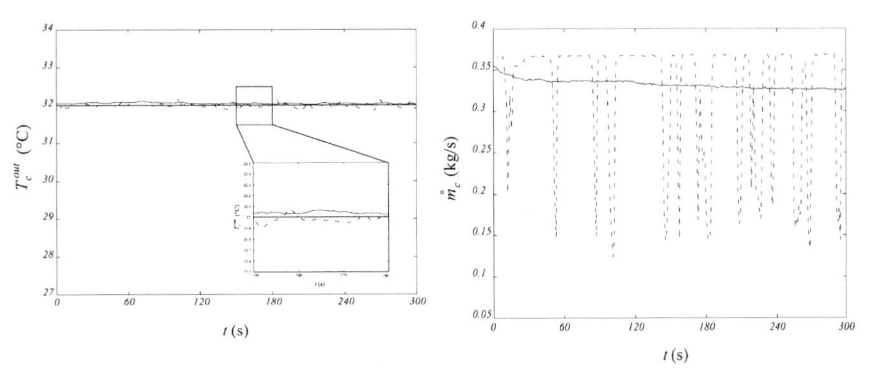

(a) Time history of T_c^{out}.

(b) Dynamics in control valve.

Fig. 13 Fuzzy vs. PID control. – Fuzzy controller; — · — PID controller.

In the second test, the fuzzy controller was compared to a standard PID controller, for which the tuning of the parameters was done using relationships from Shinskey [61]. The objective was to maintain T_c^{out} at 32°C in response to perturbations in T_h^{in}. The results are shown in Figs. 13(a) and 13(b). It is observed that both controllers are able to bring the system to the control point, and though the PID controller seems more accurate, it shows some oscillations, with a magnitude ±0.1°C, around the control point.

These oscillations in temperature arise from the dynamics in the control valve, as the PID controller continuously opens and closes the valve to keep T_c^{out} at the target value. On the other hand, once the set point has been achieved, the fuzzy controller makes very small adjustments in the amount of voltage supplied to the valve. A main advantage offered by fuzzy logic to the control of energy systems is that it provides convenient links between feedback control and human intuition regarding how a system could be controlled.

4.3 Other Applications in Energy Systems

Fuzzy logic has been successfully applied to a number energy systems, especially to perform feedback control [62, 63, 64]. However, other applications include the forecast of energy consumption [65], the simulation of thermal processes in micro-manufacture [66], energy management in fuel cells [67], supervision and planning in environmental and energy-renewable systems [68, 69], as examples of some of the more recent investigations.

5 Cluster Analysis

Cluster analysis (CA) is also known as clustering or numerical taxonomy. It is a SC methodology that classifies data into groups (or clusters), such that elements drawn from the same group are as similar to each other as possible, while those assigned to different groups are dissimilar. In contrast to the ANN supervised classification by learning processes, CA is often described as unsupervised pattern classification as the clusters are obtained solely from the data, under the hypothesis that such data contain enough information inherent to the phenomenon of interest, so that the parameters that characterize it can be established.

The application of CA has increased substantially in recent years in a variety of disciplines like biology, marketing, medicine, manufacturing, image processing, and astronomy, among others, where the common thread is some kind of classification or feature extraction from experimental data. Clustering methodologies are discussed in detail in the texts by Everitt et al. [9], and Abonyi and Feil [70]. Though several techniques in CA have been developed, and excellent reviews are available [3], this section concentrates on the application of two of the most promising in science and engineering; i.e., fuzzy C-means and Gaussian-mixtures.

5.1 Description of Fuzzy C-Means

The fuzzy C-means (FCM) algorithm is by far the most widely used clustering algorithm in practice [71, 72]. Based on the theory of fuzzy-sets [51], this technique was introduced by Dunn [73] to address the inability of the well-known K-means method [74] to accurately classify data with some type of overlap. The technique assigns to each data point a degree of membership to the different clusters, rather than assigning the point to a sole group whose centroid is the nearest. Detailed discussion of the FCM is given in the monograms by Hoppner et al. [75] and Bezdek et al. [76]. A brief description of the scheme is provided next.

Starting with a set of N measurements, e.g. $\mathrm{X} = \{\mathrm{x}_1, \mathrm{x}_2, \cdots, \mathrm{x}_N\} \subset \mathbb{R}^q$, the FCM algorithm assigns the data into a number of predetermined K groups, i.e., $\mathrm{C} = \{\mathrm{c}_1, \mathrm{c}_2, \cdots, \mathrm{c}_K\} \subset \mathbb{R}^q$ through values of membership functions u_{ij} ($i = 1, 2, \cdots, K, j = 1, 2, \cdots, N$), which provide the degree to which the data point x_j belongs to the fuzzy cluster i. The data classification into the chosen K groups is achieved by minimizing the fuzzy objective function

$$J_m(\mathrm{U},\mathrm{C}) = \sum_{i=1}^{K} \sum_{j=1}^{N} (u_{ij})^m \|\mathrm{x}_j - \mathrm{c}_i\|^2, \tag{4}$$

that provides a weighted measure of the similarity between the data and the clusters. At the beginning of each run, the values of u_{ij} are randomly assigned, and the cluster centroids computed from

$$\mathrm{c}_i = \frac{\sum_{j=1}^{N} (u_{ij})^m \mathrm{x}_j}{\sum_{j=1}^{N} (u_{ij})^m}, \quad i = 1, 2, \cdots, K. \tag{5}$$

The values u_{ij} are then updated with

$$u_{ij} = \begin{cases} \left[\sum_{k=1}^{K} \left(\frac{\|\mathrm{x}_j - \mathrm{c}_i\|}{\|\mathrm{x}_j - \mathrm{c}_k\|} \right)^{\frac{2}{(m-1)}} \right]^{-1}; & \mathrm{x}_j \neq \mathrm{c}_i, \\ 1; & \mathrm{x}_j = \mathrm{c}_i, \end{cases} \tag{6}$$

within the constraints $u_{ij} \in [0, 1]$ and $\sum_{i=1}^{K} u_{ij} = 1$. The updated values of $u_{ij}$are used in to compute J_m. The process to calculate the centroids, update the partition matrix, and evaluate the objective function, is repeated until some criterion based on convergence or a maximum number of iterations is achieved. It should be noted that the fuzzy partition exponent m appearing in Eqs. (4)–(6), may have values in $1 < m < \infty$. The larger the value of m, the less crisp the data partition into the specified groups [77]; a value $m = 1$ corresponds to a fully-crisp partition. This is a validity problem under active research [78].

5.2 Description of Gaussian-Mixture Models

Gaussian mixtures are frequently used to classify data into groups when the relationships among the data points are unknown. A main advantage offered by the method is the possibility of finding the number of groups as part of the solution. A clustering technique based on Gaussian mixtures assumes that the data can be grouped into a number of K clusters, each described by a Gaussian probability density distribution. Once the number of groups is known, the geometrical features (structures) of the groups and corresponding data classification are determined on the basis of a maximum likelihood criterion. Excellent descriptions of the technique, along with its mathematical background, are provided in several monographs [9, 79, 80, 81, 82]. The following is a brief account of the method.

Given a set of N experimental data, $\mathrm{X} = \{\mathrm{x}_1, \mathrm{x}_2, \cdots, \mathrm{x}_N\} \subset \mathbb{R}^q$, the probability distribution of each measurement x_j may be described by a linear combination of K mixture components as

$$p(\mathrm{x}_j|\Theta) = \sum_{k=1}^{K} p(\mathrm{x}_j|\omega_k, \theta_k)\, p(\omega_k), \qquad j = 1, \cdots, N \tag{7}$$

where $p(\omega_k)$ is the probability that group ω_k occurs in the sample data, and $p(\mathrm{x}_j|\omega_k, \theta_k)$ is the conditional probability of x_j belonging to cluster ω_k, modeled by cluster-specific multivariate Gaussian distribution

$$p(\mathrm{x}_j|\omega_k, \theta_k) = \frac{1}{(2\pi)^{q/2}\|\Sigma_k\|^{1/2}} \exp\left[-\frac{1}{2}(\mathrm{x}_j - \mu_k)^T \Sigma_k^{-1}(\mathrm{x}_j - \mu_k)\right]. \tag{8}$$

The parameters $\theta_k = \{\mu_k, \Sigma_k\}$, i.e., the mean vector μ_k and the covariance matrix Σ_k, characterize the shape of each component density, and $\Theta = \{(p(\omega_k), \theta_k) : k = 1, \cdots, K\}$ denotes the set of parameters of the mixture model.

The unknowns in Eqs. (7) and (8) are the number of clusters K, the parameters of the Gaussian distributions θ_k and the mixing proportions $p(\omega_k)$, all of which can be computed from the data. A number of methods have been proposed to estimate the model parameters Θ, for a prescribed K, and the set of N observations, using the well-known maximum likelihood (ML) estimation approach along with the expectation-maximization (EM) iterative algorithm [83].

The result is a maximum likelihood estimate of Θ, given as

$$\begin{aligned} \widehat{\Theta} &= \left\{\left(\hat{p}(\omega_k), \hat{\mu}_k, \hat{\Sigma}_k\right) : k = 1, \cdots, K\right\} = \\ &\operatorname{argmax}_\Theta \sum_{j=1}^{N} \log\left(\sum_{k=1}^{K} p(\mathrm{x}_j|\omega_k, \theta_k)\, p(\omega_k)\right), \end{aligned} \tag{9}$$

where now it is possible to assign the datum x_j to the group ω_k according to its maximum posterior probability, i.e., the probability that data point x_j belongs to the group ω_k. The criterion is

$$\hat{p}(\omega_k)p(\mathrm{x}_j|\omega_k, \hat{\theta}_k) > \hat{p}(\omega_l)p(\mathrm{x}_j|\omega_l, \hat{\theta}_l), \quad \text{for all } k \neq l; \quad l = 1, \cdots, K. \tag{10}$$

The number of clusters K, necessary for the classification, can be estimated based on the minimum description length (MDL) criterion [84] (although several other criteria are available in the literature [85]). The MDL is a penalized function of the negative logarithm of the maximum likelihood, i.e.,

$$MDL(K,\theta) = -\sum_{j=1}^{N} \log\left(\sum_{k=1}^{K} p\left(x_j|\omega_k,\theta_k\right) p(\omega_k)\right) + \frac{1}{2} Kq \log(Nq) \quad (11)$$

that provides a trade-off between the data representation and the model complexity. Minimization of Eq. (11) with respect to K gives the number of clusters that provide a good description of the data provided by the simplest Gaussian-mixtures model. Additional details about the algorithm are in [86], and the references therein. The next subsections show the application of CA in two different energy systems.

5.3 Application to Data Classification of Thermodynamic Properties

The application of the FCM technique is illustrated by its use to classify thermodynamic properties of fluids. This process is necessary to develop models for design and selection of engineering systems. For the analysis, a total of $N_S = 150$ data sets corresponding to pressure p, volume v, and temperature T of water were taken from the literature [87], and equally divided among the liquid (L), liquid-vapor (LV) and superheated vapor states (SV), as shown in Fig. 14. Additional information is in Avila and Pacheco-Vega [88].

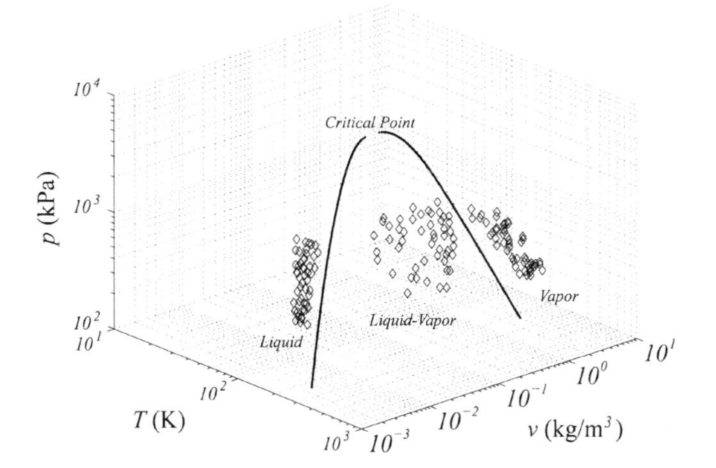

Fig. 14 Thermodynamic-property p-v-T data.

To apply the FCM technique in a classification problem, the number of clusters has to be assigned *a priori*. In this case, from the physics of the phenomenon the number of selected groups was $K = 3$. It is also convenient to normalize the variables to avoid grouping errors due to differences in units and scales, since the FCM technique computes the similarity/dissimilarity within the data based on

Euclidean distances. Though several schemes exist in the literature [89], and have been successfully applied in different settings, for purposes of this study the variables were normalized by the values of the thermodynamic quantities at the critical point. This allows larger generality, since data of different fluids are placed under the same baseline by the principle of corresponding states. The scaling is given mathematically as

$$p_i^r = \frac{p_i}{p_{cr}}; \quad v_i^r = \frac{v_i}{v_{cr}}; \quad T_i^r = \frac{T_i}{T_{cr}}, \tag{12}$$

where $i = 1, \cdots, N_S$, "*cr*" refers to the values at the critical point, "*r*" are the *reduced* thermodynamic quantities. It may also be noted that success of the normalization in the classification process largely depends on the natural structure of the data. In some situations, normalizing the variables is sufficient, e.g. [90, 91], whereas for others it may be necessary to apply some type of transformation; this application corresponds to the latter case. For this data the function that provides the sharper separation among the different data structures is the logarithmic transformation.

After taking logarithms to the scaled data, with $N_S = 150$ and $K = 3$, the results obtained are shown qualitatively in Figs. 15(a)–15(d), and quantitatively in Table 4. From the figures and the table it is clear that the FCM algorithm was able to achieve a correct classification with only one datum being misplaced from the liquid-vapor state into the subcooled liquid region, and two data points from the superheated and into the liquid-vapor state. The table indicates that 100% of the liquid-state data are grouped into the appropriate Cluster I, whereas 98% of the liquid-vapor data are placed in Cluster II and 96% of the superheated vapor data in Cluster III; i.e., $M_I = 51$, $M_{II} = 51$ and $M_{III} = 48$.

Table 4 Fraction of thermodynamic data classified in different groups.

Condition \ Group	I	II	III
Subcooled liquid	50	0	0
Liquid-vapor	1	49	0
Superheated vapor	0	2	48

The results of this example confirm the usefulness of the FCM clustering technique to identify and classify the characteristic information of a system directly from the experimental data, particularly for cases where the complexity of the phenomenon/system is substantial. In this regard, it is important to note that factors like the normalization and transformation influence the quality of the pattern identification and must be considered to ensure a correct classification.

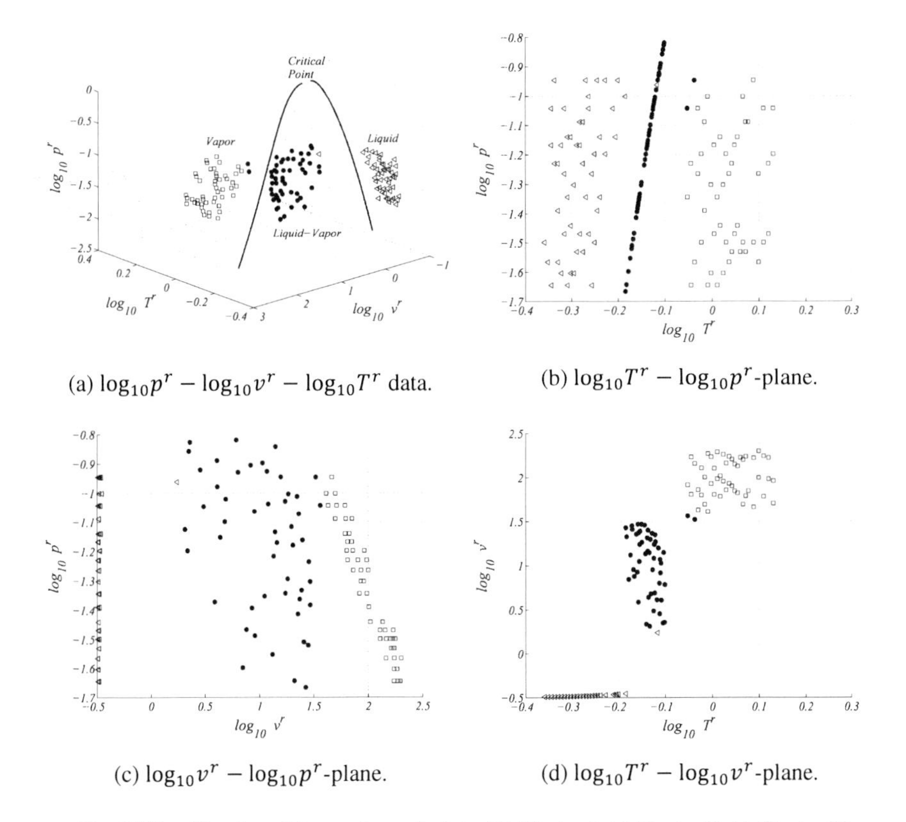

(a) $\log_{10}p^r - \log_{10}v^r - \log_{10}T^r$ data.

(b) $\log_{10}T^r - \log_{10}p^r$-plane.

(c) $\log_{10}v^r - \log_{10}p^r$-plane.

(d) $\log_{10}T^r - \log_{10}v^r$-plane.

Fig. 15 Classification of thermodynamic data. (Δ) Cluster I; (•) Cluster II; (▫) Cluster III.

5.4 Application to Classification of Performance Data in Heat Exchangers

A second illustration of the cluster analysis is the use of Gaussian mixture models to extract the regimes of operation from condensing heat exchanger data. This is the same thermal system previously analyzed in Sections 2.2 and 3.2 with ANNs and evolutionary algorithms. The information reported here has its basis on the published database [16], which was separated into dry-surface conditions, dropwise condensation and film condensation using visualization techniques.

The complexity of the classification process can be shown by looking at the function describing the system performance; i.e., $\dot{Q} = (T_w^{in}, Re_D, T_{a,db}^{in}, T_{a,wb}^{in}, \delta)$, which resembles a smooth manifold in a six-dimensional parameter space. Sections of this manifold are presented in matrix form in Fig. 16, where the top row shows the relationship between the Reynolds number Re_D versus $T_{a,db}^{in}$, $T_{a,wb}^{in}$, T_w^{in},

δ, and $\dot{Q}$. The bottom row pictures $\dot{Q}$ versus the other five variables involved. As can be seen, the figure does not present a definitive number of groups in which the data can be classified. For some planes there are two well-formed groups whereas for others five regions can be identified. Details are in Pacheco-Vega and Avila [92].

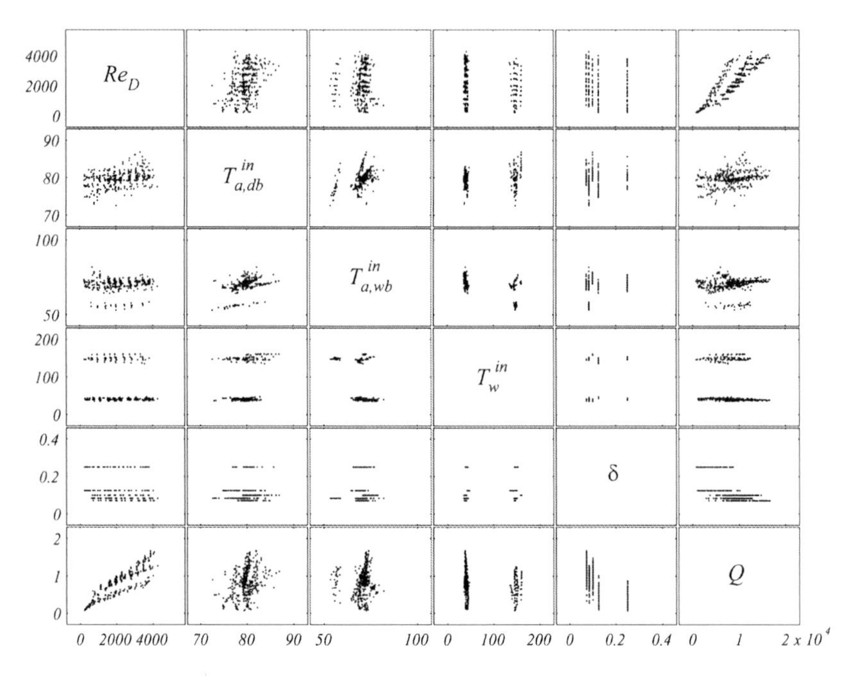

Fig. 16 Representation of heat exchanger data through planes.

The total $N = 327$ experimental runs were used in conjunction with the Gaussian-mixtures agglomerative clustering technique described before [86], to first determine the regimes of operation and then to classify the data. The convergence of the algorithm is shown in Fig. 17, where it is observed that MDL criterion achieves its minimum value at the correct number of clusters, $K = 3$, in accordance to the known physical phenomena. The classification results are presented in Figs. 18(a) and 18(b), on the plane Re_D vs. T_w^{in}. Also included are those obtained visually by McQuiston [16]. A quantitative comparison of these results is shown in Table 5. As observed, the two methods agree completely in the dry-surface data, which were all assigned to group I.

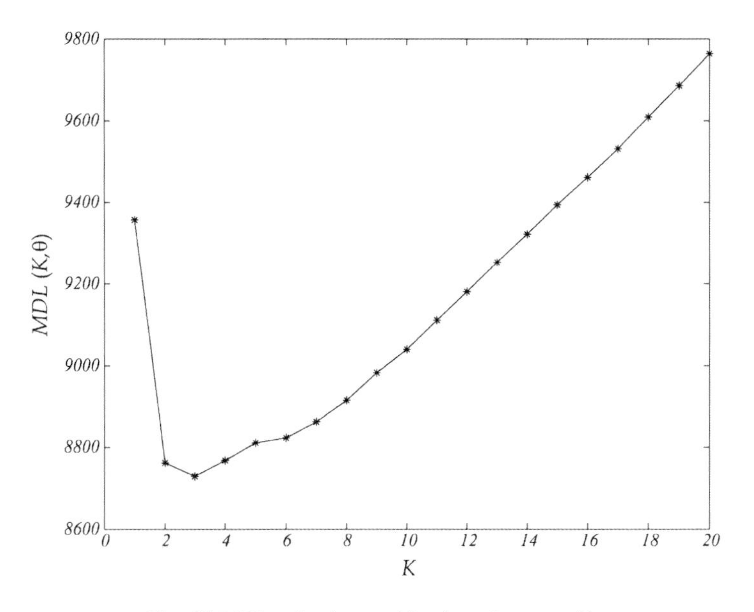

Fig. 17 MDL-criterion vs. Number of groups K.

However, the agreement in the allocation of the data corresponding to humid conditions into groups II and III is less crisp. Each of these groups contain data, in different proportions, that were visually classified as either drop or film condensation. The apparent discrepancy raises the question as to which classification is the correct one. This issue was addressed by conducting an independent data-classification via ANNs, details of which are reported in [92]. Results from the ANN-based discrimination methodology, not included in this chapter due to space limitations, agree very well with those of the Gaussian-mixtures clustering; being the best case 100% and the worst 85%.

Table 5 Classification of heat exchanger data.

Condition \ Group	I	II	III
Dry surface	100%	0	0
Drop condensation	0	35.89%	64.11%
Flim condensation	0	25.21%	74.79%

This section has shown the usefulness of cluster analysis for data classification of complex physical phenomena occurring in thermal systems.

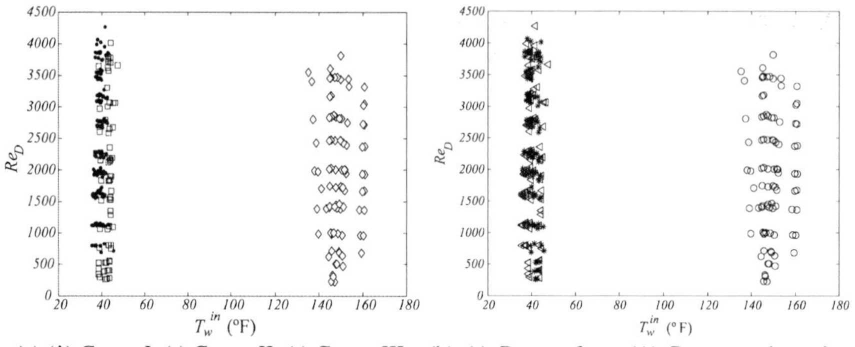

(a) (◊) Group I, (▫) Group II, (•) Group III. (b) (▫) Dry surface, (Δ) Drop condensation, (*) Film condensation.

Fig. 18 Heat exchanger data classification. (a) Algorithmic; (b) Visual [16].

5.5 Other Applications in Energy Systems

Although cluster-analysis techniques have been successfully applied to many fields, their applications to energy systems are scarce. Examples in thermal and fluids engineering include the classification of thermodynamic [88] and heat exchanger data [18, 92], turbulent flows [93] and local-wind patterns for renewable energy systems [94], environmental data for energy planning [95], energy-performance- [96] and geophysical-data [97], are among the few investigations.

6 Neuro-Fuzzy Hybrid Technique

A main advantage provided by soft computing stems from the fact that its constituent methodologies are for the most part complementary and synergistic rather than competitive, and can be combined to improve even more the quality of their individual results. There is a number of examples in several fields, including environmental engineering, medicine, planning, management and manufacturing, among others, where two or more SC techniques have been combined. A particular case is the adaptive-network-based fuzzy inference system (ANFIS), where neural networks and fuzzy logic are combined to take advantage of their individual features in modeling complex systems. ANFIS is discussed in detail by Jang [98], who developed the technique. A brief description, and subsequent application to dynamic modeling, is provided next.

6.1 Description of ANFIS

The adaptive-network-based fuzzy inference system (ANFIS), proposed by Jang [98], is a hybrid model where the nodes in the different layers of a feed-forward network handle fuzzy parameters. This is equivalent to a fuzzy inference system with distributed parameters. At the core, the technique splits the representation of prior knowledge into subsets in order to reduce the search space, and uses the backpropagation algorithm to adjust the fuzzy parameters. The resulting system is an adaptive neural network functionally equivalent to a first-order Takagi-Sugeno [99] inference system, where the input-output relationship is linear. A typical schematic of an ANFIS architecture, in the context of a heat exchanger, for three inputs ($N_v = 3$) $\mathbf{x} = [x_1, x_2, x_3]$ (e.g. $x_1 = \dot{m}_c, x_2 = T_h^{in}$, and $x_3 = t$), is shown in Fig. 19, where each layer performs a particular task in the fuzzy inference system. The output may be, for instance, $y = T_c^{out}$.

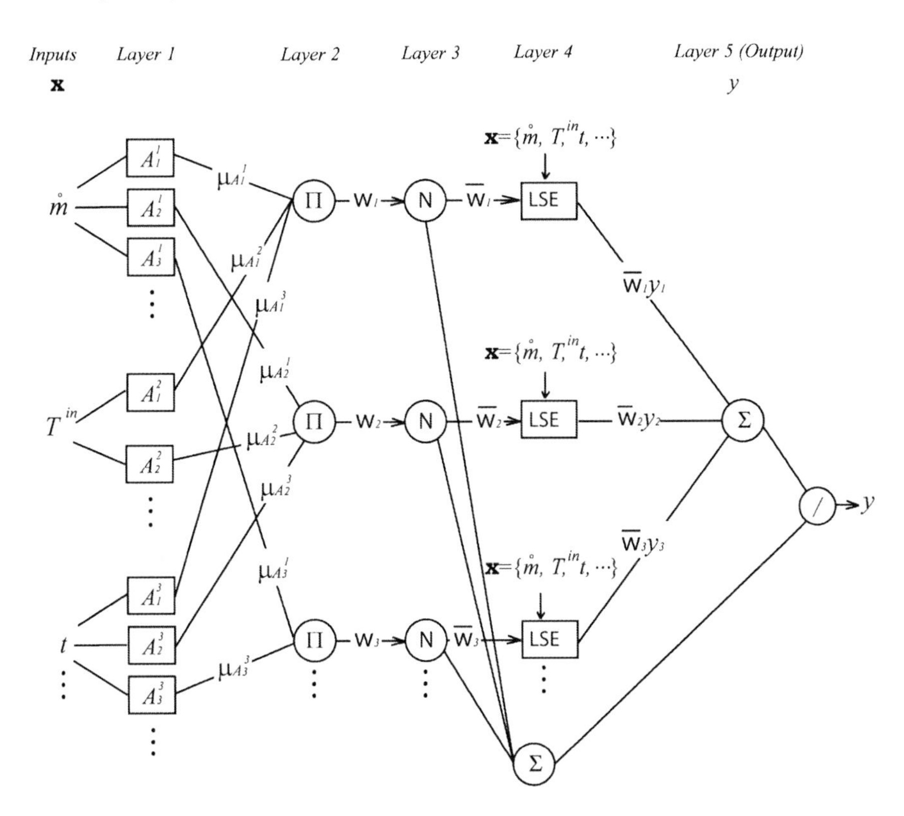

Fig. 19 Typical structure of the ANFIS scheme.

In reference to Fig. 19, an outline of the procedure [98] is:

1. In layer 1 the numerical inputs x_k, are fuzzified by computing their membership in each fuzzy set A_j^k. The output of node j, of each variable in the first layer, is

$$\mu_{A_j^k}(x_k) = \left[1 + \left(\frac{x_k - c_j}{a_j}\right)^{2b_j}\right]^{-1}, \tag{13}$$

where $\mu_{A_j^k}(x_k)$ is a bell-shaped membership function parameterized by a_j, b_j, and c_j, which adjust its geometrical structure, and are determined directly from the experimental data.

2. Layer 2 provides a weighted output (firing strength of a specific rule). The output of each node is given as the product

$$\omega_j = \prod_{k=1}^{K} \mu_{A_j^k}(x_k). \tag{14}$$

3. Layer 3 computes the normalized firing strength of a particular rule as

$$\bar{\omega}_j = \frac{\omega_j}{\sum_{r=1}^{J} \omega_r}. \tag{15}$$

4. The adaptive nodes in layer 4 calculate the weighted rule outputs based on the consequent parameters as

$$\bar{\omega}_j \cdot y_j = \bar{\omega}_j \cdot \left(p_{j0} + p_{j1} \cdot x_1 + \cdots + p_{jk} \cdot x_k + p_{jK} \cdot x_K\right), \tag{16}$$

where $\bar{\omega}_j$ is a normalized firing strength from Layer 3, and p_{jk} is the consequent parameter set of each node.

5. In layer 5, the local output functions are aggregated to form the total averaged-output for the ANFIS system ([100]), as

$$y = \frac{\sum_{j=1}^{J}(\bar{\omega}_j \cdot y_j)}{\sum_{r=1}^{J} \bar{\omega}_r}. \tag{17}$$

Note that for all the variables in a problem, ANFIS optimizes both the number of membership functions and the corresponding rules using a subtractive clustering algorithm [101].

6.2 Dynamic Model of a Heat Exchanger

The application of this hybrid technique is illustrated by its use for system identification and dynamic modeling of the heat exchanger described in Section 4.2. Details are in Ruiz-Mercado et al. [102]. The objective is to find expressions for the two outlet fluid temperatures as functions of their inlet temperatures, mass flow rates and time, i.e.,

$$\{T_c^{out}, T_h^{out}\} = \{f_1, f_2\}\left(\dot{m}_h, \dot{m}_c, T_h^{in}, T_c^{in}, t\right), \tag{18}$$

where $\left(\dot{m}_h, \dot{m}_c, T_h^{in}, T_c^{in}, t\right)$ are the hot- and cold-fluid mass flow rates, inlet and outlet temperatures, and time. For all the quantities, the experimental measurements were collected in the test facility described in Section 4.2, under the following operating conditions: constant values of $\dot{m}_h = 0.25$ kg/s, a set of four step-values for $\dot{m}_c = \{0.05, 0.1, 0.15, 0.215, 0.29\}$ kg/s, a nearly constant $T_c^{in} = 28$°C, and a linear variation of T_h^{in} from 75°C to 65°C, all in the range $t \in [15,50]$ min.

In building the model 70% of the data were initially used for training purposes, whereas the other 30% were taken aside for testing. The final model, however, was built with 100% of the data available, allowing the best possible model over the widest parameter range [18, 20]. The identification results for T_c^{out}, are illustrated in Figs. 20 and 21, where the fuzzy sets and their membership functions for $t, \dot{m}_h, \dot{m}_c, T_h^{in}$, and T_c^{in} are: $\mu_t = \{\mu_{VL}, \mu_L, \mu_M, \mu_H, \mu_{VH}\}$, $\mu_{\dot{m}_h} = \{\mu_L, \mu_H\}$, $\mu_{\dot{m}_c} = \{\mu_{VL}, \mu_L, \mu_M, \mu_H, \mu_{VH}\}$, $\mu_{T_h^{in}} = \{\mu_{VL}, \mu_L, \mu_M, \mu_H, \mu_{VH}\}$ and $\mu_{T_c^{in}} = \{\mu_{VL}, \mu_L, \mu_M, \mu_H, \mu_{VH}\}$.The linguistic labels in the fuzzy sets are: VL for "very low," L for "low," M for "mean," H for "high," and VH for "very high."

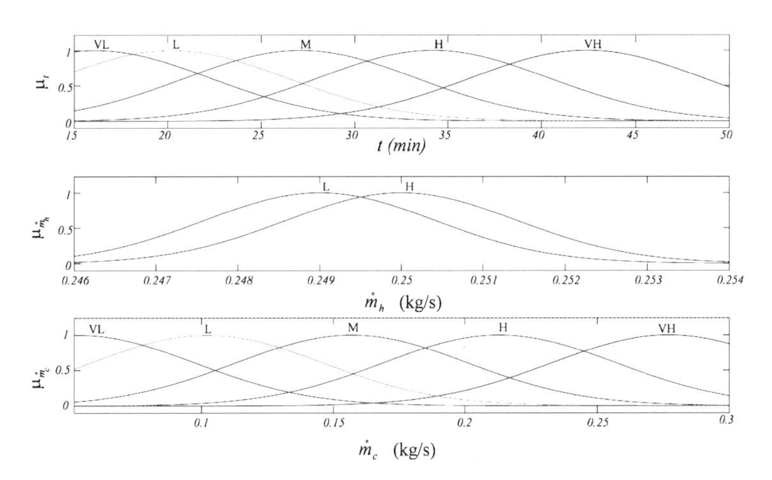

Fig. 20 Fuzzy sets for $t, \dot{m}_h$, and $\dot{m}_c$.

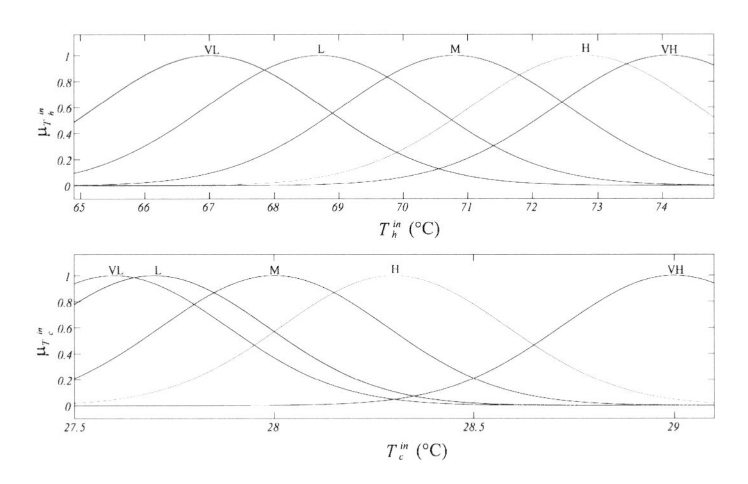

Fig. 21 Fuzzy sets for T_h^{in}, and T_c^{in}.

With these fuzzy sets, five rules were built by the algorithm. The antecedents of these rules are provided in Table 6,

Table 6 Antecedents of fuzzy rules for $T_c^{out} = f_1(\dot{m}_h, \dot{m}_c, T_h^{in}, T_c^{in}, t)$.

t	$\dot{m}_h$	$\dot{m}_c$	T_h^{in}	T_c^{in}	T_c^{out}
VH	H	VH	VL	L	y_1
M	H	M	M	M	y_2
H	L	H	L	VL	y_3
L	H	L	H	H	y_4
VL	H	VL	VH	VH	y_5

whereas the explicit representation of the consequents conforming the output for T_c^{out}, are given as

$$\begin{aligned}
y_1 &= T_{c,1}^{out} = -0.05t - 1.8\dot{m}_h - 15.7\dot{m}_c + 0.05T_h^{in} + 0.35T_c^{in} + 29.76,\\
y_2 &= T_{c,2}^{out} = -0.03t - 31.3\dot{m}_h - 29.3\dot{m}_c - 0.07T_h^{in} - 0.6T_c^{in} + 74.14,\\
y_3 &= T_{c,3}^{out} = -0.003t - 3.8\dot{m}_h - 23.8\dot{m}_c + 0.16T_h^{in} + 0.58T_c^{in} + 16.2,\\
y_4 &= T_{c,4}^{out} = -0.002t - 1.2\dot{m}_h - 16.6\dot{m}_c - 0.03T_h^{in} + 0.58T_c^{in} + 32.55,\\
y_5 &= T_{c,5}^{out} = 0.19t - 87.6\dot{m}_h - 71.2\dot{m}_c + 0.36T_h^{in} + 7.32T_c^{in} - 167.02,
\end{aligned} \tag{19}$$

From these local values, the final expression for the outlet temperature of the cold water (not provided here since it is extremely lengthy) was obtained. The procedure to build the TS fuzzy model for T_h^{out}, is similar to the one illustrated here.

The accuracy of the model to predict both the cold- and hot-water outlet temperatures was assessed based on the input conditions shown in Figs. 22(a) and 22(b). For the test, $\dot{m}_h = 0.25$ kg/s and $T_c^{in} = 28$°C whereas T_h^{in} changed from 72°C to 68°C in a linear fashion. The mass flow rate followed a step function from $\dot{m}_c = 0.15$ kg/s to 0.22 kg/s.

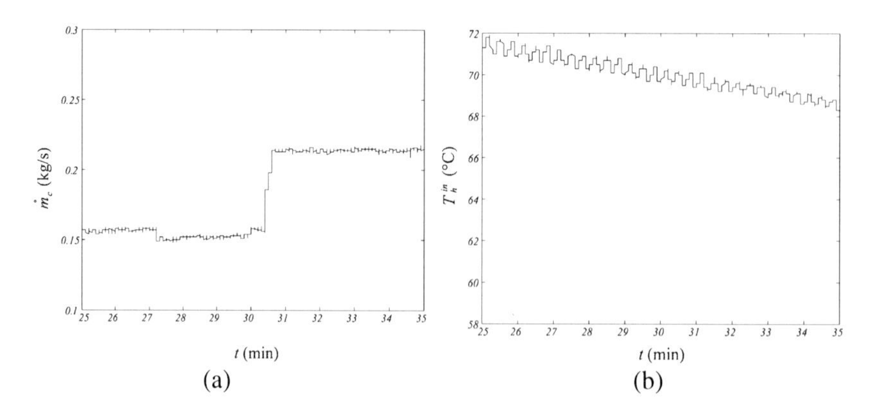

Fig. 22 Time history of input variables (a) $\dot{m}_c$ and (b) T_h^{in}.

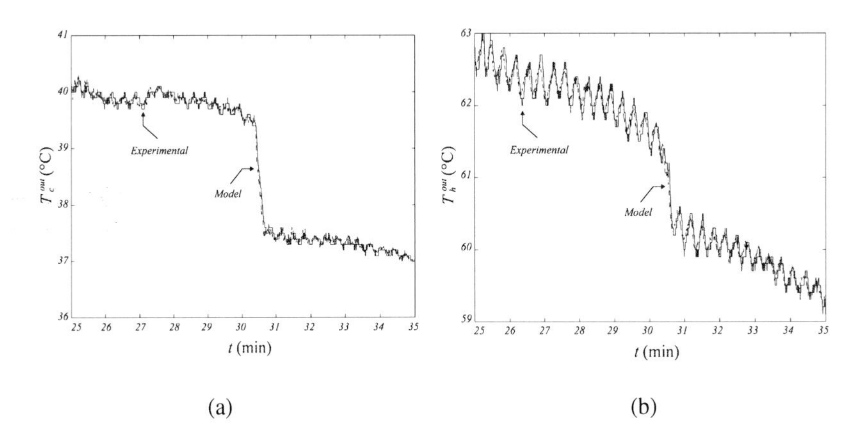

Fig. 23 Prediction of (a) T_c^{out} and (b) T_h^{out}. (—) Measurements; (— · —) TS fuzzy model.

The results from the model and the experimental measurements are shown for both T_c^{out} and T_h^{out}, respectively, in Figs. 23(a) and 23(b). In both cases the model is seen to accurately estimate the dynamic behavior of the system. In the case of T_c^{out} it can be seen that, as $\dot{m}_c$ suddenly increases, the response from both the heat exchanger and the fuzzy model is a decrease in T_c^{out}. A similar trend is obtained for the case of T_h^{out}, indicating that the ANFIS-based model perfectly resembles the characteristics of the physical system.

6.3 *Other Applications in Energy Systems*

Many other applications of hybrid methodologies, particularly the neuro-fuzzy system, to energy systems include forecasting natural gas [103] or building energy consumption [104], the modeling of photovoltaic power systems [105], control of humidity and temperature in air conditioning systems [106], performance prediction of fuel cells [107], or the analysis of component degradation in nuclear power plants [108].

7 Concluding Remarks

This chapter has reviewed several soft computing (SC) techniques used for complex systems. Due to limitations of space, the methodologies have been described here only in outline. The purpose has been to show their usefulness in a range of energy-systems-related applications, for which modeling, prediction and control by other means may be difficult or even impossible to do. It appears that soft computing may be an attractive choice to solve problems in instances where the complexity of the system is the limiting factor, as often happens in real-world applications.

Acknowledgments

The author is grateful to Professors Mihir Sen and K.T. Yang for sharing their knowledge, and for the contributions that his colleagues: G. Diaz, R.L. McClain, K. Peters, W. Cai, and G. Torres-Chavez, and his students: L.E. Vilchiz-Bravo, C. Ruiz-Mercado, and G. Avila, have made to gain a better understanding about the subject. Financial support to the students by the Mexican Council of Science and Technology (CONACyT), and to the different projects by both the Mexican Secretary of Education (SEP-PROMEP) and the National Science Foundation (NSF), is greatly appreciated.

References

[1] Sen, M., Yang, K.T.: Applications of artificial neural networks and genetic algorithms in thermal engineering. In: Kreith, F. (ed.) CRC Handbook of Thermal Engineering, Section 4.24, pp. 620–661. CRC Press, Boca Raton (1999)
[2] Sen, M., Goodwine, B.: Soft computing in control. In: Gad-el-Hak, M. (ed.) The MEMS Handbook, 2nd edn., pp. 16.1–16.35. CRC Press, Boca Raton (2006)
[3] Jain, A.K., Murty, M.N., Flynn, P.J.: Data clustering: A review. ACM Computing Surveys 31(3), 264–323 (1999)
[4] Schalkoff, R.J.: Artificial Neural Networks. McGraw-Hill, Boston (1997)
[5] Haykin, S.: Neural Networks, A Comprehensive Foundation. Prentice-Hall, Upper Saddle River (1999)
[6] Goldberg, D.E.: Genetic Algorithms in Search, Optimization and Machine Learning. Addison-Wesley, Reading (1989)
[7] Koza, J.R.: Genetic Programming Paradigm, On the Programming of Computers by Means of Natural Selection. MIT-Press, Cambridge (1992)
[8] Chen, G., Pham, T.T.: Introduction to Fuzzy Sets, Fuzzy Logic and Fuzzy Control Systems. CRC Press, New York (2000)
[9] Everitt, B.S., Landau, S., Morven, L.: Cluster Analysis, 4th edn. Arnold, New York (2001)
[10] Jang, J.S.R., Sun, C.T., Mizutani, E.: Neuro-Fuzzy and Soft Computing: A Computational Approach to Learning and Machine Intelligence. Prentice-Hall, Englewood Cliffs (1997)
[11] Tettamanzi, A.: Soft Computing: Integrating Evolutionary, Neural, and Fuzzy Systems. Springer, Berlin (1997)
[12] Karray, F.O., De Silva, C.W.: Soft Computing and Intelligent Systems Design: Theory, Tools and Applications. Addison Wesley, Upper Saddle River (2004)
[13] Zeng, P.: Neural computing in mechanics. Appl. Mech. Rev. 51(2), 173–197 (1998)
[14] Rumelhart, D.E., Hinton, G.E., Williams, R.J.: Learning internal representations by error propagation. In: Parallel Distributed Processing: Explorations in the Miscrostructure of Cognition, pp. 8.318–8.362. MIT Press, Cambridge
[15] McQuiston, F.C.: Heat, mass and momentum transfer in a parallel plate dehumidifying exchanger. ASHRAE Transactions 82(2), 87–106 (1976)
[16] McQuiston, F.C.: Heat, mass and momentum transfer data for five plate-fin-tube heat transfer surfaces. ASHRAE Transactions 84(1), 266–293 (1978)

[17] McQuiston, F.C.: Correlation of heat, mass and momentum transport coefficients for plate-fin-tube heat transfer surfaces with staggered tubes. ASHRAE Transactions 84(1), 294–309 (1978)
[18] Pacheco-Vega, A., Diaz, G., Sen, M., Yang, K.T., McClain, R.L.: Heat rate predictions in humid air-water heat exchangers using correlations and neural networks. ASME J. Heat Transfer 123(2), 348–354 (2001)
[19] Gray, D.L., Webb, R.L.: Heat transfer and friction correlations for plate finned-tube heat exchangers having plain fins. In: Tien, C.L., Carey, V.P., Ferrel, J.K. (eds.) Proceedings of the Eighth International Heat Transfer Conference, New York, NY, vol. 6, pp. 2745–2750 (1986), Hemisphere
[20] Pacheco-Vega, A., Sen, M., Yang, K.T., McClain, R.L.: Neural network analysis of fin-tube refrigerating heat exchanger with limited experimental data. Int. J. Heat Mass Transfer 44(4), 763–770 (2001)
[21] Stone, M.: Cross-validatory choice and assessment of statistical predictions. J. R. Stat. Soc. B36, 111–133 (1974)
[22] Yang, K.T.: Artificial neural networks (ANNs): A new paradigm for thermal science and engineering. ASME J. Heat Transfer 130(093001), 1–19 (2008)
[23] Mellit, A., Kalogirou, S.A.: Applications of artificial neural networks in energy systems: A review. Energy Convers. and Manage. 40, 1073–1087 (1999)
[24] Kalogirou, S.A.: Artificial neural networks in renewable energy systems: a review. Renewable and Sustainable Energy Reviews 5, 373–401 (2001)
[25] Sen, M., Yang, K.T.: A review of multiphase flow and heat transfer with artificial neural networks. In: Proceedings of the 2003 ASME International Mechanical Engineering Congress and Exposition, IMECE2003-41761, Washington, DC, (November 2003)
[26] Ghajar, A.J., Tam, L.M., Tam, S.C.: Improved heat transfer correlation in the transition region for a circular tube with three inlet configurations using artificial neural networks. Heat Transfer Engineering 25(2), 30–40 (2004)
[27] Hosoz, M., Ertunc, H.M., Bulgurcu, H.: Performance prediction of a cooling tower using artificial neural network. Energy Convers. and Manage. 48(4), 1349–1359 (2007)
[28] Najafi, G., Ghobadian, B., Tavakoli, T., Buttsworth, D.R., Yusaf, T.F., Faizollahnejad, M.: Performance and exhaust emissions of a gasoline engine with ethanol blended gasoline fuels using artificial neural network. Applied Energy 86(5), 630–639 (2009)
[29] Ridluan, A., Manic, M., Tokuhiro, A.: EBaLM-THP–A neural network thermohydraulic prediction model of advanced nuclear system components. Nuclear Engineering and Design 239(2), 308–319 (2009)
[30] Sozen, A., Arcaklioglu, E., Menlik, T.: Derivation of empirical equations for thermodynamic properties of a ozone safe refrigerant (R404a) using artificial neural network. Expert Systems With Applications 37(2), 1158–1168 (2010)
[31] Holland, J.H.: Adaptation in Natural and Artificial Systems. University of Michigan Press, Ann Arbor (1975)
[32] Michalewicz, Z.: Genetic Algorithms + Data Structures = Evolution Programs. Springer, Heidelberg (1992)
[33] Sette, S., Boullart, L.: Genetic programming: principles and applications. Engineering Applications of Artificial Intelligence 14(1), 727–736 (2001)
[34] Cai, W., Pacheco-Vega, A., Sen, M., Yang, K.T.: Heat transfer correlations by symbolic regression. Int. J. Heat Mass Transfer 49(23–24), 4352–4359 (2006)

[35] Pacheco-Vega, A., Sen, M., Yang, K.T.: Simultaneous determination of in- and over-tube heat transfer correlations in heat exchangers by global regression. Int. J. Heat Mass Transfer 46(6), 1029–1040 (2003)

[36] Pacheco-Vega, A., Sen, M., Yang, K.T., McClain, R.L.: Genetic-algorithm-based predictions of fin-tube heat exchanger performance. In: Lee, J.S. (ed.) Proceedings of the Eleventh International Heat Transfer Conference, vol. 6, pp. 137–142. Taylor & Francis, New York (1998)

[37] Gosselin, L., Tye-Gingras, M., Mathieu-Potvin, F.: Review of utilization of genetic algorithms in heat transfer problems. Int. J. Heat Mass Transfer 52(9-10), 2169–2188 (2009)

[38] Osman, M.S., Abo-Sinna, M.A., Mousa, A.A.: A solution to the optimal power flow using genetic algorithm. Applied Mathematics and Computation 155(2), 391–405 (2004)

[39] Lu, L., Cai, W., Xie, L., Li, S., Soh, Y.C.: HVAC system optimization–in-building section. Energy and Buildings 37(1), 11–22 (2005)

[40] Ooka, R., Komamura, K.: Optimal design method for building energy systems using genetic algorithms. Building and Environment 44(7), 1538–1544 (2009)

[41] Bourouni, K., MBarek, T.B., Taee, A.A.: Design and optimization of desalination reverse osmosis plants driven by renewable energies using genetic algorithms. Renewable Energy 36(3), 936–950 (2011)

[42] Jahedi, G., Ardehali, M.M.: Genetic algorithm-based fuzzy-pid control methodologies for enhancement of energy efficiency of a dynamic energy system. Energy Convers. and Manage. 52(1), 725–732 (2011)

[43] Ravagnani, M.A.S.S., Silva, A.P., Arroyo, P.A., Constantino, A.A.: Heat exchanger network synthesis and optimisation using genetic algorithm. Applied Thermal Engineering 25(7), 1003–1017 (2005)

[44] Varun, Siddhartha: Thermal performance optimization of a flat plate solar air heater using genetic algorithm. Applied Energy 87(5), 1793–1799 (2010)

[45] Dufo-Lopez, R., Bernal-Agustin, J.L.: Design and control strategies of pv-diesel systems using genetic algorithms original. Solar Energy 79(1), 33–46 (2005)

[46] McKay, B., Willis, M., Barton, G.: Steady-state modelling of chemical process systems using genetic programming. Computers Chem. Engng. 21(9), 981–996 (1997)

[47] Lee, D.-G., Kim, H.-G., Baek, W.-P., Chang, S.H.: Critical heat flux prediction using genetic programming for water flow in vertical round tubes. Int. Comm. Heat Mass Transfer 24(7), 919–929 (1997)

[48] Zdaniuk, G.J., Luck, R., Chamra, L.M.: Linear correlation of heat transfer and friction in helically-finned tubes using five simple groups of parameters. Int. J. Heat Mass Transfer 11(13–14), 3548–3555 (2008)

[49] Chakraborty, U.K.: Static and dynamic modeling of solid oxide fuel cell using genetic programming. Energy 34(6), 740–751 (2009)

[50] Lee, Y.S., Tong, L.I.: Forecasting energy consumption using a grey model improved by incorporating genetic programming. Energy Convers. and Manage. 52(1), 147–152 (2011)

[51] Zadeh, L.A.: Fuzzy sets. Information & Control 8, 338–353 (1965)

[52] Zadeh, L.A.: Fuzzy algorithms. Information & Control 12, 94–102 (1968)

[53] Zadeh, L.A.: Fuzzy logic and approximate reasoning. Synthese 30, 407–428 (1975)

[54] Isermann, R.: On fuzzy logic applications for automatic control, supervision, and fault diagnosis. IEEE Transactions on Systems, Man and Cybernetics: Part A-Systems and Humans 28(2), 221–235 (1998)

[55] Dote, Y., Ovaska, S.J.: Industrial applications of soft computing: A review. Proceedings of the IEEE 89(9), 1243–1265 (2001)

[56] Mordeson, J.N., Nair, P.S.: Fuzzy Mathematics: An Introduction for Engineers and Scientists. Physica-Verlag, New York (1998)

[57] Klir, G.J., Yuan, B.: Fuzzy Sets and Fuzzy Logic: Theory and Applications. Prentice-Hall, Englewood Cliffs (1995)

[58] Pacheco-Vega, A., Ruiz-Mercado, C., Peters, K., Vilchiz-Bravo, L.: On-line fuzzy-logic-based temperature control of a concentric-tube heat exchanger facility. Heat Transfer Engineering 30(14), 1208–1215 (2009)

[59] Mamdani, E.H.: Application of fuzzy algorithms for control of simple dynamic plant. Proceedings of the IEEE 121(12), 1585–1588 (1974)

[60] Ruiz Mercado, C.: Control of a Concentric-Tubes Heat Exchanger with Fuzzy Logic (in Spanish). MS Thesis, Universidad Autonoma de San Luis Potosi, San Luis Potosi, Mexico (2005)

[61] Shinskey, F.G.: Process Control Systems: Application, Design, and Tuning. McGraw-Hill, New York (1996)

[62] Caputo, A.C., Pelagagge, P.M.: Fuzzy control of heat recovery systems from solid bed cooling. Applied Thermal Engineering 20, 49–67 (2000)

[63] Shahnawaz-Ahmed, S., Shah-Majid, M., Novia, H., Abd-Rahman, H.: Fuzzy logic based energy saving technique for a central air conditioning system. Energy 32(7), 1222–1234 (2007)

[64] Altas, I.H., Sharaf, A.M.: A novel maximum power fuzzy logic controller for photovoltaic solar energy systems. Renewable Energy 33(3), 388–399 (2008)

[65] Lau, H.C.W., Cheng, E.N.M., Lee, C.K.M., Ho, G.T.S.: A fuzzy logic approach to forecast energy consumption change in a manufacturing system. Expert Systems with Applications 34(3), 1813–1824 (2008)

[66] Xie, H., Mahajan, R.L., Lee, Y.-C.: Fuzzy logic models for thermally based microelectronic manufacturing processes. IEEE Transactions on Semiconductor Manufacturing 8(3), 219–226 (1995)

[67] Gao, D., Jin, Z., Lu, Q.: Energy management strategy based on fuzzy logic for a fuel cell hybrid bus. Journal of Power Sources 186(1), 311–317 (2008)

[68] Courtecuisse, V., Sprooten, J., Robyns, B., Petit, M., Francois, B., Deuse, J.: A methodology to design a fuzzy logic based supervision of hybrid renewable energy systems. Mathematics and Computers in Simulation 81(2), 208–224 (2008)

[69] Li, Y.F., Li, Y.P., Huang, G.H., Chen, X.: Energy and environmental systems planning under uncertainty-An inexact fuzzy-stochastic programming approach. Applied Energy 87(10), 3189–3211 (2010)

[70] Abonyi, J., Feil, B.: Cluster Analysis for Data Mining and System Identification. Birkhauser Verlag AG, Berlin (2007)

[71] Baraldi, A., Blonda, P.: A survey of fuzzy clustering algorithms for pattern recognition-Part I. IEEE Trans. Sys., Man, Cyber.-Part B: Cybernetics 9(6), 778–785 (1999)

[72] Baraldi, A., Blonda, P.: A survey of fuzzy clustering algorithms for pattern recognition-Part II. IEEE Trans. Sys., Man, Cyber.-Part B: Cybernetics 9(6), 786–801 (1999)

[73] Dunn, J.C.: A fuzzy relative of the ISODATA process and its use in detecting compact well-separated clusters. Cybernetics and Systems: An International Journal 3(3), 32–57 (1973)

[74] Duda, R., Hart, P.: Pattern Classification and Scene Analysis. Wiley Interscience, New York (1973)

[75] Hoppner, F., Klawonn, F., Kruse, R., Runkler, T.: Fuzzy Cluster Analysis: Methods for Classification, Data Analysis and Image Recognition. Wiley and Sons, Baffins Lane (1999)

[76] Bezdek, J.C., Keller, J.M., Krishnapuram, R., Pal, N.R.: Fuzzy Models and Algorithms for Pattern Recognition and Image Processing. Springer, New York (2005)

[77] Bezdek, J.C.: Pattern Recognition with Fuzzy Objetive Function Algorithms. Plenum Press, New York (1981)

[78] Bouguessa, M., Wang, S.R., Sun, H.J.: An objective approach to cluster validation. Pattern Recognition Letters 27(13), 1419–1430 (2006)

[79] Webb, A.: Statistical Pattern Recognition. John Wiley & Sons, LTD, Chichester (2002)

[80] Richards, J.A., Jia, X.: Remote Sensing Digital Image Analysis: An Introduction. Springer, Berlin (1999)

[81] Mitchell, T.: Machine Learning. McGraw-Hill, New York (1997)

[82] Jain, A.K., Dubes, R.C.: Algorithms for Clustering Data. Prentice Hall, Englewood Cliffs (1988)

[83] Dempster, A.P., Laird, N.M., Rubin, D.B.: Maximum likelihood from incomplete data via EM algorithm. J. Royal Statist. Soc. B 39(1), 1–38 (1977)

[84] Rissanen, J.: A universal prior for integers and estimation by minimum description length. Annals of Statistics 11(2), 417–431 (1983)

[85] Fonseca, J.R.S., Cardoso, M.G.M.S.: Mixture-model cluster analysis using information theoretical criteria. Intelligent Data Analysis 11(2), 155–173 (2007)

[86] Chen, S., Bouman, C.A., Lowe, M.J.: Clustered components analysis for functional MRI. IEEE Transactions on Medical Imaging 23(1), 85–98 (2004)

[87] Wagner, W., Pruß, A.: The IAPWS formulation 1995 for the thermodynamic properties of ordinary water substance for general and scientific use. J. Physical and Chemical Reference Data 31(2), 387–535 (2002)

[88] Avila, G., Pacheco-Vega, A.: Fuzzy C-means-based classification of thermodynamic property data: A critical assessment. Numerical Heat Transfer, Part A 56(11), 880–896 (2009)

[89] Han, J., Kamber, M.: Data Mining: Concepts and Techniques. Morgan Kaufmann Publishers, San Francisco (2006)

[90] Kim, S.Y., Lee, J.W., Bae, J.S.: Effect of data normalization on fuzzy clustering of DNA microarray data. BMC Bioinformatics 7(Article number 134) (2006)

[91] Aruga, R., Mirti, P., Zelano, V.: Influence of transformation and scaling of archaeometric data on clustering and visual-display. Analusis 18(10), 597–598 (1990)

[92] Pacheco-Vega, A., Avila, G.: Classification of condensing heat exchangers performance data by Gaussian mixtures. In: Proceedings of the ASME 2009 Heat Transfer Summer Conference, San Francisco, CA (July 2009), HT2009-88627

[93] Vernet, A., Kopp, G.A.: Classification of turbulent flow patterns with fuzzy clustering. Engineering Applications of Artificial Intelligence 15(3-4), 315–326 (2002)

[94] Gomez-Muñoz, V.M., Porta-Gandara, M.A.: Local wind patterns for modeling renewable energy systems by means of cluster analysis techniques. Renewable Energy 25(2), 171–182 (2002)

[95] Di Piazza, A., Di Piazza, M.C., Ragusa, A., Vitale, G.: Environmental data processing by clustering methods for energy forecast and planning. Renewable Energy 36(3), 1063–1074 (2011)

[96] Santamouris, M., Mihalakakou, G., Patargias, P., Gaitani, N., Sfakianaki, K., Papaglastra, M., Pavlou, C., Doukas, P., Primikiri, E., Geros, V., Assimakopoulos, M.N., Mitoula, R., Zerefos, S.: Using intelligent clustering techniques to classify the energy performance of school buildings. Energy and Buildings 39(1), 45–51 (2007)

[97] Paasche, H., Tronicke, J.: Cooperative inversion of 2D geophysical data sets: A zonal approach based on fuzzy c-means cluster analysis. Geophysics 72(3), A35–A39 (2007)

[98] Jang, J.S.R.: ANFIS: Adaptive network based fuzzy inference system. IEEE Transactions on Systems, Man, and Cybernetics 23(3), 665–685 (1993)

[99] Takagi, T., Sugeno, M.: Fuzzy identification of systems and its applications to modeling and control. IEEE Transactions on Systems, Man and Cybernetics 15, 116–132 (1985)

[100] Lin, C.T., Lee, G.S.G.: Neural Fuzzy System: A Neuro-Fuzzy Synergism to Intelligent Systems. Prentice Hall, Upper Saddle River (1996)

[101] Chiu, S.L.: Fuzzy model identification based on cluster estimation. Journal of Intelligent & Fuzzy Systems 2(3), 267–278 (1994)

[102] Ruiz-Mercado, C., Pacheco-Vega, A., Torres-Chavez, G.: A Takagi-Sugeno fuzzy dynamic model of a concentric-tubes heat exchanger. Chemical Product and Process Modeling 4(2), 1–22 (2009)

[103] Kaynar, O., Yilmaz, I., Demirkoparan, F.: Forecasting of natural gas consumption with neural network and neuro fuzzy system. Energy Education Science and Technology Part A–Energy Science and Research 26(2), 221–238 (2011)

[104] Li, K., Su, H.: Forecasting building energy consumption with hybrid genetic algorithm hierarchical adaptive network-based fuzzy inference system. Energy and Buildings 42(11), 2070–2076 (2010)

[105] Mellit, A., Kalogirou, S.A.: Anfis-based modelling for photovoltaic power supply system: A case study. Renewable Energy 36, 250–258 (2011)

[106] Soyguder, S., Alli, H.: An expert system for the humidity and temperature control in HVAC systems using ANFIS and optimization with fuzzy modeling approach. Energy and Buildings 41(8), 814–822 (2009)

[107] Viral, Y., Ingham, D.B., Pourkashanian, M.: Performance prediction of a proton exchange membrane fuel cell using the ANFIS model. International Journal of Hydrogen Energy 34(22), 9181–9187 (2009)

[108] Ferreira-Guimaraes, A.C., Cunha-Cabral, D., Franklin-Lapa, C.M.: Adaptive fuzzy system for degradation study in nuclear power plants' passive components. Progress in Nuclear Energy 48(7), 655–663 (2006)

Use of Soft Computing Techniques in Renewable Energy Hydrogen Hybrid Systems

Gabriele Zini, Simone Pedrazzi, and Paolo Tartarini

University of Modena e Reggio Emilia,
Via Vignolese 905, 41125 Modena
{gabriele.zini,simone.pedrazzi,paolo.tartarini}@unimore.it

Abstract. Soft computing techniques are important tools that significantly improve the performance of energy systems. This chapter reviews their many contributions to renewable energy hydrogen hybrid systems, namely those systems that consist of different technologies (photovoltaic and wind, electrolyzers, fuel cells, hydrogen storage, piping, thermal and electrical/electronic control systems) capable as a whole of converting solar energy, storing it as chemical energy (in the form of hydrogen) and turning it back into electrical and thermal energy.

Fuzzy logic decision-making methodologies can be applied to select amongst renewable energy alternative or to vary a dump load for regulating wind turbine speed or find the maximum power point available from arrays of photovoltaic modules. Dynamic fuzzy logic controllers can furthermore be utilized to coordinate the flow of hydrogen to fuel cells or employed for frequency control in micro-grid power systems.

Neural networks are implemented to model, design and control renewable energy systems and to estimate climatic data such as solar irradiance and wind speeds. They have been demonstrated to predict with good accuracy system power usage and status at any point of time. Neural controls can also help in the minimization of energy production costs by optimal scheduling of power units.

Genetic or evolutionary algorithms are able to provide approximate solutions to several complex tasks with high number of variables and non-linearities, like optimal operational strategy of a grid-parallel fuel cell power plant, optimization of control strategies for stand-alone renewable systems and sizing of photovoltaic systems.

K. Gopalakrishnan et al. (Eds.): Soft Comput. in Green & Renew. Ener. Sys., STUDFUZZ 269, pp. 37–64.
springerlink.com

Particle swarm optimization techniques are applied to find optimal sizing of system components in an effort to minimize costs or coping with system failures to improve service quality.

These techniques can also be implemented together to exploit their potential synergies while, at the same time, coping with their possible limitations.

This chapter covers soft computing methods applied to renewable energy hybrid hydrogen systems by providing a description of their single or mixed implementation and relevance, together with a discussion of advantages and/or disadvantages in their applications.

1 Introduction

A *renewable energy hydrogen hybrid system* (REHHS) is a collection of systems of different technologies (energy conversion, electrolyzers, fuel cells, hydrogen storage, piping, thermal and electrical/electronic systems) capable, as a whole, of converting energy from a renewable energy source into electrical energy, storing it as chemical energy (in the form of hydrogen) and turning it back into electrical and/or thermal energy when needed [Romm 2004].

The importance of renewable energy systems is manifold and due to growing costs of fossil fuels and their environmental impact, together with the need to exploit renewable energy sources and securing energetic independence for every nation [Penner 2006].

The main drawback is the unevenness and dependability of many forms of renewable energies which can, in principle, hamper their widespread adoption.

Notwithstanding, hydrogen energy storage can provide a solution to the problem. Hydrogen can indeed represent a potential viable energy vector for stationary and non-stationary applications, posing a viable alternative to the dominance of oil and other carbon-based energy vectors.

Many publications address research in this area; see for instance Hammache and Bilgen [1987], Hollenberg et al [1995], Ulleberg and Morner [1997], Kolhe et al [2003], Maclay et al [2007]. A review of solar and wind hydrogen system models can be found in Deshmukh and Boehm [2008].

2 Modelling

2.1 General System Description

A general form of a REHHS is outlined in Fig. 1.

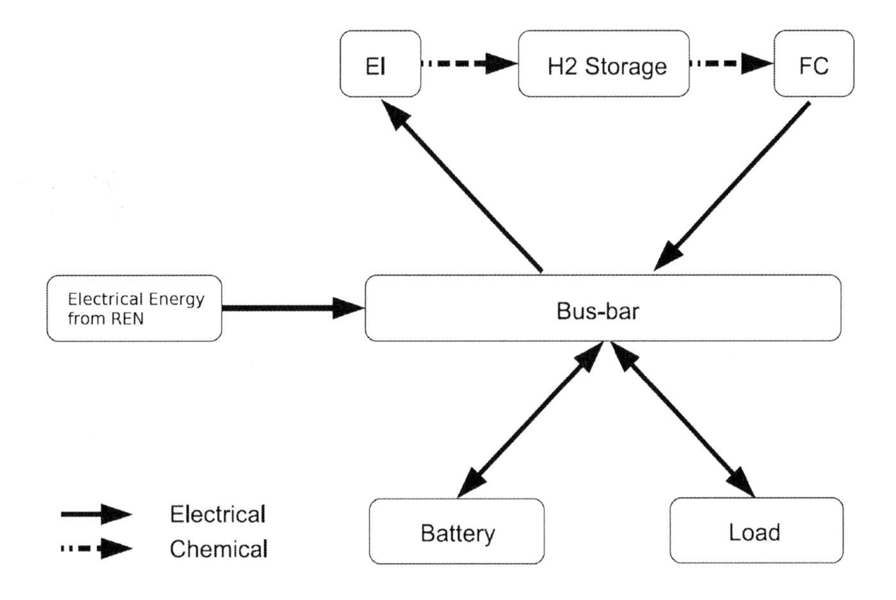

Fig. 1 System schematic (adapted from Zini and Tartarini [2010]).

The energy from a renewable energy source (REN) is converted to electricity by means of different conversion mechanisms that are characteristic of the energy form itself. For instance, if solar photons are to be directly converted into electricity, photovoltaic cells are used for the task. If mechanical energy from wind is the chosen form of energy, aerodynamical blades (rotors) are employed to convert the wind stream into mechanical energy (kinetic energy in the turbine rotation) and finally by asynchronous generators into electrical energy. Solar thermal energy is converted by Rankine cycles into electricity. Other renewable energy sources can also be successfully used (i.e. energy from waves, tides, water, biomasses, biofuels, ...).

In any case, the electrical energy originated from the renewable energy source is managed by means of electrical conditioning, protection and control circuits of well consolidated technology: the output from the conversion system is connected to a DC bus-bar, the power electrical distribution which acts as the backbone of the hybrid system. The electrolyzer receives current from the bus-bar and produces hydrogen and oxygen that are compressed and stored. When the control logic switches on the fuel cell depending upon load requests, electrical energy is converted from stored gases and enters the bus-bar. Energy is then supplied to the load by a DC/AC inverter connected to the bus-bar. Finally, a battery keeps the bus-bar always charged within a controlled power range to guarantee smooth functioning and adequate quality of service of the overall hybrid system.

In the next paragraph, a description of the main sub-system modelling will be highlighted in order to facilitate the understanding of the issues related to the control of these highly integrated energy systems.

2.2 *Renewable Energy Modelling*

2.2.1 Solar Photovoltaic

The photovoltaic system modules capture portions of the solar radiation reaching the earth's surface: the direct (G_b) and the diffuse (G_d) [Liu and Jordan 1963, Garg 1982, Duffie and Beckman 2006]. The total solar radiation is indeed expressed by:

$$G_T = G_b R_b + G_d R_d + (G_b + G_d) R_r \tag{1}$$

where R_b is the tilt factor for direct radiation, R_d for diffuse radiation and R_r for reflected radiation. They are given by the following equations:

$$R_b = \frac{\sin\delta\sin(\varphi-\beta)+\cos\delta\cos\omega\cos(\varphi-\beta)}{\sin\varphi\sin\delta-\cos\varphi\cos\delta\cos\omega},$$

$$R_d = \frac{1+\cos\beta}{2}, \tag{2}$$

$$R_r = \rho\frac{1-\cos\beta}{2}$$

where β is the tilt angle of the photovoltaic modules, φ is the location latitude, $\omega =$ (12-t) π/12 is the hour angle in radians as a function of time t (in hours), ρ is the ground reflectivity (albedo), and δ is the declination angle expressed by:

$$\delta = 23.45°\sin\left[\frac{360°(284+n)}{365}\right] \tag{3}$$

with n the Julian day of the year.

G_b and G_d represent respectively the direct and the diffuse solar radiation on a flat surface at an hour angle ω and are expressed as a function of the daily average total direct solar radiation energy for a horizontal surface (H_{b0}) and the daily average value of the diffuse solar radiation energy (H_{d0}), all expressed in Wh/m^2:

$$G_b = \frac{180°}{24}\frac{\sin\delta\sin\varphi+\cos\delta\cos\omega\cos\varphi}{\omega_s\sin\varphi\sin\delta-\cos\varphi\cos\delta\cos\omega_s}H_{b0} \tag{4}$$

$$G_d = \frac{180°}{24}\frac{\cos\omega-\cos\omega_s}{\sin\omega_s-\omega_s\cos\omega_s}H_{d0} \tag{5}$$

where $\omega_s = cos^{-1}(-tan\varphi\ tan\delta)$ is the sunrise hour angle.

The site data (β, φ, ω, δ, H_{b0}, H_{d0}) that are needed to compute G_T are available in specific databases.

The photovoltaic sub-system converts solar radiation energy into electrical energy; to reach a satisfactory trade-off between model complexity and precision, its behavior is profiled with the single-diode model [Duffie and Beckman 2006].

From Kirchhoff's laws, the *I-V* relationship of the equivalent circuit can be written as:

$$I = I_L - I_0\left(\exp\frac{V + IR_s}{a} - 1\right) - \frac{V + IR_s}{R_{sh}} \tag{6}$$

where I_L is the photo current generated when the diode is radiated by solar energy, I_0 is the diode reverse saturation current, R_s is the series resistance, R_{sh} is the shunt resistance. The term a is set equal to (NKT/q) where K is the Boltzmann constant, T the temperature, q the electron charge constant and N is a parameter that depends on diode technology.

A set of several different parameters has been measured at standard conditions (irradiation at 1000 W/m^2, temperature at 25°C, air mass at 1.5) for a number of commercially available photovoltaic modules. These parameters are used to solve the model equations and ensure a very good realism in photovoltaic system simulation since the input data directly derive from the assessment of real-world photovoltaic components.

A maximum power point tracker (MPPT) embedded in the system, constantly attempts at finding the tension and current values ($V_{mp} - I_{mp}$) that maximize power conversion constantly while coping with variable irradiation and meteo conditions.

The photovoltaic equation system is:

$$\begin{cases} I_{mp} = I_L - I_0\left(\exp\dfrac{V_{mp} + I_{mp}R_s}{a} - 1\right) - \dfrac{V_{mp} + I_{mp}R_s}{R_{sh}} \\ \left.\dfrac{dP}{dV}\right|_{P=P_{mp}} = 0 = I_{mp} + V_{mp}\dfrac{-\dfrac{I_0}{a}\exp\dfrac{V_{mp} + I_{mp}R_s}{a} - \dfrac{1}{R_{sh}}}{1 + \dfrac{I_0 R_s}{a}\exp\dfrac{V_{mp} + I_{mp}R_s}{a} + \dfrac{R_s}{R_{sh}}} \end{cases} \tag{7}$$

which gives the operating maximum power output values of tension and current used as input to the downstream sub-system.

By taking into proper account the MPPT efficiency η_{mppt} and the boost-converter efficiency η_{bc}, the current to the bus-bar is given by:

$$I_{pv \to bus} = (\eta_{mppt}\eta_{bc}V_{mp}I_{mp})/V_{bus} \tag{8}$$

2.2.2 Wind Energy Modelling

A wind speed profile in an average site can be modelled after the Weibull probability density function [Zini and Tartarini 2010]:

$$h(v) = \left(\frac{k}{c}\right)\left(\frac{v}{c}\right)^{k-1}\exp\left(-\frac{v}{c}\right)^k \tag{9}$$

with form factor k ranging between 1.5 and 2.5 and a scale factor c between 5 and 10 m/s. The form factor defines the shape of the function while the scale factor accounts for the wind speed distribution.

The power contained in a wind stream of speed v passing through a surface A is:

$$P_{wind} = \frac{\rho}{2} A v^3 = \frac{\rho}{2} \frac{D^2 \pi}{4} v^3 \tag{10}$$

where D is the rotor diameter and ρ is the air density, function of temperature, humidity and air pressure.

Many losses occur along the path of the conversion of the kinetic energy in the wind stream to the electrical energy supplied to the load.

To start with aerodynamic losses that reduce the portion of kinetic energy transferred to the rotor: P_{wind} must hence be multiplied by a power coefficient c_p, usually ranging between 0.4 and 0.5 and function of the *tip speed* and wind speed ratio, where tip speed is the ratio between the rotational speed of the tip of the blade and the actual velocity of the wind. c_p must be carefully set up in the design phase since optimal functioning occurs when rotor speed is controlled to maintain c_p at design values with respect to wind speed.

The kinetic energy from the rotation of the turbine is then transferred to a mast, which is connected by means of mechanical gears (introducing a conversion efficiency η_{mech}) to an electric asynchronous generator which converts kinetic energy into AC electrical energy (with efficiency η_{asyn}). Since DC current is needed to supply the load, a further conversion step that converts AC to DC is needed, with a reduction of efficiency due to a coefficient $\eta_{AC\text{-}DC}$.

P_{aero} power provided by the aero-generator to the system is hence given by:

$$P_{aero} = (c_p \eta_{mech} \eta_{asyn} \eta_{AC-DC}) P_{wind} \tag{11}$$

Since wind turbines can undergo damages if they are allowed to rotate over certain wind speeds, accurate control logics must make sure that the rotor is limited to speeds under a pre-determined *cut-out speed*. Additionally, to avoid rotation at ineffective low wind velocities, rotation must be prevented when wind speeds occur below a *cut-in speed*. Between the cut-in and the cut-out speeds, a maximum power tracking control logic is in place to ensure maximum power conversion by regulating the rotation speed in order to maintain a constant c_p.

This range of limitation of useful speeds further reduces the converted power available to satisfy load demands and overall system efficiency.

2.2.3 Other Renewable Energy Models

Different other renewable energy sources can be employed. A description of all diverse sources is out of the scope of this chapter, but as far as they can be converted to electrical energy, the outlined hybrid system maintains its engineering structure and design.

Of course, depending on any particular source and load characteristics, final system results and behaviour will be significantly different.

2.3 Hydrogen Loop

2.3.1 Electrolyzer

The electrolyzer is the sub-system capable of converting electrical energy into chemical energy by separating water into its basic components, oxygen and hydrogen.

The electrolyzer receives P_{el} from the bus-bar through a buck converter with conversion efficiency η_{bc}:

$$P_{el} = V_{el} I_{el} = \eta_{bc} P_{bc} \tag{12}$$

where V_{el} and I_{el} are the electrolyzer voltage and current. Voltage can be written as:

$$V_{el} = V_{el,0} + C_{1,el} T_{el} + C_{2,el} \ln\left(\frac{I_{el}}{I_{el,0}}\right) + \frac{R_{el} I_{el}}{T_{el}} \tag{13}$$

where T_{el} is the electrolyte temperature; $V_{el,0}$, $C_{1,el}$, $C_{2,el}$, $I_{el,0}$, R_{el} are constants that are determined empirically and account for polarization effects and electrolyzer technology. T_{el} can be calculated from [Ulleberg 2003]:

$$\dot{Q}_{gen} = \dot{Q}_{store} + \dot{Q}_{loss} + \dot{Q}_{cool} \tag{14}$$

where $\dot{Q}_{gen}$ is the internal heat generation, $\dot{Q}_{store}$ is the thermal energy storage, $\dot{Q}_{loss}$ is the heat loss to the surrounding ambient, and $\dot{Q}_{cool}$ is the heat transferred to the auxiliary cooling system. T_{el} is obtained by substitution in (13).

Equations (12) and (13) yield the electrolyzer operating point I_{el}-V_{el}, hence the hydrogen production flow rate becomes:

$$\dot{n}_{H_2} = \eta_{F,el} \frac{N_{cells} I_{el}}{zF} \tag{15}$$

where $\eta_{F,el}$ is the Faraday's efficiency of the electrolyzer, N_{cells} is the number of the cells in series, z is the number of moles of electrons for moles of water (z=2) and F is the Faraday's constant.

2.3.2 Hydrogen Storage

Many technologies provide hydrogen storage capabilities, like compression, liquefaction or adsorption in materials like carbon nano-structures or hydrides.

In case a polytropic compression is considered, its power is given by [Deshmukh and Boehm 2008]:

$$P_{comp} = \frac{\dot{n}_{gas} L_{comp}}{\eta_{comp}} = \frac{\dot{n}_{gas}}{\eta_{comp}} \frac{mRT_{in,c}}{m-1} \left[1 - \left(\frac{p_{out}}{p_{in}}\right)^{\frac{m-1}{m}}\right] \tag{16}$$

where $\dot{n}_{gas}$ is gas flow, η_{comp} is the compressor efficiency, m is the polytropic coefficient, R is the universal gas constant, $T_{in,c}$ is the inlet gas compressor temperature, $p_{in,c}$ and $p_{out,c}$ are the inlet and outlet compressor pressure and gases are considered ideal.

From the definition of polytropic compression and from the equation of state of ideal gases:

$$T_{in,s} = T_{in,c}\left(\frac{p_{in,c}}{p_{out,c}}\right)^{\frac{1-m}{m}} \tag{17}$$

where $T_{in,s}$ is the storage tank inlet gas pressure. From the ideal gas equation of state and by integration of hydrogen/oxygen flows:

$$p_s = \left[n_{ini} + \int_0^t (\dot{n}_{in,s} - \dot{n}_{out,s})d\tau\right]\frac{RT_s}{V} \tag{18}$$

where n_{ini} is the number of moles in the initial conditions; $\dot{n}_{in,s}$ and $\dot{n}_{out,s}$ are inlet and outlet gas flows in the tank, T_s is the temperature of the storage tank and V is the storage tank volume. By balancing the thermal flows in the compressor:

$$\dot{Q}_{gen} = \dot{Q}_{store} + \dot{Q}_{loss}\text{, or } C_{in}(T_{in,s} - T_s) = C_s\frac{dT_s}{dt} + \frac{(T_s - T_a)}{R_t} \tag{19}$$

where C_{in} is the thermal capacity of inlet gas, T_a is the ambient temperature, R_t thermal resistance of the tank.

Integrating Eq. 19:

$$T_s = T_{ini} + \int_0^t \left[\frac{C_{in}(T_{in,s} - T_s)}{C_s} + \frac{(T_a - T_s)}{\tau_t}\right]d\tau \tag{20}$$

where T_{ini} is the temperature of the tank in the initial condition, C_{in} is the heat capacity of gas, C_s is the heat capacity of the tank, and $\tau_t = R_t\, C_t$ is the thermal time constant of the tank.

The pressure p_s in the storage tank can finally be obtained by solving the system of equations (17), (18) and (20).

2.3.3 Fuel Cell

The fuel cell converts the chemical energy stored in hydrogen into electrical energy, reversing the conversion operated by the electrolyzer at the beginning of the hydrogen loop.

The boost-converter connects the electrical energy from the fuel cell to the busbar with an efficiency η_{bo}:

$$P_{bo} = \eta_{bo} P_{fc} = V_{fc} I_{fc} \tag{21}$$

where P_{fc}, V_{fc}, and I_{fc} are fuel cell output power, voltage and current. Voltage is given by:

$$V_{fc} = V_{fc,0} + C_{1,fc} T_{fc} + C_{2,fc} \ln\left(\frac{I_{fc}}{I_{fc,0}}\right) + \frac{R_{fc} I_{fc}}{T_{fc}} \tag{22}$$

where $V_{fc,0}$, C_{1fc}, C_{2fc} (V), $I_{fc,0}$ and R_{fc} are constants determined experimentally and accounting for the fuel cell technology adopted and T_{fc} is the fuel cell operating temperature [Ulleberg 1998].

From (21) and (22) derive I_{fc} which substituted in:

$$\dot{n}_{H_2} = \eta_{F,fc} \frac{N_{fc} I_{fc}}{zF} \tag{23}$$

yields the hydrogen rate of fuel cell consumption $\dot{n}_{H_2}$, where η_{Ffc} is the fuel cell Faraday's efficiency, N_{fc} is the number of stack cells, z is the number of moles of electrons for moles of water (z=2), and F is the Faraday's constant.

2.4 Battery

Batteries can be employed to smooth out the potential discontinuities between on/off cycles of the different sub-systems but are not meant to provide principal energy storage, which is performed by the hydrogen storage unit.

Knowing the current, the voltage is given by:

$$U_B(t) = (1 + \alpha t) U_{B,0} + R_i I_B(t) + K_i Q_R(t) \tag{24}$$

where α is the self-discharge rate, $U_{B,0}$ is the open circuit voltage at time 0, R_i is the internal resistance, K_i is the polarization coefficient, and Q_R is the rate of accumulated charge.

The total energy stored in the batteries $E(t)$ is given by:

$$E(t) = E_{in} + \int_0^t U_B(\tau) I_B(\tau) d\tau \tag{25}$$

where E_{in} is the initial charge of the batteries.

The state of charge of the batteries (SOC) is defined using the following equation:

$$SOC = \frac{E(t)}{E_{\max}} \tag{26}$$

where E_{max} is the total battery capacity. This is expressed as a percentage and gives the portion of storage available for use.

2.5 Example of Deterministic Control Logic

A control logic is needed to coordinate the interactions between the various components of the overall system. A deterministic control flow can be based on the bus-bar equation that balances the electrical currents entering (positive sign) or exiting the bus-bar (negative sign) [Pedrazzi et al 2010]:

$$I_{ren\_en \to bus} - I_{bus \to el} - I_{bus \to comp} + I_{fc \to bus} - I_{bus \to load} \pm I_{bat \leftrightarrow bus} = 0 \qquad (27)$$

If the renewable energy conversion system does not convert enough power to supply the load ($I_{ren\_en \to bus} < I_{bus \to load}$), the fuel cell is switched on by providing a positive $I_{fc \leftrightarrow bus}$. In case the battery SOC reaches a minimum (usual values are around 20-25%). If the energy available from the renewable source is higher than the energy needed by the load ($I_{ren\_en \to bus} > I_{bus \to load}$), the fuel cell is disconnected and batteries charged (with a negative $I_{bat \leftrightarrow bus}$, from bus-bar to battery). When the batteries reach a maximum SOC (i.e. 85%), they are switched-off and the electrolyzer and the compressor are activated to yield and store new hydrogen.

Soft computing techniques can be considered to provide more apt and optimized algorithms to improve overall system efficiency. The next chapters will describe several contributions to the subject as per available scientific literature.

3 Fuzzy Logic

Fuzzy Logic (FL) is a soft computing technique based on approximate reasoning typical of human cognition. FL models are empirically-based, relying on experience rather than technical understanding of the system. The knowledge is interpreted as a collection of elastic or, equivalently, fuzzy constraint on a collection of variables [Robert 1995, Machado and Rocha 1992] allowing partial set membership rather than complete membership or non membership [Zadeh 1965, 1972, 1973].

FL incorporates simple “IF X AND Y THEN Z” rules useful to solve control problem rather than attempting to rigorously model a system from a mathematical point of view.

FL requires some numerical parameters in order to operate such as what is considered significant error and significant rate-of-change-of-error, but exact values of these numbers are usually not critical. Originally, FL was conceived as a better method for sorting and handling data, but has proven to be an excellent choice for many control system applications since it is particularly apt at mimicking human control logic. By using descriptive language to deal with input data more like a human operator, FL techniques are very robust and forgiving of operator and data input, and often work out of the box with little or no tuning at all [Yager 1987].

Fuzzy systems are also suitable for uncertain or approximate reasoning, especially for the system with a mathematical model that is difficult to derive. FL allows decision making with estimated values under incomplete or uncertain information [Mellit and Kalogirou 2008]. These techniques have been successfully applied in a number of applications like, computer vision, decision making and system design including some soft computing techniques training schemes, to program controllers for cement kilns, braking systems, elevators, washing machines, hot water heaters, air conditioners, video cameras, rice cookers and photocopiers [Lakhmi and Martin 1998].

FL can be successfully employed to control, design and simulate renewable energy hydrogen hybrid system in order to choose the best renewable source alternatives, to design and/or control sub-systems parameters and overall system coordination.

3.1 Control of FL in REHHS

Bilodeau and Agbossou [2006] used a dynamic fuzzy logic controller (FLC) to control a virtual stand-alone renewable energy system with hydrogen storage very similar to the one depicted in Fig. 1.

The use of fuzzy logic is appropriate since it caters for the use of multiple input variables without increasing design complexity, since the desired behaviour can easily be described in words and the entire range of inputs can be defined by using a minimum set of rules. Also, there is no need for historical data, this being an important advantage over other types of soft computing techniques such as neural networks and genetic algorithms.

The goals of the controller are the reduction of energy transfer from the short-term storage to the long-term storage and vice-versa, and prevention of excessive use of batteries.

The controller has two input variables: the *net power flow* dP between the power provided by the sources and the power consumed by the load, and the battery *state of charge* (*SOC*).

The FLC output variable is the power set point P; the boost converter and the fuel cell are switched on when the output is positive and, conversely, the buck converter and the electrolyzer are turned on when the output is negative.

The following FLC rules are chosen to define the system behaviour:

- IF $dP < 0$ THEN $P > 0$
- IF $dP = 0$ THEN $P = 0$
- IF $dP > 0$ AND $SOC > 0.55$ THEN $P < 0$
- IF $SOC < 0.5$ THEN $P > 0$
- IF $SOC > 0,55$ THEN $P < 0$

The implication operator is MIN and the aggregation operator is MAX; the output is defuzzified using the centroid method.

Conclusions show that the FLC is able to correctly control the system, avoiding that batteries be discharged under the SOC, maintain a good quality of service to the load while obtaining a certain amount of hydrogen storage. Overall system efficiency can be increased by fuzzy logic over conventional deterministic control logics.

3.2 Load Control in a Wind-Hydrogen REHHS

A small hydrogen stand alone power system with a wind turbine FLC has been presented by Khan and Iqbal [2009]. A FLC is applied to control a dump load that regulates the wind turbine speed. From the estimated rotor torque, optimal rotor speed can be calculated for the whole operating range. The reference speed is compared with the actual rotor speed and the FLC adjusts the dump load R_{dump} to achieve optimal variable speed operation.

The fuzzification is performed using five Gaussian membership functions for each of the parameters. The inputs are named with the following linguistic functions depending on the magnitude of deviation from zero: NB (negative big), NM (negative medium), ZR (zero), PM (positive medium) and PB (positive big). Similarly, the output variable is named: VL (very low), LO (low), MD (medium), HI (high) and VH (very high).

Comparing the fuzzified inputs and determining their degree of fulfilment for each of the given rules, an output of each rule is found. These outputs are defuzzified by the centroid method.

Another load control algorithm for a wind-hydrogen REHHS has been developed by Miland et al [2005]. A Distributed Intelligent Load Controller (DILC) based on fuzzy logic algorithm balances the flow of active power in the system and its control system frequency. The system voltage is maintained within the limits specified in international standards by a synchronous compensator.

The maximum load enabled by the DILC is 30 kW, exceeding the turbine maximum power even in the strongest winds. The DILC engages to maintain the system frequency at 50 Hz or within the specified limits. The fuzzy controller has two inputs, frequency and rate of change of frequency, used to compute the control outputs.

The rules for the fuzzy controller were selected using knowledge of stand-alone power system dynamics. A self-tuning feature is embedded in the fuzzy controller, aiming at creating a flexible controller capable of good performance over a wide range of systems and different operating conditions. The tuning is achieved by automatically modifying the input membership functions of the fuzzy controller, as a result of the controller monitoring its own frequency control performance.

3.3 Frequency Control in a Micro-grid Power System

Li et al [2005] submitted a study on the frequency stability of a wind-operated micro-grid hydrogen power system for the control of a micro-turbine with the fuel cell and electrolyzer hybrid system.

The authors apply self-tuning fuzzy proportional-integral (FPI) controllers to deal with real-time frequency fluctuations and sudden real power imbalances. This controller readjusts the PID gains in real-time to improve the process output response during system operations. Therefore, the fuzzy-logic-based self-tuning or self-organizing PI controller may prevail over the deterministic PI controller.

The control system consists of an adaptive PI controller and a fuzzy self-tuning mechanism that adjusts the PI parameters.

3.4 MPPT in a Hybrid PV/Fuel Cell System

El-Shatter et al [2002] applied fuzzy regression modelling (a fuzzy variation of classical regression analysis [El-Shatter et al 1997, Kahraman 2006] to maximum power point tracking (MPPT). The system comprises a PV array, an electrolyzer, an hydrogen storage tank and a PEM fuel cell stack.

The flows of power and hydrogen are set by a controller according to the state of the system. The fuzzy regression modelling input parameters are the solar irradiation and the panel surface temperature, while the outputs of the model are the MPP voltage and current.

3.5 Selection among Renewable Energy Alternatives

Fuzzy axiomatic design (AD) and fuzzy analytic hierarchy process (AHP) have been compared by Kahraman et al [2009] for the best selection among renewable energy alternatives in Turkey.

Biomass, geothermal, hydropower, solar and wind are the energy alternatives analyzed on the basis of technological, environmental, socio-political and economic criteria.

The first technique employed is a modified fuzzy AHP method applied to work out the priority weights of energy alternatives. In a typical AHP method, experts have to give a definite number within a 1 – 9 range to the pair-wise comparison so that the priority vector can be computed.

Often, experts cannot compare two factors due to the lack of adequate information. In this case, a classical AHP method has to be discarded due to the existence of fuzzy or incomplete comparisons, so a fuzzy AHP method designed by Zeng et al [2007] is preferred, where fuzzy aggregation is used to create group decisions and defuzzication is finally engaged to transform the fuzzy scales into crisp scales for the computation of the priority weights.

The group preference of each factor is calculated by applying fuzzy aggregation operators, i.e. fuzzy multiplication and addition operators. The application of this method reveals that wind energy is the most appropriate renewable energy alternative for Turkey in the future and the ranking of energy alternatives is determined as follows:

1. Wind.
2. Solar.
3. Biomass.
4. Geothermal.
5. Hydropower.

Another technique that can be applied is the fuzzy AD method to provide a design framework for engineers. The primary goal of AD is to provide a thinking process to create a new design and/or to improve the existing design.

To improve a design, the axiomatic approach uses two axioms named as *independence axiom* and *information axiom*. The independence axiom states that the independence of functional requirements (FR) must always be maintained, whereas FR are defined as the minimum set of independent requirements that characterize the design goals.

The information axiom states that the design having the smallest information content is the best one among those that satisfy the independence axiom. This facilitates the selection of proper alternatives that have minimum information content. Such axiom is used in order to select the best alternatives when there is more than one design that satisfies independence axiom. The authors extend the information axiom under fuzzy environment and the new methodology is used for decision-making problems under fuzzy environment.

Again, wind energy is selected as the most suitable alternative with respect to predetermined FR.

4 Artificial Neural Networks

Artificial neural networks (ANN) are soft computing techniques that mimic the operations of biological neural systems as a collection of small individually interconnected processing units called *neurons* [Mellit and Kalogirou 2008]. Such algorithms have been applied successfully in several fields of mathematics, engineering, medicine, economics, meteorology, psychology, neurology, and many others [Kalogirou 2001].

An important feature of ANN is the automatic learning about a given problem domain achievable through the training phase. ANN can work with many numeric variables that would be difficult to deal with by other means. It is a black box approach that does not require sophisticated mathematical knowledge by the user.

The method is robust even in the presence of noise in the input data and can present a high degree of accuracy when used to generalize over a set of raw and unstructured data [Mellit and Kalogirou 2008].

Nevertheless, ANN are not suitable when there is scarcity of appropriate data since ANN needs a set of data which should be large enough to be representative of the observed system behaviour [Ghaboussi et al 1991].

A basic neuron model [Yu and Jenq-Neng 2001] consists of two parts: the net function z and the activation function $f(z)$.

The net function (Eq. 28) determines how the *network inputs* x_i are combined inside the neuron, each input is weighted by w_i known as *synaptic weights*, while the *bias* θ is used to define an offset or threshold:

$$z = \sum_{i=1}^{N} w_i x_i + \vartheta \tag{28}$$

The output y of the neuron is related to the network inputs via the sigmoid activation function shown in Eq. 29:

$$y = f(z) = \frac{1}{1 + e^{-z}} \tag{29}$$

Other types of net and activation functions have been proposed [Yu and Jenq-Neng 2001]. Neurons can be interconnected in layers (such as an input layer, an output layer and one or more hidden layers). The configuration of the interconnections can be described efficiently with a graph consisting of nodes (neurons) and arcs (synaptic links).

To be effective, ANN algorithms must undergo a training and testing period. In the training (or learning) phase, weights are adjusted when data pass between artificial neurons along the connections. A learning rule is used to find a set of weights such that the error is minimum [Shahin et al 2001]. Once ANN have been trained, the testing phase is initiated with new patterns presented to ANN for prediction or classification [Yu and Jenq-Neng 2001].

Several architectures and algorithms of ANN have been developed for engineering problem solving. Details regarding the theory and mathematics behind the most widely used ANN are available in Mellit and Kalogirou [2008].

This section presents how ANN can be implemented to estimate climatic data such as solar irradiance or to model, design and control renewable energy systems.

4.1 Estimation of Solar Irradiance

A neural network model (NNM) for predicting global solar irradiance (GSI) distribution on horizontal surfaces has been developed by Zervas et al [2008].

The GSI is predicted using a Radial Basis Function (RBF) neural network, a 2-layer network where the learning process is performed in two different stages [Mellit and Kalogirou 2008]. The training approach is divided in two phases. In the first phase, the input data set x_n is employed to determine the first layer weights (unsupervised stage). In the second phase, the first layer weights are fixed while the second layer weights are evaluated. The second stage is supervised as both input and target data are required. Final optimization is achieved through a classic least squares approach [Mellit and Kalogirou 2008].

Zervas et al [2008] have used the RBF neural network to predict the parameters of a Gaussian function employing meteo data and daylight duration as input variables. This function approximates the GSI daily distribution.

Fuzzy logic has been applied to neural networks, in particular to the RBF training algorithm [Sarimveis et al 2002] by exploiting databases containing local observations of the input variables and the parameters of the Gaussian function over a year long period of time. A correction methodology for the two tails of the Gaussian function further improves the accuracy of the model. The obtained soft computing model is capable of providing reliable future predictions of the daily GSI distributions on horizontal surfaces, given only the meteo and the daylight duration.

4.2 Simulation and Control of a PEM Fuel Cell System

Hatti and Tioursi [2009] developed a dynamic neural network model and controller for a PEM fuel cell power system. The neural network model has been derived using data collected from single cells.

The dynamic neural network controller is based on an input – output non-linear mapping principle, and trains the neural controller offline or online by Levenberg–Marquardt back-propagation method and Bayesian algorithms.

After the training process, the network was ready to generate I-V characteristics for a broad range of conditions.

The aim of the controller is to regulate the active, reactive and steady state power by acting on parameters like voltage and phase angle. For steady state power adjustment, power changes must be followed by a proper hydrogen flow rate adjustment.

In the training phase, proportional-integral controllers are used to generate the modulation index and phase angle as reference for the dynamic neural network controller. It is possible to show that the use of dynamic neural network controller provides viable results.

4.3 Control of REHHS with a Diesel Generator

Al-Alawi et al [2007] developed a predictive artificial-neural-network-based prototype controller for the optimum operation of an integrated hybrid renewable energy-based water and power supply system. The integrated system consists of photovoltaic modules, a diesel generator, a battery bank for energy storage and a reverse osmosis desalination unit. The electrical load is from typical households and a desalination plant.

The ANN controller needs to take decisions, based on predictive information, over on/off cycles of the diesel generator under light loads and high solar radiation levels while maintaining high efficiency. The key objectives are to reduce fuel dependency, greenhouse gas emissions, and engine wear and tear due to incomplete combustion.

The designed ANN consists of four input nodes, representing time in a 24-hour period, power from PV panels, power from battery, and power from inverter. The two output nodes represent the power from diesel generator and the generator on/off status.

After adjustments to the parameters, the network is capable of converging to a threshold of 0.00001. The statistical analysis of the results indicates that the R2 value for the testing set of 186 cases is 0.979. This indicates that ANN-based model can predict the power usage and generator status at any point of time with high accuracy.

5 Genetic Algorithms

Soft computing programming stems from ideas that can be traced back to the 50's by scholars who were investigating machine learning by taking advantage of analogies with human behaviour [Fogel 1999].

Genetic algorithms (GA), also called evolutionary algorithms, were later developed and successfully applied to real life optimization problems [Goldberg 1989, Bäck 1996].

In GA, the programming starts with a set of data (called genes or genotype) encoded in potential solutions (called individuals or phenotypes) to an optimization problem [Schmitt Lothar 2001]. Individuals are initially generated randomly, and are allowed to generate new individuals by inheritance/cross-over, selection and mutation of the original genotypes. For each new generation of individuals, the fitness of every individual in the population is checked against predetermined criteria, and only fit individuals are randomly selected and allowed to generate new individuals to form a new population until a defined fitness level has been reached. The basic algorithm steps are: *initialization*, *selection*, *reproduction* and *termination* of the experiment.

With these algorithms an optimal solution is not necessarily always found; hence, often, the number of iterations are capped to save computing time. GA can also be successfully used in combination with deterministic algorithms to produce an initial set of potential solutions that will be fine-tuned by other methodologies [Eiben and Smith 2007].

Applications in energy systems have been numerous [Miranda et al 1998]. In the following paragraphs, uses in REHHS are presented and discussed.

5.1 Efficient Design and Control

As already discussed, REHHS are complex systems characterized by a large number of variables that increase the complexity of sizing and controlling [Dufo-Lòpez et al 2007, Bernal-Agustìn and Dufo-Lòpez 2009]. Evolutionary algorithms present the advantage of having low computational requirements yielding good solutions in a reasonable timeframe.

In this example, the REHHS is an hybrid of a PV and Wind conversion system, a diesel generator with a battery set and an hydrogen loop to provide electrical energy to a load. The problem consists in searching for the components and control strategies that minimize the lowest total net present cost of the system.

The GA consists of two sub-GA.

The main sub-GA is devoted to finding the best system design, the optimal control strategy for every combination of the system design as per the outcome of the main GA.

The genotype of the main algorithm is a vector of 11 integers:

- number of PV panels in parallel;
- type of PV panel;
- number of wind turbines;
- type of wind turbine;
- type of hydro turbine;
- number of batteries in parallel;
- type of battery;
- type of AC generator;

- type of fuel cell;
- type of electrolyser;
- type of inverter.

The fitness function of the ith iteration is given by:

$$fitness_{MAIN_i} = \frac{(N_m + 1) - i}{\sum_j [(N_m + 1) - j]}, j = 1 \ldots N_m \tag{30}$$

where j is the rank in the population (1 for the best individual, N_m for the worst). The fitness function determines the probability of selecting an individual on which to apply the genotype modification process.

The secondary algorithm is devised to find the control strategy that minimizes, for each configuration provided by the main algorithm, the cost function. Its genotype consists of twelve control variables [Dufo-Lòpez et al 2007]:

- minimum power of the AC generator recommended by the manufacturer;
- minimum power of the fuel cell recommended by the manufacturer;
- minimum state of charge of the battery recommended by the manufacturer;
- AC generator critical power limit;
- batteries SOC set point for the AC generator;
- fuel cell critical power limit;
- batteries SOC set point for the fuel cell;
- set point for the amount of H2 stored in the tank
- power below which it is more economical to store energy in the batteries than in the H2 tank;
- intersection point of the cost of supplying energy with the batteries and the cost of supplying energy with the fuel cell;
- intersection point of the cost of supplying energy with the fuel cell and the cost of supplying energy with the AC generator.

Similarly to the main algorithm, the fitness function of the ith iteration is given by:

$$fitness_{SEC_i} = \frac{(N_{sec} + 1) - i}{\sum_j [(N_{sec} + 1) - j]}, j = 1 \ldots N_{sec} \tag{31}$$

where j is the rank in the population (1 for the best individual, N_{sec} for the worst).

Elitism is the rule common to both algorithms; best individuals are not lost from one generation to the following. Gene evolution is performed by means of either a uniform or a non-uniform mutation.

Non–uniform mutation is carried out by changing a randomly selected gene e by the following:

$$e' = \begin{cases} e + \Delta(N_{gen\_main}, HL - e) & \text{if a random binary number is 0} \\ e + \Delta(N_{gen\_main}, e - LL) & \text{if a random binary number is 1} \end{cases} \tag{32}$$

where HL and LL are the higher limit and the lower limit of the gene *e*. The function $\Delta(N_{gen\_main}, y)$ produces a value in the interval [0,y] so that the probability of $\Delta(N_{gen\_main}, y)$ approaching 0 increases as N_{gen_main} increases. This operator initially searches solutions globally, then locally in the last generations in order to increase the likelihood of generating individuals with genotypes closer to its successor. It is written as:

$$\Delta\left(N_{gen\_main}, y\right) = y \cdot \left(1 - r^{\left(\frac{1-N_{gen\_main}}{N_{gen\_main\_max}}\right)^{b}}\right) \tag{33}$$

where r is a random number in [0,1], $N_{gen\_main\_max}$ is the maximum generation number, and b is a parameter determining the degree of dependency upon the generation number.

Uniform mutation is achieved by applying to a random gene a value obtained by applying a uniform distribution of probability in the interval [LL,HL] where LL and HL respectively are the lower and the higher limit for the same gene.

The parameter ranges of the genotype for the individuals in the main algorithm are outlined in Table 1.

Table 1 Genotype of main algorithm (adapted from Bernal-Agustìn and Dufo-Lòpez [2009]).

Parameter	Values
PV module power	0, 50, 125 W_p
Number of modules in parallel	[1,20] (4 in series)
Wind power	0, 275, 640, 1760 W
Number of wind turbines in parallel	[1,3]
Battery capacities	0,.43, 200, 462 Ah
Number of batteries	[1,6] (4 in series)
Diesel generators:	0, 1.9, 3, 4.5, 5.5 kVA
FC power	0, 1, 2, 3, and 5 kW
Electrolyzer power	0, 1, 2, 3, 4.2 kW
Inverters	3, 4.5, 5.5 kVA

Each gene has an associated cost. The total cost depends on final gene selections and is the function that must be minimized by the algorithm.

5.1.1 Deterministic Evaluation

A deterministic enumerative algorithm has been applied on all the possible combinations (6 480 000) of different genes configurations. The result that minimizes the net present cost of the system consists of 4 serial and 20 parallel PV panels of

125 W_p, 4 serial and 2 parallel batteries of 462 Ah, a diesel generator of 3 kVA, three wind turbines of 1 760 W and an inverter of 4.5 kVA.

The minimum net present cost is 222 468.10 € (with the exclusion of the hydrogen loop components). The enumerative method has lasted 1 day and 9 h for evaluation of all the possible combinations on the same hardware set.

5.1.2 GA Evaluation

The design of the GA algorithm has considered the following:

- Number of generations: 10, 20, 30, 40, 50, 100.
- Number of population (Nm): 100, 200, 300, 400, 500, 1000.
- Crossing rate (CRm): 50%, 70%, 90%.
- Mutation rate (MRm): 0.1%, 0.5%, 1%.
- Mutation typology: uniform and non-uniform.

Every optimization has been performed three times. The runs have lasted from 9 seconds to 28 minutes, and nearly all optimizations that lasted more than 1 minute have reached the same solution yielded by the deterministic algorithm. Figures 2 and 3 show a percentage of optimizations that have reached the global optimal solution with different mutation rates and typologies. It is possible to conclude that convergence improves with uniform mutation and higher mutation rates. This can be particularly true when the number of generations or population size is low.

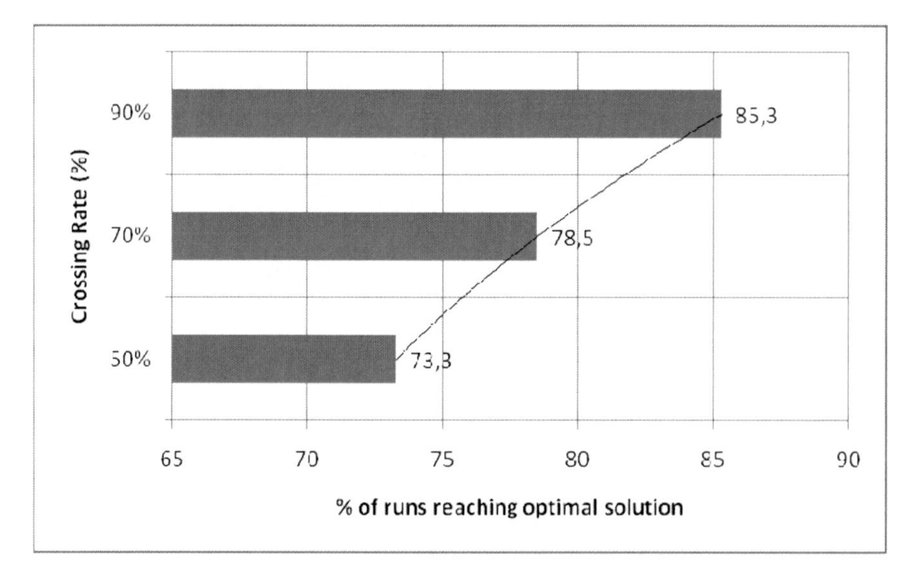

Fig. 2 Percentage of projects reaching the global solution by different mutation rates (adapted from Bernal-Agustìn and Dufo-Lòpez [2009]).

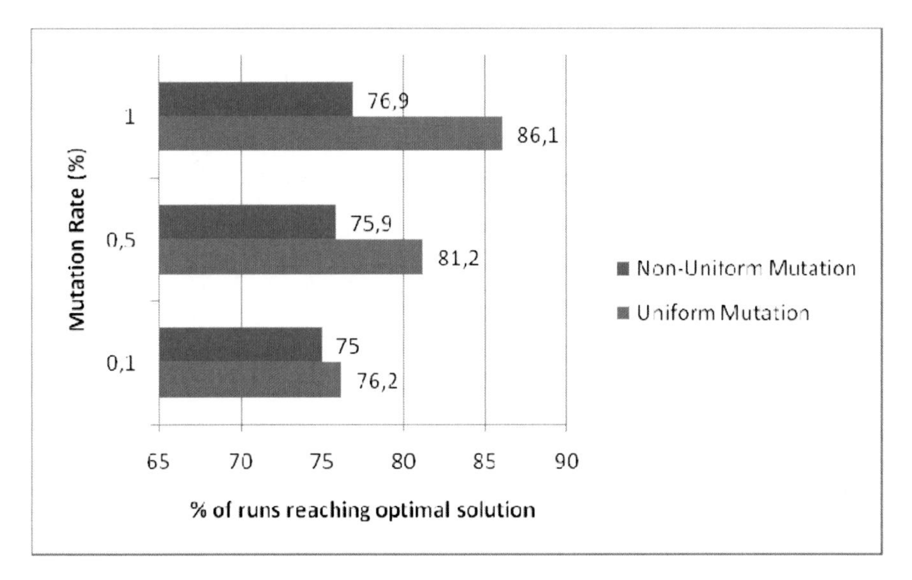

Fig. 3 Optimal solution with varying mutation rates and mutation typologies (adapted from Bernal-Agustìn and Dufo-Lòpez [2009]).

The evolution of 6 optimization runs for a design of 20 generations, 200 population and a 90% crossing rate shows that by increasing the crossing rate, the convergence becomes faster. After the 13th generation, the solution is not significantly improved. With a 90% crossing rate, the size of the population can be reduced while still having a close-to-100% probability of reaching a global optimal solution.

To reach a high probability of obtaining the optimal solution, the algorithm could be designed with the following settings:

- number of generations higher than 15;
- population size higher than 0.003% of the number of total combinations;
- crossing rate 90%;
- mutation rate 1%;
- uniform mutation.

5.2 *Sizing Optimization of a Stand-Alone Lighting System*

A practical application of a combination of a GA with a Simplex Algorithm (SA) is described in Lagorse et al [2009] where system parameter and cost optimization is performed for a stand-alone photovoltaic hydrogen hybrid system supplying a street lighting load.

Since power production depends on variable weather conditions, deterministic algorithms can be difficult to be modelled; on the contrary, a heuristic global search like GA is easily defined to look for a set of individuals whose genotype will become the initial data set that will be further refined by the SA.

The genotype of the individuals is set to be the following:

- PV power (P_{pv});
- Battery capacity (Q_b);
- FC power (P_{fc});
- PV tilt angle (β);
- Minimum SOC (SOC_{min});
- Maximum SOC (SOC_{max}).

The function to be optimized is an economic cost function. System parameters are to be determined by evaluation of penalties related to excess or shortage of energy produced or stored.

GA is designed as:

- 15 individuals per generation;
- 100 generations;
- *Roulette* selection scheme;
- 47% mutation probability.

The fitness is defined as a percentage, so that, if the total cost tends to zero, fitness tends to 100%. Table 2 outlines the values of the genotype of the best individual as evaluated by the GA.

Table 2 Genotype of best individual obtained by GA (adapted from Lagorse et al [2009]).

Variable	Value
β	53.3°
P_{pv}	94.6 W_p
Q_b	2.56 kWh
P_{fc}	282.6 W
SOC_{min}	36.1%
SOC_{max}	90%
Cost	9631.51 €
Fitness	80.74%

The final values obtained by GA are the initial values of the SA. Table 3 shows the results after 75 iterations of the SA.

Table 3 Genotype of best individual obtained by SA applied on population from GA (adapted from Lagorse et al [2009]).

Variable	Value
β	50.6°
P_{pv}	148 W_p
Q_b	2.537 kWh
P_{fc}	128 W
SOC_{min}	42%
SOC_{max}	58%
Cost	6515.53 €
Fitness	67.42%

REHHS hydrogen consumption is 195 kWh/year with a 6 year fuel cell useful life. Storage, using a traditional compression technique, necessitates a 1650 l tank at a pressure of 20 MPa.

GA is employed in combination with a traditional deterministic algorithm, the strength of soft computing techniques in finding solution through heuristics joining forces with a more traditional algorithm to increase final solution accuracy. By examining the results in Tables 3 and 4, a significant difference is evident between the solution given by the GA and the following refinement from the SA. Overall computing time is decreased by the synergy between a combined heuristic and deterministic approach.

6 Particle Swarm Optimization

Kennedy and Eberhart [1995] introduced the concept of *Particle Swarm Optimization* (PSO), where each possible solution in the design space is called *particle* and the overall set of particles is called a *swarm*. Each particle moves with an adaptable velocity within the search space and retains in its memory the best position it has encountered. The best position attained by all individuals of the swarm is communicated to all the particles [Parasopoulos and Vrahatis 2004, Jarboui et al 2008].

If the search space is n-dimensional, then a particle i can be represented by a position vector $X_i = [x_{i1}, \dots ,x_{in}]^T$ and a velocity vector $V_i = [v_{i1}, \dots ,v_{in}]^T$. Particles adjust their positions in the $k+1$ iteration according to the following equations:

$$X_{k+1}^{i} = X_{k}^{i} + V_{k+1}^{i} \tag{34}$$

$$V_{k+1}^{i} = w_k V_k^i + c_1 r_1 \left(P_k^i + X_k^i\right) + c_2 r_2 \left(P_k^g + X_k^i\right) \tag{35}$$

where subscript k represents a time increment, P_k^i represents the best possible position for particle i at time k, P_k^g corresponds to the global best position in the swarm at time k; r_1 and r_2 represent uniform random numbers between 0 and 1. c_1

and c_2 are respectively, the *cognitive* and the *social* parameters. If $c_1 > c_2$, the cognitive parameter overcomes the social parameter implying that each single particle will strive to find the position that represents its own best personal position; in the opposite case when the social overcomes the cognitive parameter, the ensemble of particles is attracted by the global best overall position. Finding a good balance between the two parameters means avoiding excess particle wandering in case the cognitive is higher than the social parameter, or reaching a global pseudo-optimal position if the social is higher than the cognitive parameter. In Eq. 35, w is the inertia component used to balance between global (higher w) or local exploration (lower w).

6.1 Optimal Sizing of a REHHS with Hydrogen from Biomass and Wind

The PSO has been applied to devise the optimal component sizing [Hakimi and Moghaddas-Tafreshi 2009] of an REHHS where hydrogen is obtained from a wind conversion source and municipal wastes.

The algorithm is performed under a certain set of assumptions over the location, the availability of wind and quantity of municipal waste. In this case, the residential area comprises a population of 2000 with a daily *pro capite* waste production of 0.6 kg, with a daily H_2 production rate of 50 kg (equivalent to 1890 kWh). Hydrogen is then stored in a pressurized tank.

The PSO is run with $c1 = c2 = 2$, $w = 0.7$, a population size of $p = 60$ and a number of iterations $g = 500$.

Inputs of the optimization procedure are the capital, replacement, O&M costs, the efficiency, the lifetime of components and lifetime of the project, and meteo data in the specific region of interest of the study.

The PSO algorithm computes, with Eqs. 34 and 35, the best position of the particle and the best position of the group by determining the objective function and comparing it with the values obtained from previous iterations. By comparing all the best positions of the particles, the best group position is finally found as the solution of the problem.

The constraints adopted in the optimization algorithm are basic physical limitations due, for instance, to hydrogen storage in tanks and performances of the many sub-systems that compose the REHHS under consideration.

Results provide a nominal power of 7.5 kW for each wind turbine, 1 kW for each electrolyzer and each fuel cell, a hydrogen tank capacity of 1 kg, with a lifetime of the project of 20 years, a reactor throughput of 750 kg/day, a hydrogen production from the reformer of 31.2 kg H_2/day, and a compressor power of 50 kW. The system cost estimation reaches the optimum after 200 generations.

6.2 Minimization of Costs with Component Outages

A similar approach has been followed by Kashefi Kaviani et al [2009] for a wind and photovoltaic REHHS. The PSO algorithm is devised to minimize the cost of the system over 20 years of operations with components that may be subject to failures.

For five different optimization experiments, the algorithm converges to the same global optimum after nearly 100 generations. It is worth noting that reaching the same optimum (or fitness value) does not mean that the solution combination is the same, because different points in the solution space may produce very similar fitness values.

The energy stored in the hydrogen tank during one year of REHHS operations oscillates daily, with a maximum around 5 700 – 5 800 kWh most frequently in the first part of the year, and a minimum around 350 – 400 kWh in the second part of the year.

The reliability of many different components can have a profound impact on system's overall reliability and economic results. Results from this study show that, for instance, inverter reliability is so impactful that it represents the most important issue for system overall reliability.

The PSO algorithm has been proved to offer very quick results that can be of help when doing first approximation studies in, say, initial design or sensitivity analysis, while deterministic methods can be used in conjunction with PSO to further refine the original approximate solution.

7 Summary and Conclusions

Fuzzy logic has already shown its potential in many different engineering and scientific applications. It comes to no surprise that also in renewable energy engineering and science it represents a very powerful instrument for control or selection of alternative possibilities. The qualitative, rather than quantitative, approach is very useful when a rigorous mathematical modelling is not practical, but this limits the possibility to improve the understanding of how a complex system really works.

Neural networks necessitate of an initial phase of training and testing, but are instrumental in providing engineers with important engines for the control of complex hybrid systems like REHHS. An important and interesting property of such soft computing algorithms is the self-tuning capability shown by trained neural nets, granting the chance to assist the functioning of the system also in far-from-normal situations and improve robustness of REHHS. Very useful when the complex inter-relations between sub-components are difficult to be precisely defined, neural networks promote a black-box approach that only considers inputs and outputs, achieving an alternative interpretation of the reality of the systems.

Advantages of evolutionary algorithms are the low computational efforts needed to obtain partial or final solutions. The heuristic quality of the algorithm means that it is not necessary to deterministically define the boundaries of the problem, but rather the definition of the genotype and fitness functions. Disadvantages entail in some cases the use of deterministic approaches in order to fine-tune the genotype of the individuals in case an optimal solution is not found.

Particle Swarm Optimization algorithms are effective in reaching approximate solutions in very small amounts of time; deterministic computations can then be applied to fine-tune the solution. Another advantage is that the combination offering global optimal solutions can be non necessarily unique, thus providing a set of possible solutions that can be chosen depending on other pre-conditions not originally used in the design of the PSO runs.

As in many other engineering and scientific fields, the use of soft-computing factually adds value and options to improve efficiency and effectiveness of problem solving techniques for renewable energy systems.

References

Al-Alawi, A., Al-Alawi, S., Islam, S.M.: Predictive control of an integrated PV-diesel water and power supply system using an artificial neural network. Renewable Energy 32, 1426–1439 (2007)

Bäck, T.: Evolutionary algorithms in theory and practice. Oxford University Press, Oxford (1996)

Bernal-Agustìn, J.L., Dufo-Lòpez, R.: Efficient design of hybrid renewable energy systems using evolutionary algorithms. Energy Conversion and Management 50, 479–489 (2009)

Bilodeau, A., Agbossou, K.: Control analysis of renewable energy system with hydrogen storage for residential applications. Journal of Power Sources 162, 757–764 (2006)

Deshmukh, S.S., Boehm, R.F.: Review of modeling details related to renewably powered hydrogen systems. Renewable and Sustainable Energy Reviews 9(12), 2301–2330 (2008)

Duffie, J.A., Beckman, W.A.: Solar engineering of thermal processes. Wiley, Chichester (2006)

Dufo-Lòpez, R., Bernal-Agustìn, J.L., Contreras, J.: Optimization of control strategies for stand-alone renewable energy systems with hydrogen storage. Renewable Energy 32, 1102–1126 (2007)

Eiben, A.E., Smith, J.E.: Introduction to evolutionary computing (Natural computing series). Springer, Berlin (2007)

El-Shatter, T.F., El-Hagry, M.T., Aboueldahab, M.E., Elkousy, A.A.: Fuzzy modeling and simulation of photovoltaic system. In: Proceedings of the 14th European Photovoltaic Solar Energy Conference, Barcelona, Spain, June 30 -July 4 (1997)

El-Shatter, T.F., Eskandar, M.F., El-Hagry, M.T.: Hybrid PV/fuel cell system design and simulation. Renewable Energy 27, 479–485 (2002)

Fogel, D.B.: Evolutionary computation toward a new philosophy of machine intelligence, 2nd edn. Wiley-IEEE Press (1999)

Garg, H.P.: Treatise on solar energy. Wiley, Chichester (1982)

Ghaboussi, J., Garrett Jr., J.H., Wu, X.: Knowledge-Based Modeling of Material Behavior with Neural Networks. ASCE J. Engrg. Mech. 117(1), 132–153 (1991)

Goldberg, D.E.: Genetic algorithms in search, optimization and machine learning. Addison-Wesley, Reading (1989)

Hakimi, S.M., Moghaddas-Tafreshi, S.M.: Optimal sizing of a stand-alone hybrid power system via particle swarm optimization for Kahnouj area in south-east of Iran. Renewable Energy 34, 1855–1862 (2009)

Hammache, A., Bilgen, E.: Photovoltaic hydrogen production for remote communities in northern latitudes. Solar & Wind Technology 2(4), 139–144 (1987)

Hatti, M., Tioursi, M.: Dynamic neural network controller model of PEM fuel cell system. International Journal of Hydrogen Energy 34, 5015–5021 (2009)

Haykin, S.: Neural networks: a comprehensive foundation, 2nd edn. Macmillan, New York (1999)

Hollenberg, J.W., Chen, E.N., Lakeram, K., Modroukas, D.: Development of a photovoltaic energy conversion system with hydrogen energy storage. Int. J. Hydrogen Energy 3(20), 239–243 (1995)

Jarboui, B., Damak, N., Siarry, P., Rebai, A.: A combinatorial particle swarm optimization for solving multi-mode resource-constrained project scheduling problems. Appl. Math. Computing 195, 299–308 (2008)

Kahraman, C., Beşkese, A., Tunç Bozbura, F.: Fuzzy Regression Approaches and Applications. Studies in Fuzziness and Soft Computing, vol. 201, pp. 589–615. Springer, Heidelberg (2006)

Kahraman, C., Kaya, I., Cebi, S.: A comparative analysis for multiattribute selection among renewable energy alternatives using fuzzy axiomatic design and fuzzy analytic hierarchy process. Energy 34, 1603–1616 (2009)

Kalogirou, S.A.: Artificial neural networks in renewable energy systems applications: a review. Renewable and Sustainable Energy Reviews 5, 373–401 (2001)

Kashefi Kaviani, A., Riahy, G.H., Kouhsari, S.M.: Optimal design of a reliable hydrogen-based stand-alone wind/PV generating system, considering compo-nent outages. Renewable Energy 34, 2380–2390 (2009)

Kennedy, J., Eberhart, R.C.: Particle swarm optimization. In: Proceedings of the 1995 IEEE International Conference on Neural Networks, Perth, Australia, vol. 4, pp. 1942–1948. IEEE Service Center, Piscataway (1995)

Khan, M.J., Iqbal, M.T.: Analysis of a small wind-hydrogen stand-alone hybrid energy system. Applied Energy 86, 2429–2442 (2009)

Kolhe, M., Agbossou, K., Hamelin, J., Bose, T.K.: Analytical model for predicting the performance of photovoltaic array coupled with a wind turbine in a stand-alone renewable energy system based on hydrogen. Renewable Energy 28, 727–742 (2003)

Lagorse, J., Paire, D., Miraoui, A.: Sizing optimization of a stand-alone street lighting system powered by a hybrid system using fuel cell, PV and battery. Renewable Energy 34, 683–691 (2009)

Lakhmi, C.J., Martin, N.M.: Fusion of neural networks, fuzzy systems and genetic algorithms: industrial applications. CRC Press LLC, Boca Raton (1998)

Li, X., Song, Y.J., Han, S.B.: Frequency control in micro-grid power system combined with electrolyzer system and fuzzy PI controller. Journal of Power Source 180, 468–475 (2008)

Liu, B.Y.H., Jordan, R.C.: The long term average performance of flat plate solar energy collectors. Solar Energy 7, 53–70 (1963)

Machado, R.J., Rocha, A.F.: A hybrid architecture for fuzzy connec tionist expert systems. In: Kandel, A., Langholz, G. (eds.) Hybrid Architectures for Intelligent Systems, CRC Press, Boca Raton (1992)

Maclay, J.D., Brouwer, J., Samuelsen, G.S.: Dynamic modeling of hybrid energy storage systems coupled to photovoltaic generation in residential applications. Journal of Power Source 163, 916–925 (2007)

Mellit, A., Kalogirou, S.A.: Artificial intelligence techniques for photovoltaic applications: A review. Progress in Energy Combustion Science 34, 574–632 (2008)

Miland, H., Glöckner, R., Taylor, P., Aaberg, R.J., Hagen, G.: Load control of a wind-hydrogen stand-alone power system. International Journal of Hydrogen Energy 31, 1215–1235 (2005)

Miranda, V., Srinivasan, D., Proenca, L.M.: Evolutionary computation in power systems. International Journal of Electrical Power & Energy Systems 20(2), 89–98 (1998)

NeuroShells (2007) An ANN development software from ward systems group. Inc., http://www.wardsystems.com

Parasopoulos, K.E., Vrahatis, M.N.: On the computation of all global minimizers through particle swarm optimization. IEEE Trans. Evol. Comput. 3, 8 (2004)

Pedrazzi, S., Zini, G., Tartarini, P.: Complete modeling and software implementation of a virtual solar hydrogen hybrid system. Energy Conversion and Management 51, 122–129 (2010)

Penner, S.S.: Steps toward the hydrogen economy. Energy 31, 33–43 (2006)

Robert, F.: Neural Fuzzy Systems. Abo Akademi University (1995)

Romm, J. (ed.): The hype about hydrogen: fact and fiction in the race to save the climate. Island Press (2004)

Sarimveis, H., Alexandridis, A., Tsekouras, G., Bafas, G.: A fast and efficient algorithm for training radial basis function neural networks based on a fuzzy partition of the input space. Ind. Eng. Chem. Res. 41, 751–759 (2002)

Schmitt Lothar, M.: Theory of generic algorithms. Theoretical Computer Science 259(1-2), 1–61 (2001)

Shahin, M.A., Jaksa, M.B., Maier, H.R.: Artificial neural network applications in geotechnical engineering. Australian Geomechanics 36(1), 49–62 (2001)

Ulleberg, Ø.: Modeling of advanced electrolyzers: a system simulation approach. Int. J. Hydrogen Energy 28, 21–33 (2003)

Ulleberg, Ø., Morner, S.O.: Trnsys simulation models for solar hydrogen systems. Solar Energy 4-6(59), 271–279 (1997)

Yager, R.R.: Fuzzy Sets and Applications: Selected Papers by L.A. Zadeh. John Wiley, New York (1987)

Yu, H.H., Jenq-Neng, H.: Handbook of neural network signal processing. CRC press, Boca Raton (2001)

Zadeh, L.A.: Fuzzy sets. Inform Control 8, 338–353 (1965)

Zadeh, L.A.: A fuzzy-set-theoretic interpretation of linguistic hedges. Cyber Net 2, 4–34 (1972)

Zadeh, L.A.: Outline of a new approach to the analysis of complex systems and decision processes. IEEE Trans. Syst. Man Cybernet 3, 28–44 (1973)

Zeng, J., An, M., Smith, N.J.: Application of a fuzzy based decision making methodology to construction project risk assessment. International Journal of Project Management 25, 589–600 (2007)

Zervas, P.L., Sarimvies, H., Palyvos, J.A., Markatos, N.G.C.: Model-based optimal control of a hybrid power generation system consisting of photovoltaic arrays and fuel cells. Journal of Power Source 181, 327–338 (2008)

Zini, G., Tartarini, P.: Wind-hydrogen energy stand-alone system with carbon storage: Modeling and simulation. Renewable Energy 35, 2461–2467 (2010)

Soft Computing in Absorption Cooling Systems

Arzu Şencan Şahin[1] and Soteris A. Kalogirou[2]

[1] Department of Mechanical Education, Technical Education Faculty,
Süleyman Demirel University, 32260 Isparta, Turkey
sencan@tef.sdu.edu.tr

[2] Cyprus University of Technology, Department of Mechanical Engineering and Materials Sciences and Engineering, P.O. Box 50329, 3603 Limassol, Cyprus
Soteris.kalogirou@cut.ac.cy

Abstract. Absorption cooling systems make sense in many applications for process water cooling. Instead of mechanically compressing a refrigerant gas, as in the conventional vapor compression process, absorption cooling uses a thermochemical process. Two different fluids are used, a refrigerant and an absorbent. Heat directly from natural gas combustion, solar energy, waste-heat source or indirectly from a boiler, drives the process.

In recent years, soft computing (SC) methods have been widely utilized in the analysis of absorption cooling systems. Soft computing is becoming useful as an alternate approach to conventional techniques. Soft computing differs from conventional (hard) computing in that, unlike hard computing, it is tolerant of imprecision, uncertainty, partial truth, and approximation.

In this chapter, the research of applying soft computing methods for absorption cooling applications is presented.

1 Introduction

Absorption cooling systems have become increasingly popular in recent years from the viewpoint of energy and environment. Absorption cooling system uses a source of heat to provide the energy needed to drive the cooling process.

Absorption cooling operates similarly to conventional electric vapor compression chillers with some very important differences. The major differences are seen in the components of the system and the refrigerant used in the cycle. Absorption systems use what is called a "thermal compressor", which uses thermal energy to operate, in place of the conventional system's compressor, which uses electricity. Absorption cooling system is shown schematically in Fig. 1.

Compared to an ordinary cooling cycle the basic idea of an absorption system is to avoid compression work. This is done by using a suitable working pair. The working pair consists of a refrigerant and a solution that can absorb the refrigerant. Usually, LiBr-H_2O is used, water is the refrigerant. As shown in Fig. 1, when the

K. Gopalakrishnan et al. (Eds.): Soft Comput. in Green & Renew. Ener. Sys., STUDFUZZ 269, pp. 65–95.
springerlink.com

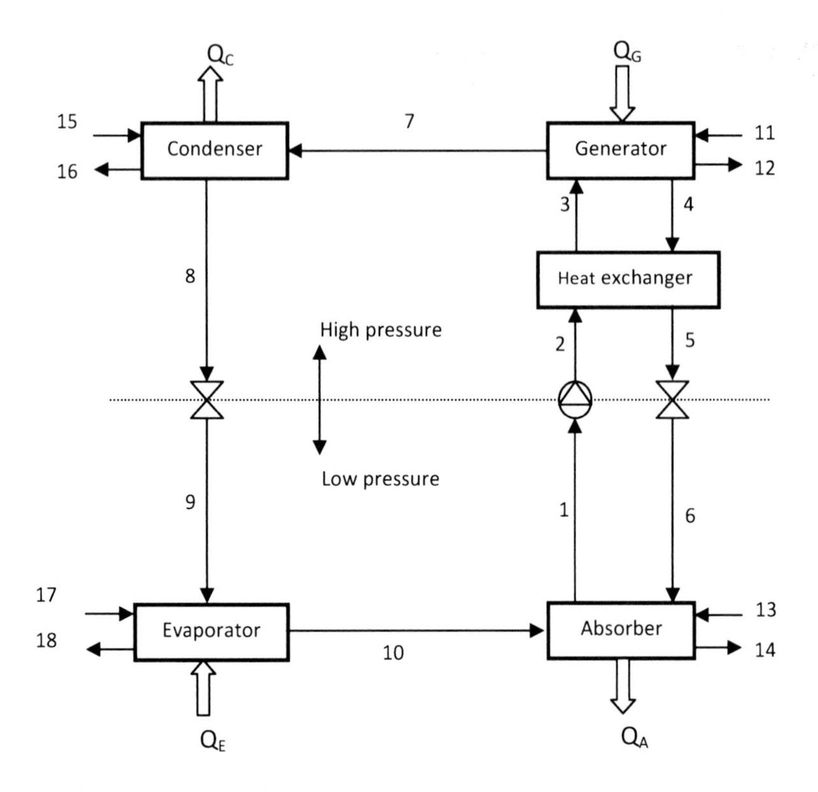

Fig. 1 Absorption cooling system

refrigerant vapour is coming from the evaporator (10) it is absorbed in a liquid (1). This liquid is pumped to higher pressure (1-2), where the refrigerant is boiled out of the solution by the addition of heat (3-7). Subsequently, the refrigerant goes to the condenser (7-8) like in an ordinary cooling cycle. Finally, the liquid with smaller amount of refrigerant returns back to the absorber (6) [Herold et al. 1996].

The basic absorption cycle employs two fluids, the absorbate or refrigerant, and the absorbent. The most commonly fluids used are water as the refrigerant and lithium bromide as the absorbent. These fluids are separated and recombined in the absorption cycle. In the absorption cycle the low-pressure refrigerant vapor is absorbed into the absorbent releasing a large amount of heat. The liquid refrigerant/absorbent solution is pumped to a high-operating pressure generator using significantly less electricity than that for compressing the refrigerant for an electric chiller. Heat is added at the high-pressure generator from a gas burner, steam, hot water or hot gases. The added heat causes the refrigerant to desorb from the absorbent and vaporize. The vapors flow to a condenser, where heat is rejected and condense to a high-pressure liquid. The liquid is then throttled though an expansion valve to the lower pressure in the evaporator where it evaporates by absorbing heat and provides useful cooling. The remaining liquid absorbent, in the generator passes through a valve, where its pressure is reduced, and then is recombined with the low-pressure refrigerant vapors returning from the evaporator so the cycle can be repeated [Herold et al. 1996, Absorption Chillers 2010].

Soft computing is a methodology tending to fuse synergically the different aspects of fuzzy logic, neural networks, evolutionary algorithms, and non-linear distributed systems in such a way as to define and implement hybrid systems which manage to come up with innovative solutions in the various sectors of intelligent control, classification, and modeling and simulating complex non-linear dynamic systems.

The basic principle of soft computing is its combined use of the new computation techniques that allow it to achieve a higher tolerance level towards imprecision and approximation, and thereby new software/hardware products can be obtained at lower cost, which are robust and better integrated in the real world. Hybrid systems derived from this combination of soft computing techniques are considered to be the new frontier of artificial intelligence [Foryuna et al. 2001].

Soft computing methodologies have been advantageous in many applications. In contrast to analytical methods, soft computing methodologies mimic consciousness and cognition in several important respects: they can learn from experience; they can universalize into domains where direct experience is absent; and, through parallel computer architectures that simulate biological processes, they can perform mapping from inputs to the outputs faster than inherently serial analytical representations [Chaturvedi 2008].

This chapter aims to present a wide view of various soft computing techniques. In addition, this chapter introduces various applications of soft computing in absorption cooling. These include system modeling and determination of working fluids properties.

2 Soft Computing Techniques

Soft Computing techniques are based on the way information processing is performed in biological systems. The complex biological information processing system enables the human beings to survive with accomplishing tasks like recognition of surrounding, making prediction, planning, and acting accordingly. Soft computing differs from conventional (hard) computing in that, unlike hard computing, it is tolerant of imprecision, uncertainty, partial truth, and approximation. The guiding principle of soft computing is: Exploit the tolerance for imprecision, uncertainty, partial truth, and approximation to achieve tractability, robustness and low solution cost. The principal constituents of Soft Computing (SC) are artificial neural network (ANN), fuzzy logic (FL), Adaptive Network based Fuzzy Inference System (ANFIS), genetic algorithm (GA) and Data Mining (DM).

2.1 Artificial Neural Networks (ANN)

According to Haykin [1994], a neural-network is a massively parallel distributed processor that has a natural propensity for storing experiential knowledge and making it available for use. It resembles the human brain in two respects: the knowledge is acquired by the network through a learning process, and inter-neuron connection strengths, known as synaptic weights, are used to store the knowledge.

Artificial neural-network (ANN) models may be used as alternative methods in engineering analyses and predictions. ANNs mimic somewhat the learning process of a human brain. They operate like a "black box" model, and require no detailed information about the system. Instead, they learn the relationship between the input parameters and the controlled and uncontrolled variables by studying previously recorded data, in a way similar to how a non-linear regression might be performed. Another advantage of using ANNs is their ability to handle large and complex systems with many interrelated parameters. They seem to simply ignore excess data that are of minimal significance, and concentrate instead on the more important inputs.

A schematic diagram of typical multilayer feed-forward neural-network architecture is shown in Fig. 2. The network usually consists of an input layer, some hidden layers and an output layer. In its simple form, each single neuron is connected to other neurons of a previous layer through adaptable synaptic weights. Knowledge is usually stored as a set of connection weights (presumably corresponding to synapse-efficacy in biological neural systems). Training is the process of modifying the connection weights, in some orderly fashion, using a suitable learning method. The network uses a learning mode, in which an input is presented to the network along with the desired output and the weights are adjusted so that the network attempts to produce the desired output. The weights, after training, contain meaningful information whereas before training they are random and have no meaning.

Figure 3 illustrates how information is processed through a single node. The node receives weighted activations of other nodes through its incoming connections. First, these are added up (summation). The result is then passed through an

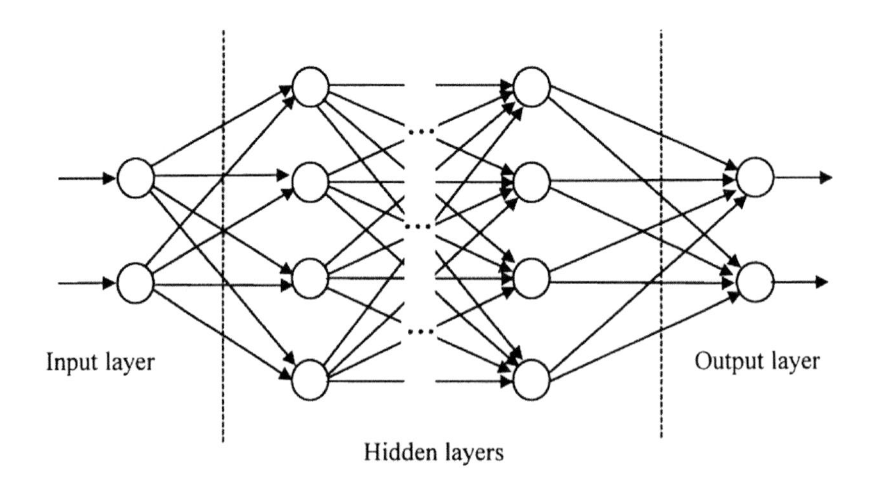

Fig. 2 Schematic diagram of a multilayer feed-forward neural-network

activation function, the outcome being the activation of the node. For each of the outgoing connections, this activation value is multiplied by the specific weight and transferred to the next node.

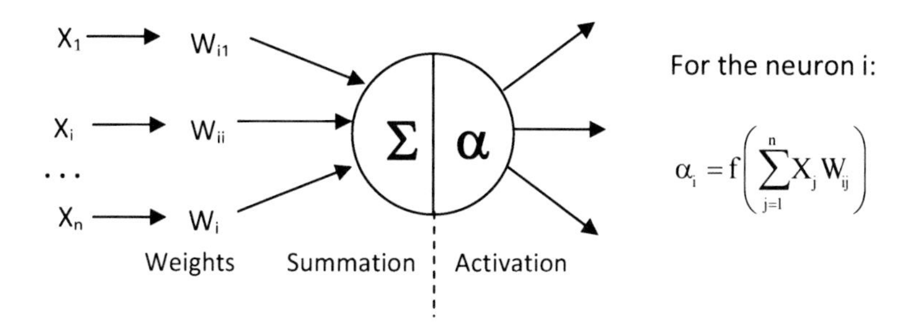

Fig. 3 Information processing in a neural network unit

A training set is a group of matched input and output patterns used for training the network, usually by suitable adaptation of synaptic weights. The outputs are the dependent variables that the network produces for the corresponding input. It is important that all the information the network needs to learn is supplied to the network as a data set. When each pattern is read, the network uses the input data to produce an output, which is then compared with the training pattern, i.e. the correct or desired output. If there is a difference, the connection weights (usually but not always) are altered in such a direction that the error is decreased. After the network has run through all the input patterns, if the error is still greater than the maximum desired tolerance, the ANN runs again through all the input patterns repeatedly until all the errors are within the required tolerances. When the training reaches a satisfactory level, the network holds the weights constant and uses the trained network to make decisions, identify patterns or define associations in new input data sets which were not used to train it. More details on neural networks can be found in [Haykin, 1994, Kalogirou 2000, 2001, Matlab 2010].

2.2 *Fuzzy Logic (FL)*

Fuzzy Logic is basically a multivalued logic that allows intermediate values to be defined between conventional evaluations like yes/no, true/false, black/white, etc. Notions like rather warm or pretty cold can be formulated mathematically and processed by computers.

Fuzzy set theory has been introduced in 1965 by Zadeh and basically it means filling with real numbers the interval between 0 and 1, allowing intermediate values between these two extremes. Fuzzy logic shares with human reasoning the ability of making use of approximate information in order to generate good decisions and precise solutions. Since Aristotle, the theory of logic stated that every proposition must either be true or false, excluding the Middle. In contrast, fuzzy logic is designed to allow computers to make use of the distinctions among data with shades of gray. It proposes making the membership function (or the values False and True) operate over the range of real numbers [0, 1]. This should not lead to confusion between the degree of truth used in fuzzy theory and probabilities, which are conceptually distinct. Boolean logic can be seen as a subset of fuzzy logic [Paulescu 2008].

A general fuzzy system, as shown in Fig. 4 has the components of fuzzification, fuzzy rule base, fuzzy output engine and defuzzification.

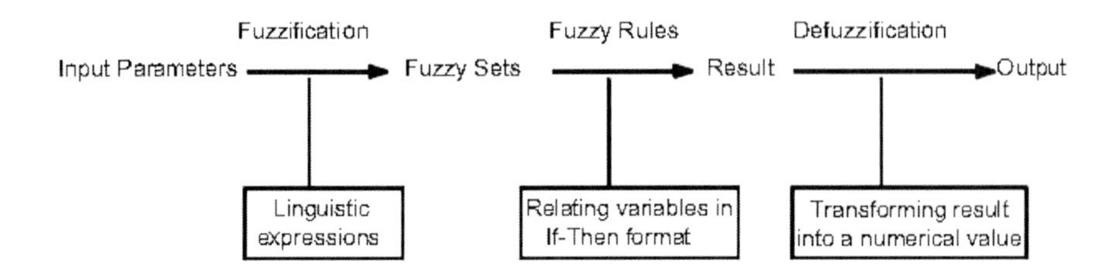

Fig. 4 Schematic of a general fuzzy system

During building up a fuzzy model, firstly, effective (input) parameters should be determined for the system. Secondly, each of the effective parameter should be divided into fuzzy sets, that is, classes with sharply defined fuzzy boundaries in which the transition from membership to non- membership is gradual rather than abrupt, on the scale from 0 to 1. These fuzzy sets are labeled with linguistic expressions like low, medium, high, etc., reflecting the variable physical condition (Fig. 4). In this way, the variable is considered not as a global quantity, but in partial groups that provide more room for the justification of sub-relationships between two or more variables on the basis of fuzzy words. Then, fuzzy rules are written between these variables in IF–THEN format based on the data and expert decision. Lastly, results are defuzzified to a specific number as an output. The purpose of defuzzification is to convert the final fuzzy set representing the overall conclusion into a real number that, in some sense, best represents this fuzzy set. More details on fuzzy logic can be found in [Paulescu 2008, Kucukali and Baris 2010, Matlab 2010; Lau et al. 2008].

2.3 Adaptive Network Based Fuzzy Inference System (ANFIS)

ANFIS is a multilayer feed-forward network consisting of nodes and directional links, which combines the learning capabilities of a neural network and reasoning capabilities of fuzzy logic. This hybrid structure of the network can extend the prediction capabilities of ANFIS beyond ANN and fuzzy logic techniques when they are used alone. Analyzing the mapping relation between the input and output data, ANFIS can establish the optimal distribution of membership functions using either a back-propagation gradient descent algorithm alone, or in combination with a least-squares method [Ertunc and Hosoz 2008]. A basic ANFIS was illustrated in Fig. 5.

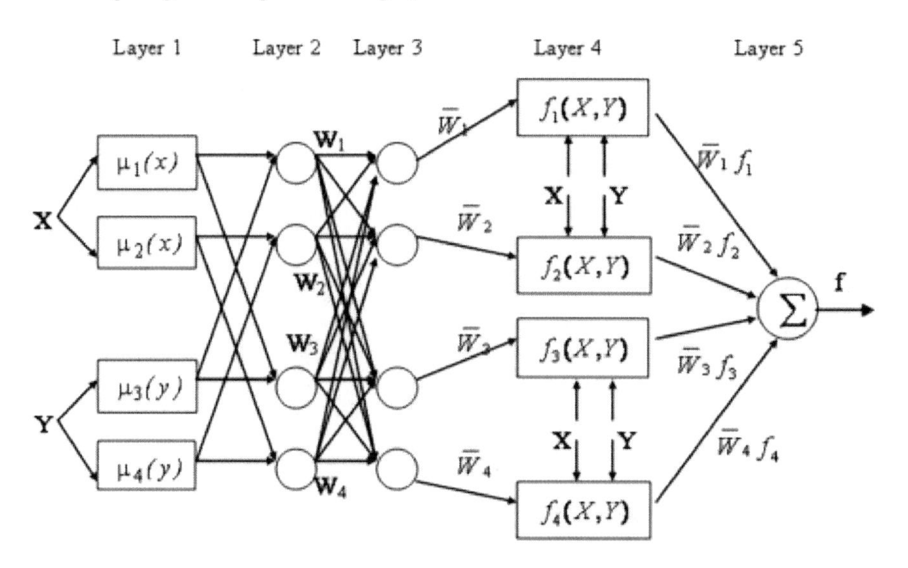

Fig. 5 A basic structure of the ANFIS [Reprinted from *Journal of Hydrology*, Vol. 367, Cobaner et al. 2009, Suspended sediment concentration estimation by an adaptive neuro-fuzzy and neural network approaches using hydro-meteorological data, pp. 52-61, Copyright 2009, with permission from Elsevier]

Depending on the types of inference operations of ''if–then rules'', most fuzzy inference systems can be classified into three types; Mamdani's system, Sugeno's system and Tsukamoto's system. Mamdani's system is the most commonly used, whereas, Sugeno's system is more compact and computationally more efficient; the output is crisp, so, without the time consuming and mathematically intractable defuzzification operation, it is by far the most popular candidate for sample-data based fuzzy modeling and it lends itself to the use of adaptive techniques [Takagi and Sugeno 1985].

In first-order Sugeno's system, a typical rule set with two fuzzy IF/THEN rules can be expressed as:

Rule: 1. If x is A_1 and y is B_1, then $f_1 = p_1x + q_1y + r_1$
Rule: 2. If x is A_2 and y is B_2, then $f_2 = p_2x + q_2y + r_2$

where A_i and B_i are the fuzzy sets, f_i is the output set within the fuzzy region specified by the fuzzy rule p_i and q_i and r_i are the design parameters that are determined during the training process [Efendigil et al. 2009]. The architecture of ANFIS is shown in Fig. 5. The functionality of nodes in ANFIS can be summarized as follows [Cobaner et al. 2009]:

Layer 1: every node i in this layer is an adaptive node, representing membership functions described by generalized bell functions, e.g.,

$$Z_{1,i} = \mu_1(X) = \frac{1}{1 + \left|(X - c_1)/a_1\right|^{2b_1}} \tag{1}$$

where X = input to the node and a_1, b_1 and c_1 = adaptable variables known as premise parameters. The outputs of this layer are the membership values of the premise part.

Layer 2: this layer consists of the nodes which multiply incoming signals and sending the product out. This product represents the firing strength of a rule.

$$Z_{2,1} = W_1 = \mu_1(x)\mu_3(y) \tag{2}$$

Layer 3: in this layer, the nodes calculate the ratio of the ith rules firing strength to the sum of all rules' firing strengths.

$$Z_{3,1} = \overline{W_1} = \frac{W_1}{W_1 + W_2 + W_3 + W_4} \tag{3}$$

Layer 4: this layer's nodes are adaptive with node functions.

$$Z_{4,1} = \overline{W_1} f_1 = \overline{W}(p_1 x + q_1 y + r_1) \tag{4}$$

where W_1 is the output of Layer 3 and $\{p_i, q_i, r_i\}$ are the parameter set. Parameters of this layer are referred to as consequent parameters.

Layer 5: this layer's single fixed node computes the final output as the summation of all incoming signals.

$$f = \sum_{i=1}^{n} \overline{W_i} f_i \tag{5}$$

More information for ANFIS can be found in related literature [Matlab 2010, Jang 1993].

2.4 Data Mining (DM)

Data mining refers to the application of a machine learning method, as well as other automatic knowledge acquisition methods, to the generation of potentially useful knowledge from the organization and analysis of raw data [Xie et al. 2006].

Data mining is a powerful technique for extracting predictive information from large databases. The automated analysis offered by data mining goes beyond the retrospective analysis of data. Data mining tools can answer questions that are too time-consuming to resolve with methods based on first principles. In data mining, databases are searched for hidden patterns to reveal predictive information in patterns that are too complicated for human experts to identify [Hoffman and Apostolakis 2003]. Data mining is applied in a wide variety of fields for prediction, e.g. stock-prices, customer behavior, production control and many others. In addition, data mining has also been applied to other types of scientific data such as bio-informatical, astronomical, and medical data [Li and Shue 2004].

A general approach consists of several steps in which the data are collected and screened for errors, and descriptors are defined. Finally, the descriptors are trained to discriminate between observed data and generated erroneous yet plausible data. A short sketch of data mining approach is given as follows [Hoffman and Apostolakis 2003]:

1. Data are collected.
2. Poor quality data are removed. This step is very important since even a small number of erroneous data can influence the result heavily.
3. The descriptors for a specific problem have to be defined. These are a set of attributes, which, for each data point, contain information relevant to the problem being addressed.
4. Decoys have to be generated. These simple decoys allow for the derivation of a first approximation of the potentials.
5. During training, the descriptors are optimized to discriminate between the actual data and the decoys.
6. Finally, the trained potentials were validated.

The process of knowledge discovery in databases can be seen in Fig. 6.

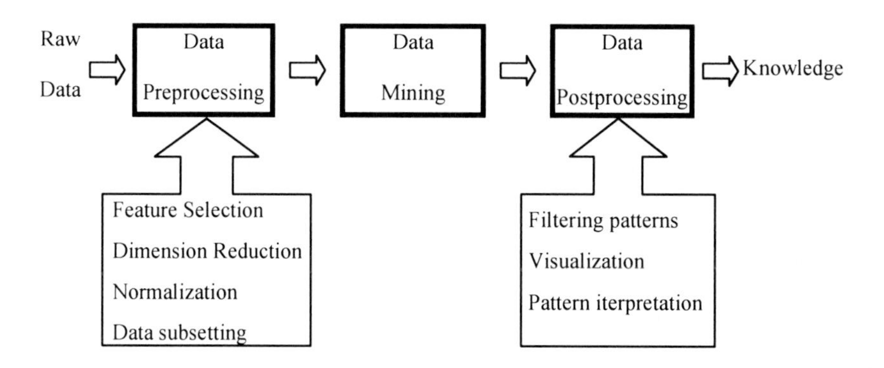

Fig. 6 The process of knowledge discovery in databases

2.5 Genetic Algorithm (GA)

GAs were first proposed by John Holland in 1975 [Holland 1975] and further developed for engineering applications by Goldberg [1989]. Evolution by natural selection is one of the most compelling themes of modern science, and it gives a revolutionary way of thinking about biological systems. This is a form of evolution known as the GA that takes place in a computer. In GAs, selection operates on strings of binary digits stored in the computer's memory, and over time, the functionality of these strings evolves in much the same way that natural populations of individuals evolve. GAs allow engineers to use a computer to evolve solutions over time instead of designing them by hand. An algorithm is a step-by-step procedure for accomplishing some specific task. Many algorithms may be readily implementable by computer programs. Thus, an algorithm is the general description of a procedure, and a program is its realisation as a sequence of instructions to a computer. Although GAs are known primarily as a problem solving method, they can also be used to study evolution itself and to model dynamic systems [Sen et al. 2001]. GA is fundamentally different from the classic optimization algorithms. A genetic algorithm is a probabilistic search technique that has its roots in the principles of genetics [Yuzgec et al. 2006].

Genetic algorithms (GA) are suitable for finding the optimum solution in problems where a fitness function is present. Genetic algorithms use a ''fitness'' measure to determine which of the individuals in the population survive and reproduce. Thus, survival of the fittest causes good solutions to progress. A GA works by selective breeding of a population of ''individuals'', each of which could be a potential solution to the problem [Kalogirou 2004].

The search procedure starts from a set of initial possible solutions that are represented by ''chromosomes"; this set is called the ''initial population". The solutions presented in a population are then chosen according to a fitness criterion and are then used in order to generate a new population of solutions. This multiplication and selection procedure permits the ''quality" of the new population to be improved with respect to the initial one. The generation and the selection procedures of new populations are repeated until a given convergence condition is satisfied.

The general form of a GA can be summarized as follows:

1. Start with a random generation of an initial population of N chromosomes.
2. Carry out a fitness evaluation, f(x), for each x-chromosome forming the population.
3. Apply a cross-over operation to the population in order to generate a new one according to the following steps [Dipama et al. 2008]:
   (a) Select two parent chromosomes according to their best fitness.
   (b) Use a cross-over probability in order to reproduce the two parents into two new chromosomes (offspring's). Note that if crossing the parents is not carried out, then the offspring's become an exact replica of their parents.
   (c) Use a mutation probability to modify the new chromosomes.
   (d) Relocate the new chromosomes in the population space.
4. Use this new population for continuing searching the best solution, i.e., continue the execution of the algorithm.
5. Carry out a test to check if a convenient convergent criterion is satisfied; if this condition is achieved stop the procedure and select the chromosome that has the best fitness as the solution of the problem.
6. If step 5 is not satisfied then go back to step 2.

One of the main advantages of GAs as opposed to other optimization techniques is their ease of use. Furthermore, easy-to-use commercial GA toolboxes are now available. One of the specificities of GAs is that they do not necessitate the calculation of the objective function gradient with respect to the design variables. This feature is particularly helpful in some cases such as for material allocation, ordering (combinatory) problems, multi-objective problems, and mixed integer nonlinear programming (MINLP) [Gosselin et al. 2009].

3 Applications of Soft Computing in Absorption Cooling Systems

Soft computing has been used by various researchers in absorption cooling system applications. This section presents an overview of these applications. The types of applications on the use of soft computing techniques in absorption cooling presented in this chapter are summarized in Table 1.

Table 1 Summary of numbers of applications presented in the absorption cooling applications

AI technique	Area	Number of applications
Artificial neural networks	Modeling of absorption systems Optimization of absorption systems Refrigerant-absorbent pairs	8 3 3
Fuzzy logic	Controller of solar air-conditioner Performance prediction	2 1
Genetic algorithms	Optimization of absorption systems	1
Data Mining	Modeling of absorption systems Refrigerant-absorbent pairs	1 1

3.1 Applications of Artificial Neural Networks

Table 2 shows a summary of applications of artificial neural networks for absorption system applications.

Sozen et al. [2003] used artificial neural networks for the analysis of ejector–absorption refrigeration systems (EARSs). ANNs method was used to determine the properties of liquid and two phase boiling and condensing of an alternative working fluid couple (methanol/LiBr), which does not cause ozone depletion for EARS. The back-propagation learning algorithm with three different variants and logistic sigmoid transfer function was used in the network. In addition, this paper presents a comparative performance study of the EARS using both analytic functions and the properties of the fluid couple predicted by the ANN. After training, it was found that the average error is less than 1.3% and R^2 values are about 0.9999.

Table 2 Summary of absorption system applications of artificial neural networks

Authors	Year	Subject
Sozen et al.	2003	Modeling of absorption systems
Sencan et al.	2007	
Sozen and Akcayol	2004	
Rosiek and Batlles	2010	
Manohar et al.	2006	
Sozen et al.	2004a	
Aly et al.	2010	
Sencan et al.	2006	
Chow et al.	2002	Optimization of absorption systems
Hernandez et al.	2009	
Colorado et al.	2010	
Sozen et al.	2004b	Refrigerant-absorbent pairs
Sozen et al.	2005	
Sencan and Kalogirou	2005	

Additionally, when the results of analytic equations obtained by using experimental data and by means of ANN were compared, deviations in coefficient of performance (COP), exergetic coefficient of performance (ECOP) and circulation ratio (F) for all working temperatures were found to be less than 1.8%, 4%, 0.2%, respectively. Deviations for COP, ECOP and F at a generator temperature of 90°C for which the COP of the system is maximum, are 1%, 2%, 0.1%, respectively.

A theoretical modeling of an absorption heat transformer for the temperature range obtained from an experimental solar pond was presented by Şencan et al. [2007]. The working fluid pair in the absorption heat transformer is aqueous ternary hydroxide fluid consisting of sodium, potassium and caesium hydroxides in the proportions 40:36:24 (NaOH:KOH:CsOH). Different methods such as linear regression (LR), pace regression (PR), sequential minimal optimization (SMO), M5 model tree, M50 rules, decision table and back propagation neural network (BPNN) are used for modeling the absorption heat transformer. The best results were obtained by the back propagation neural network model. A new formulation based on the BPNN is presented to determine the flow ratio (FR) and the coefficient of performance (COP) of the absorption heat transformer. Figure 7 shows the architecture of the BPNN used for the flow ratio and COP prediction. As seen from the figure, the evaporator temperature, absorber temperature, condenser temperature and generator temperature are the input data and the flow ratio (FR) and COP of the AHT are the actual outputs.

Sozen and Akcayol [2004] used artificial neural network for the performance analysis of a solar-driven ejector-absorption refrigeration system (EARS) with an aqua/ammonia working fluid. The use of artificial neural networks has been proposed to determine the performance parameters as functions of only the working temperature, under various working conditions. Thus, this study is considered to

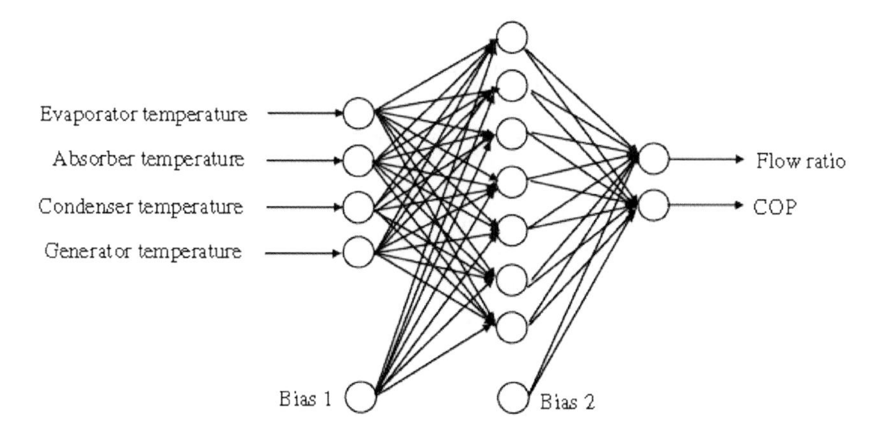

Fig. 7 BPNN model used for flow ratio and COP prediction [Şencan et al. 2007]

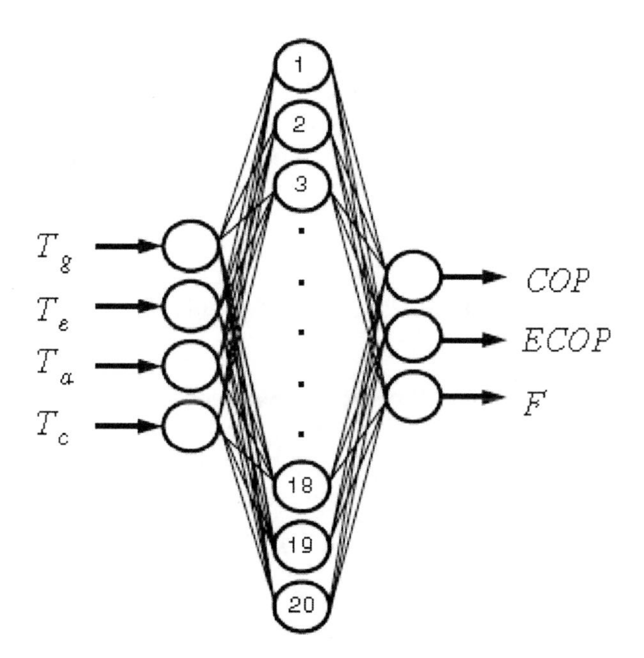

Fig. 8 ANN architecture used for the estimation of COP, ECOP and F [Reprinted from *Applied Energy Journal*, Vol. 79, Sozen and Akcayol, 2004, Modelling (using artificial neural-networks) the performance parameters of a solar-driven ejector-absorption cycle, pp. 309-325, Copyright 2004, with permission from Elsevier]

be helpful in predicting the performance of an EARS prior being set up in an environment where the temperatures are known. The statistical coefficient of multiple determination (R^2–value) is equal to 0.976, 0.9825, 0.9855 for the coefficient of performance (COP), exergetic coefficient of performance (ECOP) and circulation ratio (F), respectively. The ANN structure employed is shown in Fig. 8. The hidden layer has twenty neurons, and the input layer has four neurons for generator temperature (T_g), evaporator temperature (T_e), absorber temperature (T_a) and condenser temperature (T_c). There are three output neurons for COP, ECOP and F.

Rosiek and Batlles [2010] used ANNs to model a solar-assisted air-conditioning system installed in the Solar Energy Research Center (CIESOL). This system consists mainly of the single-effect LiBr-H_2O absorption chiller fed by water provided from either solar collectors or hot water storage tanks. The present work describes only solar cooling systems based on absorption chiller and powered by solar collectors. The experimental data were collected during the cooling period of 2008. The ANN was used with the main goal of predicting the efficiency of the chiller and global system using the lowest number of input variables. The configuration 7-8-4 (7 inputs, 8 hidden and 4 output neurons) was found to be the optimal topology. The results demonstrate the accuracy of ANN predictions with a Root Mean Square Error (RMSE) of less than 1.9% and practically null deviation, which can be considered very satisfactory. Figure 9 presents the configuration of the two-layer back propagation network selected in this work. The input layer includes the entering generator temperature (T_{eg}), leaving generator temperature (T_{lg}), entering evaporator temperature (T_{ee}), leaving evaporator temperature (T_{le}), incident radiation intensity (I), leaving flat-plate collector temperature (T_{out}) and collector's mass flow rate (m_c). The hidden layer has seven nodes, and the output layer includes four neurons representing the coefficient of

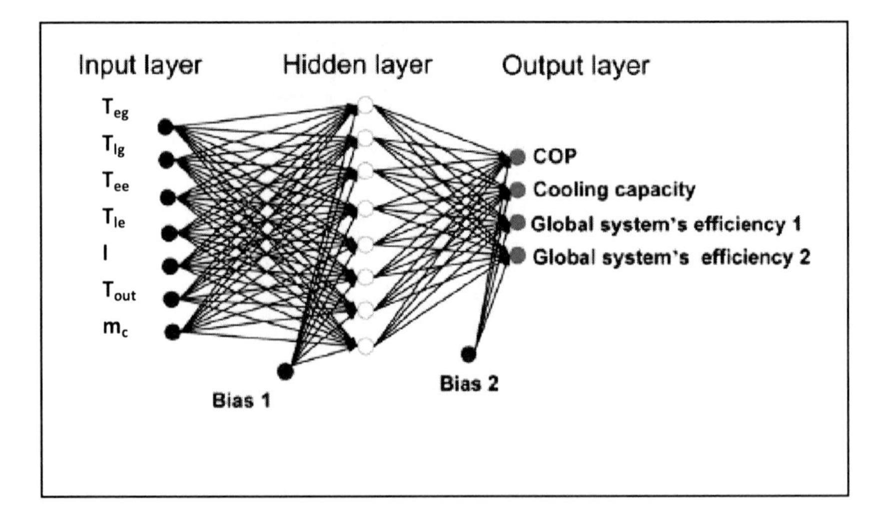

Fig. 9 ANN architecture used for the absorption system [Reprinted from *Renewable Energy Journal*, Vol. 35, Rosiek and Batlles, 2010, Modelling a solar-assisted air-conditioning system installed in CIESOL building using an artificial neural network, pp. 2894-2901, Copyright 2010, with permission from Elsevier]

performance (COP), cooling capacity (Q_{cool}), global system efficiency 1 (η_{s1}) and global system efficiency 2 (η_{s2}). The global system efficiency 1 (η_{s1}) is defined as the quotient between the cooling capacity and the useful collectors' array energy. The global system efficiency 2 (η_{s2}) is defined as the quotient between the cooling capacity and the incidence energy on the collectors' array.

Manohar et al. [2006] carried out the modeling of a double effect absorption chiller using steam as heat input. The modeling is based on the artificial neural network technique with 6-6-9-1 configuration (i.e., it includes two hidden layers with 6 and 9 neurons in each). The neural network is a fully connected feed forward configuration using the back propagation learning algorithm. The model predicts the chiller coefficient of performance (COP) based on the time, chilled water inlet (C_{hi}) and outlet (C_{ho}) temperatures, cooling water inlet (C_{wi}) and outlet (C_{wo}) temperatures and steam pressure (stpr). The network was trained with one year of experimental data and predicts the performance within ±1.2% of the actual values. Figure 10 shows the schematic diagram of the four-layer, feed forward ANN used for modeling the absorption system.

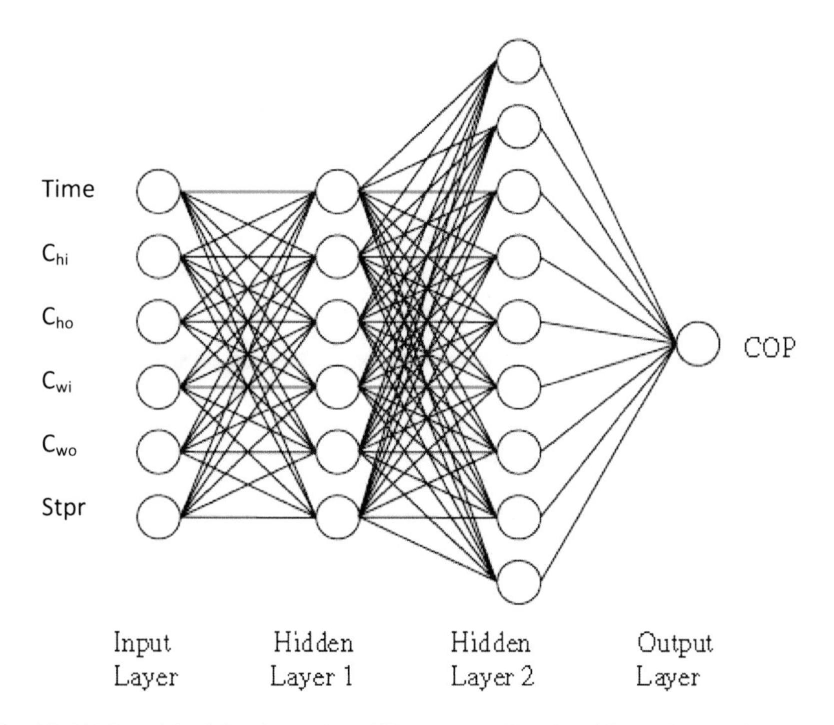

Fig. 10 ANN model of the absorption chiller system [Reprinted from *Energy Conversion and Management Journal*, Vol. 47, Manohar et al., 2006, Modelling of steam fired double effect vapour absorption chiller using neural network, pp. 2202-2210, Copyright 2006, with permission from Elsevier]

Sozen et al. [2004a] used artificial neural networks to determine the properties of the liquid and two-phase boiling and condensation of an alternative working fluid couple (methanol/LiCl), which does not cause ozone depletion. A comparative performance study of the ejector absorption heat pump (EAHP) is performed between the analytic functions and the values predicted by the ANN for the properties of the couple. The back propagation learning algorithm with three different variants and logistic sigmoid transfer function were used in the network. In order to train the neural network, limited experimental measurements were used as training and test data. The input layer consists of three neurons; temperature, pressure and concentration of the couples. Specific volume is in the output neuron. After training, it was found that the maximum error was less than 3%, the average error was less than 1.2% and the R^2 values were about 0.9999. Additionally, the comparison of the results between analytic equations obtained by using experimental data and the ANN show that the deviations of the refrigeration effectiveness of the system for cooling (COP_r), exergetic coefficient of performance of the system for cooling ($ECOP_r$) and circulation ratio (F) for all working temperatures were less than 1.7%, 5.1%, and 1.9%, respectively. Deviations for COP_r, $ECOP_r$ and F at a generator temperature of 90°C (cut off temperature), at which the coefficient of performance of the system is maximum, are 0.9%, 1.8%, and 0.1%, respectively. When this system was used for heating, similar deviations were obtained. In Fig. 11, the selected neural network architecture is shown schematically.

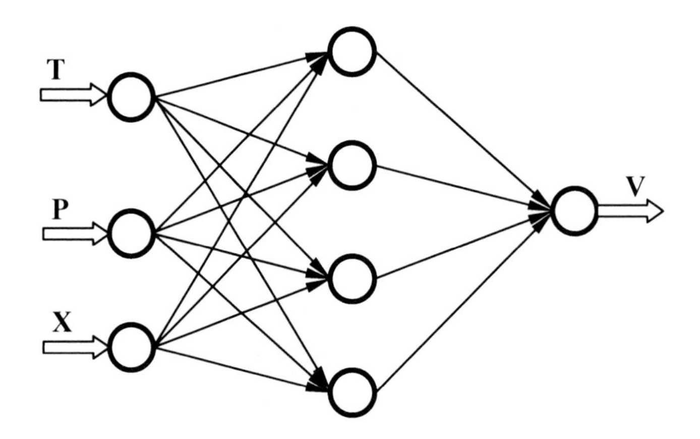

Fig. 11 Neural network architecture for thermal properties of methanol/LiCl [Sozen et al. 2004a]

Aly et al. [2011] carried out an investigation on the performance of lithium chloride (LiCl) absorption cooling system using an artificial neural network model. Using the proposed model, the effect of system design parameters, namely; regenerator length and air flow rate on the performance of the system are investigated. Additionally, the variation of the thermo-physical parameters along the regenerator length is presented.

Sencan et al. [2006] carried out the thermodynamic analysis of absorption systems using an artificial neural network. ANN is used for the determination of the thermodynamic properties of LiBr–water and LiCl–water solutions. The ANN is successfully applied to determine the enthalpy values of both solutions. The R^2-values in both cases were about 0.999, which can be considered as very satisfactory. In this study, in order to calculate the enthalpy values, mathematical formulations were derived from the ANN model. In addition, performance analysis of absorption systems operating with LiBr–water and LiCl–water solutions is carried out. Enthalpy values of both solutions were obtained using simple equations derived from the ANN models. Figure 12 shows the architecture of the ANN used for the LiBr–water solution enthalpy prediction. In this, the temperature and concentration are the input parameters and enthalpy of the solution is the actual output. Configuration 2-8-1 appeared to be the most optimal topology for this application. Figure 13 shows the architecture of the ANN used for the LiCl–water solution enthalpy prediction. Again the temperature and concentration are the input parameters and enthalpy of the solution is the actual output. Configuration 2-4-1 appeared to be the most optimal topology for this application.

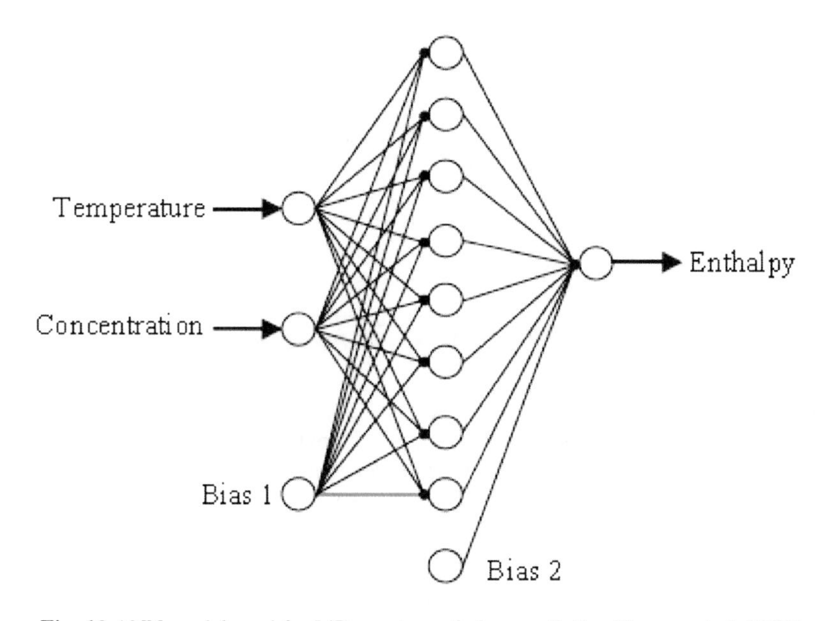

Fig. 12 ANN model used for LiBr–water enthalpy prediction [Sencan et al. 2006]

Chow et al. [2002] investigated the concept of integrating a neural network and a genetic algorithm in the optimal control of an absorption chiller system. Based on a commercial absorption unit, a neural network was used to model the system characteristics. A genetic algorithm is also employed as a global optimization tool. Figure 14 gives a brief outline of the optimization plan.

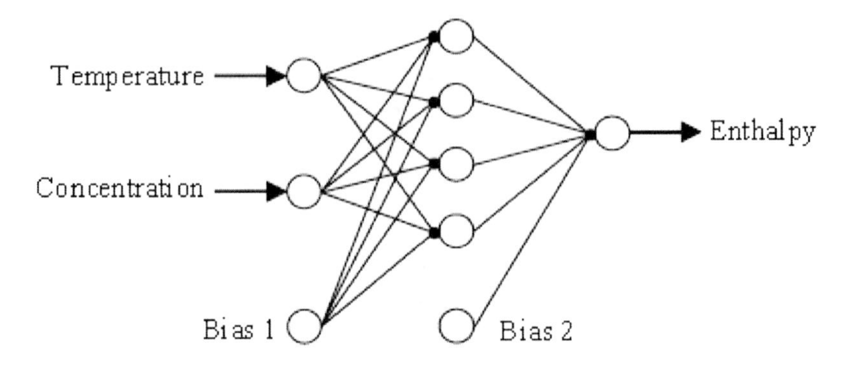

Fig. 13 ANN model for LiCl–water enthalpy prediction [Sencan et al. 2006]

Artificial neural network inverse (ANNi) is applied to calculate the optimal operating conditions on the coefficient of performance (COP) for a water purification process integrated to an absorption heat transformer with energy recycling [Hernandez et al. 2009]. An artificial neural network (ANN) model is developed to predict the COP which was increased with energy recycling. This ANN model takes into account the input and output temperatures for each one of the four components (absorber, generator, evaporator, and condenser), as well as two pressures and LiBr- H_2O concentrations. For the network, a feedforward with one hidden layer, Levenberg–Marquardt learning algorithm, hyperbolic tangent sigmoid transfer function in the hidden layer and linear transfer function in the output layer were used. The best fit of the training dataset was obtained with three neurons in the hidden layer. Simulations and experimental data test were in good agreement for the validation dataset (R> 0.99). This ANN model can be used to predict the COP when the input variables (operating conditions) are known. Input variables are input-temperature in the absorber that comes from generator ($T_{in.GE-AB}$), input-temperature in the absorber that comes from evaporator ($T_{in.EV-AB}$), output-temperature in the absorber towards generator ($T_{out.AB-GE}$), input-temperature in the generator that comes from absorber ($T_{in.AB-GE}$), output-temperature in the generator towards condenser ($T_{out.GE-CO}$), output-temperature in the generator towards absorber ($T_{out.GE-AB}$), input-temperature of the condenser that comes from generator ($T_{in.CO}$), output-temperature in the condenser towards evaporator ($T_{out.CO}$), input-temperature in the evaporator that comes from condenser ($T_{in.EV}$), output-temperature in the evaporator towards absorber ($T_{out.EV-AB}$), pressure in absorber (P_{AB}), pressure in generator (P_{GE}), LiBr concentration in the absorber inlet ($X_{in.AB}$), LiBr concentration in the absorber outlet ($X_{out.AB}$), LiBr concentration in the generator inlet ($X_{in.GE}$) and LiBr concentration in the generator outlet ($X_{out.GE}$). However, to control the COP of the system, a strategy is developed to estimate the optimal input variables when a COP is required from ANNi. An optimization

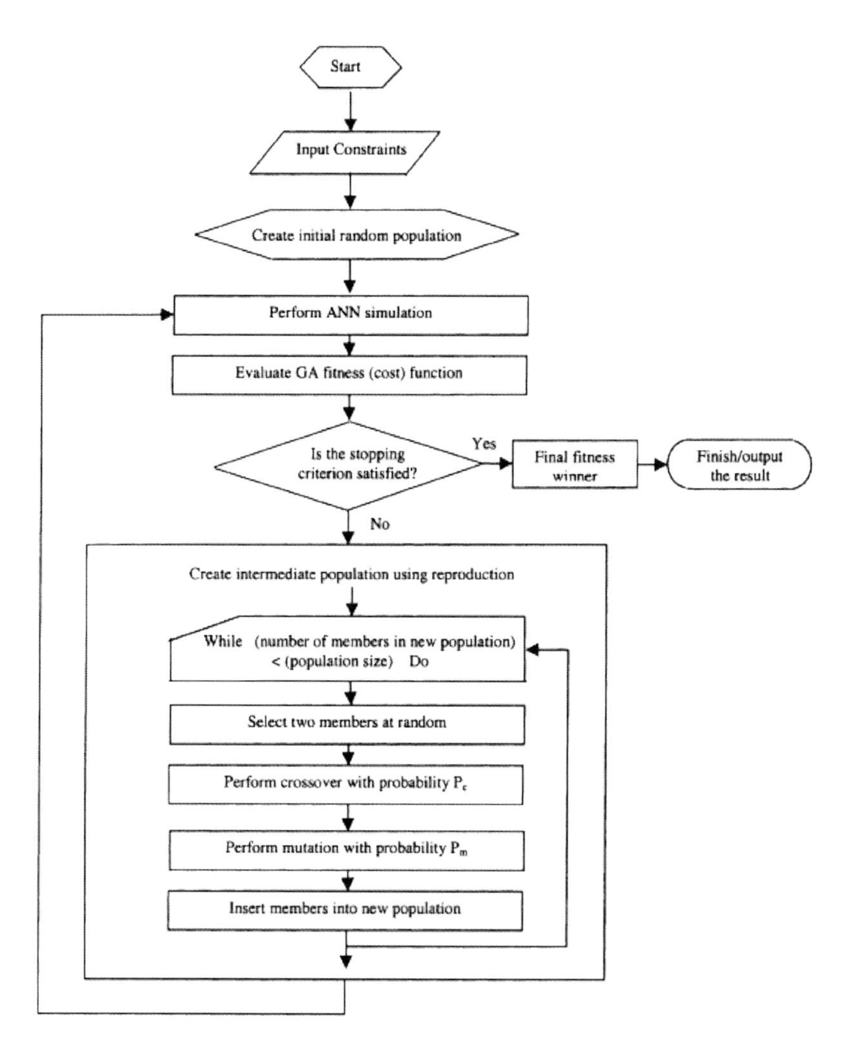

Fig. 14 Optimization process based on ANN and GA [Reprinted from *Energy and Buildings Journal*, Vol. 34, Chow et al., 2002, Global optimization of absorption chiller system by genetic algorithm and neural network, pp. 103-109, Copyright 2002, with permission from Elsevier]

method (the Nelder–Mead simplex method) is used to fit the unknown input variables resulted from the ANNi. The neural network model shown in Fig. 15 with three neurons in the hidden layer (51 weights and four biases) was found to be the most efficient in predicting the COP values of the water purification process integrated to an absorption heat transformer with energy recycling.

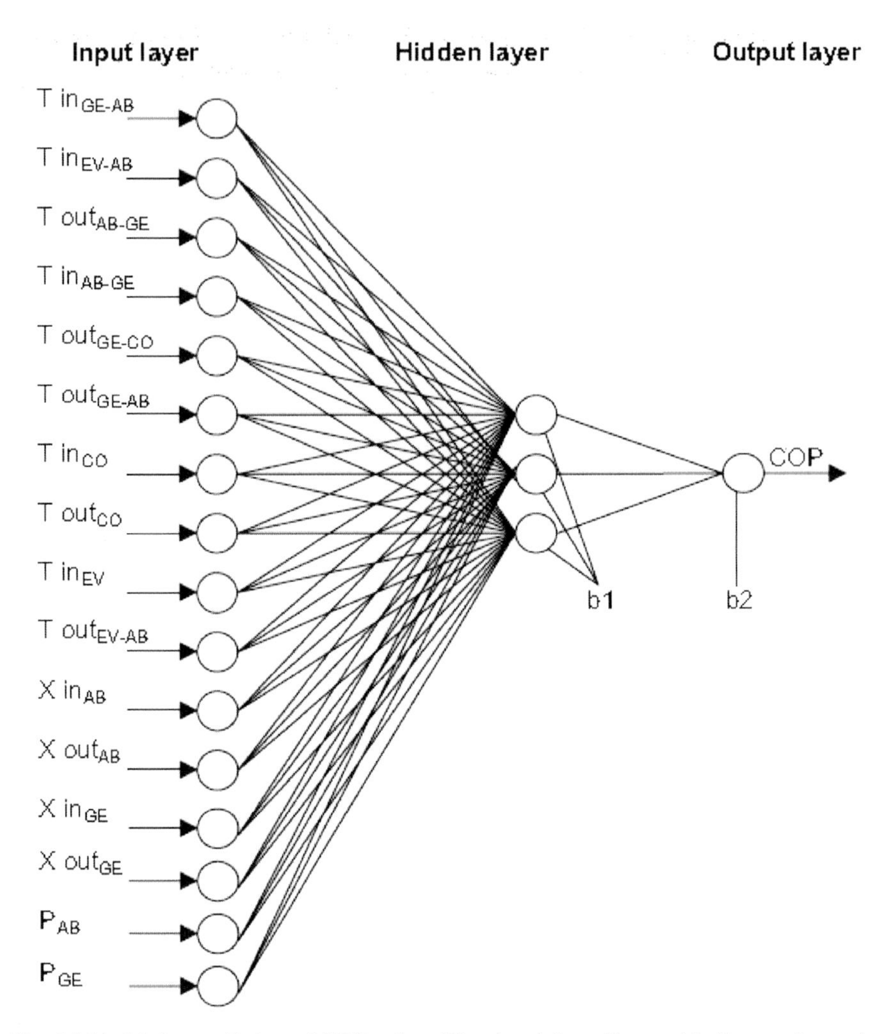

Fig. 15 Model for prediction of COP values [Reprinted from *Renewable Energy Journal*, Vol. 34, Hernandez et al., 2009, Optimum operating conditions for a water purification process integrated to a heat transformer with energy recycling using neural network inverse, pp. 1084-1091, Copyright 2009, with permission from Elsevier]

Colorado et al. [2011] determined the optimal operation conditions of a single-stage heat transformer by means of an artificial neural network inverse. Analysis based on first and second law of thermodynamics together with the direct and artificial neural networks inverse (ANNi) have been used to develop a methodology to decrease the total irreversibility of an experimental single-stage heat transformer. With the proposed methodology it is possible to calculate the optimal input parameters that should be used in order to operate the heat transformer with lower irreversibilities. The mathematical validation of ANNi was carried out with

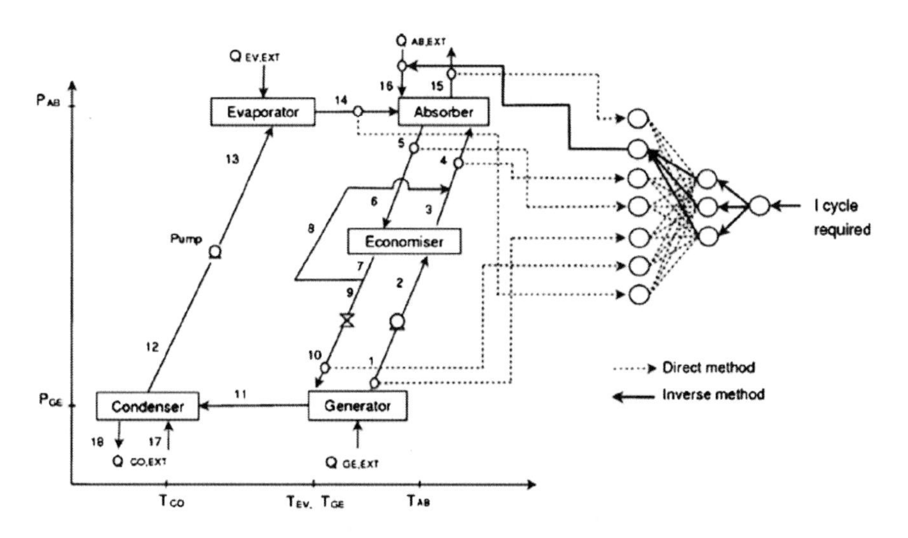

Fig. 16 Application of neural network inverse in an absorption heat transformer [Reprinted from *Applied Energy Journal*, Vol. 88, Colorado et al., 2011, Optimal operation conditions for a single-stage heat transformer by means of an artificial neural network inverse, pp. 1281-1290, Copyright 2011, with permission from Elsevier]

a comparison between the total cycle irreversibility (I_{cycle}) obtained thermodynamically and the I_{cycle} determined by using the ANNi. The results showed a mean discrepancy of 0.9% of the I_{cycle} values. Figure 16 shows a schematic diagram of the strategy to connect the ANNi to the absorption heat transformer.

Sozen et al. [2004b] also used artificial neural networks to determine the properties of liquid and two phase boiling and condensation of two alternative refrigerant/absorbent couples (methanol–LiBr and methanol–LiCl). The back-propagation learning algorithm and logistic sigmoid transfer function were used in the network. Variants of the algorithm used in the study are scaled conjugate gradient (SCG), Pola–Ribiere conjugate gradient (CGP), and Levenberg–Marquardt (LM). In order to train the neural network, limited experimental measurements were used as training and test data. In input layer, various temperatures are used in the range of 298–498K (with 25K increase), pressures (0.1–40MPa) and concentrations of 2, 7, and 12% of the couples. The specific volume is the output parameter. After training, it was found that the maximum error is less than 3%, average error is about 1% and R^2 value is 99.999.

Finally, in a similar work Sozen et al. [2005] used a new approach based on artificial neural networks to determine the properties of liquid and two phase boiling and condensation of two alternative refrigerant/absorbent couples (methanol/LiBr and methanol/LiCl). In order to train the neural network, the same measurements as in Sozen et al. [2004b] were used as training data and test data. The input and output data were also the same as in previous model [Sozen et al. 2004b]. In this work back-propagation learning algorithm with three different variants, namely scaled conjugate gradient (SCG), Pola–Ribiere conjugate gradient (CGP), and

Levenberg–Marquardt (LM), and logistic sigmoid transfer function were used in the network so that the best approach is selected. The most suitable algorithm is found to be SCG with 8 neurons in the hidden layer. For this topology, it is found that after the training, the maximum error is less than 3%, the average error is about 1% and R^2 value is 99.999. As a result, formulation was given to determine the properties of liquid and two phase boiling and condensation of two alternative refrigerant/absorbent couples.

Sencan and Kalogirou [2005] used artificial neural networks (ANN) to determine the thermodynamic properties of two alternative refrigerant/absorbent couples (LiCl - H_2O and LiBr + $LiNO_3$ + LiI + LiCl - H_2O). These pairs can be used in absorption heat pump systems. In order to train the network, limited experimental measurements were used as training and test data. Two feedforward ANNs were trained, one for each pair, using the Levenberg-Marquardt algorithm. The inputs of the network are concentration (X) and temperature (T), and the output is vapor pressure. The training and validation was performed with good accuracy. The correlation coefficient obtained when unknown data were used to the networks was 0.9997 and 0.9987 for the two pairs respectively which is very satisfactory. The present methodology proved to be much better that the linear multiple regression analysis. Using the weights obtained from the trained network a new formulation is presented for the determination of the vapor pressures of the two refrigerant/absorbent couples. In Table 3 a comparison is presented between the actual vapor pressure and vapor pressure predicted with the equations derived from ANN for LiCl-water fluid couple. In Table 4 a comparison is presented between the actual vapor pressure and vapor pressure predicted with the equations derived from ANN for LiBr + $LiNO_3$ + LiI + LiCl-water fluid couple.

Table 3 Comparison between actual vapor pressure and vapor pressure obtained with equations derived from ANN for LiCl-water fluid couple [Sencan and Kalogirou 2005]

X (%)	T (°C)	Actual vapor pressure (mmHg)	ANN predicted vapor pressure (mmHg)	Error	Percentage difference (%)*
12.907	40	46.79	46.80	-0.01	-0.02
19.265	60	107.33	107.36	-0.03	-0.02
22.768	40	34.31	34.30	0.01	0.03
22.768	80	227.87	228.00	-0.13	-0.05
26.456	60	79.67	79.61	0.06	0.07
29.788	60	66.55	66.53	0.02	0.03
33.692	50	31.22	31.21	0.01	0.03
36.976	40	14.04	13.93	0.11	0.78
36.976	90	166.21	166.46	-0.25	-0.15
40.756	70	52.27	52.07	0.20	0.38
44.186	40	8.01	7.96	0.05	0.62
44.186	80	67.48	68.05	-0.57	-0.84

* Percentage difference (%) = (Error/ Actual vapor pressure) *100.

Table 4 Comparison between actual vapor pressure and vapor pressure obtained with equations derived from ANN for LiBr + $LiNO_3$ + LiI + LiCl-water fluid couple [Sencan and Kalogirou 2005]

X (%)	T (K)	Actual vapor pressure (kPa)	ANN predicted vapor pressure (kPa)	Error	Percentage difference (%)*
50	350.71	12.35	11.95	0.40	3.24
50	386.47	54.78	54.13	0.65	1.19
50	405.01	98.92	98.45	0.47	0.48
51.8	353.53	12.24	12.41	-0.17	-1.39
51.8	400.66	81.51	79.77	1.74	2.13
55	334.21	3.81	3.96	-0.15	-3.94
55	367.18	19.11	18.55	0.56	2.93
58	361.84	11.00	10.99	0.01	0.09
58	399.04	52.09	53.39	-1.30	-2.49
60	361.06	8.77	8.83	-0.06	-0.68
60	387.59	27,66	27,68	-0.02	-0.07

* Percentage difference (%) = (Error/ Actual vapor pressure) *100.

3.2 Applications of Fuzzy Logic

Table 5 shows a summary of fuzzy logic applications for absorption systems.

Table 5 Summary of absorption system applications of fuzzy logic

Authors	Year	Subject
Lygouras et al. Lygouras et al.	2007 2008	Controller of solar air-conditioner
Sozen et al.	2004c	Performance prediction

The implementation of a variable structure fuzzy logic controller for a solar powered air conditioning system and its advantages were investigated by Lygouras et al. [2007]. Two DC motors are used to drive the generator pump and the feed pump of the solar air-conditioner. Two different control schemes for the DC motors rotational speed adjustment are implemented and tested: the first one is a pure fuzzy controller, its output being the control signal of the DC motor driver. A 7x7 fuzzy matrix assigns the controller output with respect to the error value and the derivative of the error. The second scheme is a two-level controller. The lower level is a conventional PID controller, and the higher level is a fuzzy controller acting over the parameters of the low level controller. Step response of the two

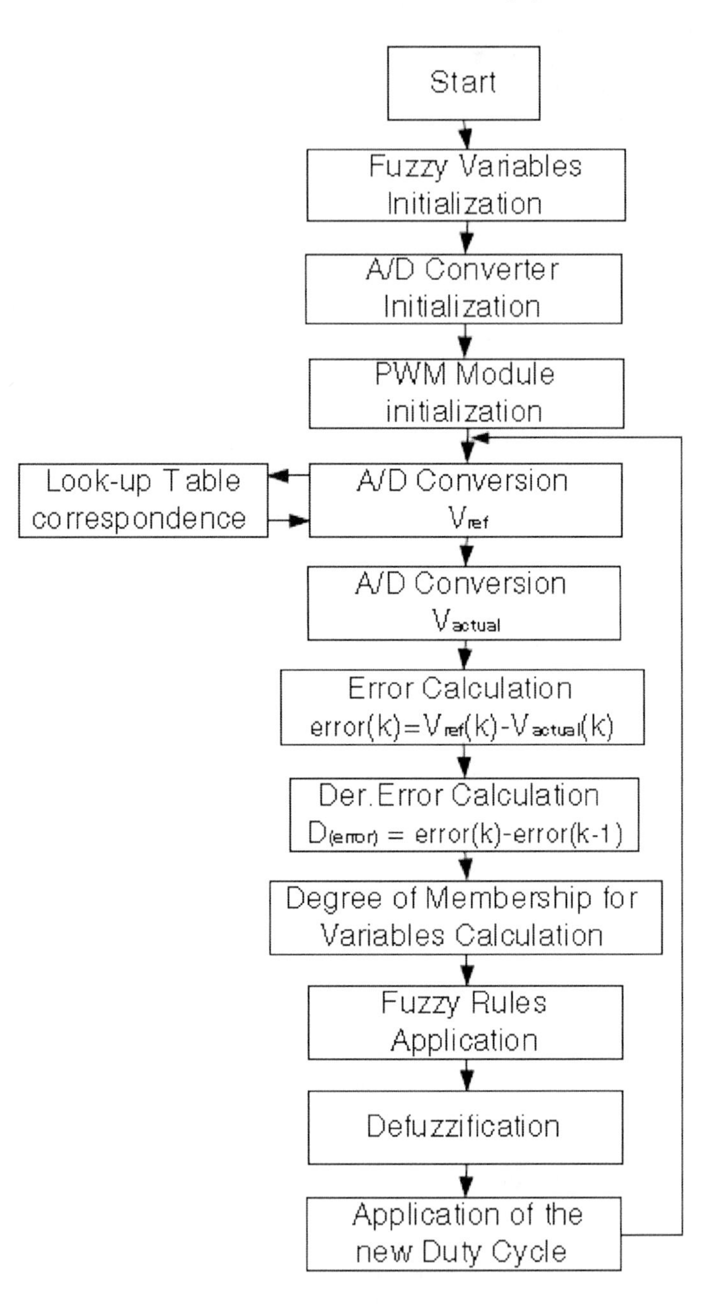

Fig. 17 Flow chart implemented in fuzzy controller software [Reprinted from *Applied Energy Journal*, Vol. 84, Lygouras et al., 2007, Fuzzy logic controller implementation for a solar air-conditioning system, pp. 1305-1318, Copyright 2007, with permission from Elsevier]

control loops are presented as experimental results. The contribution of this design is that in the control system, the fuzzy logic is implemented through software in a common, inexpensive, 16-bit microcontroller, which does not have special abilities for fuzzy control. In Fig. 17, the flow chart implemented in fuzzy controller software is shown.

The design and implementation of a Two-Input/Two-Output (TITO) variable structure fuzzy-logic controller for a solar-powered air-conditioning system was also described by Lygouras et al. [2008]. Two DC motors are used to drive the generator pump and the feed pump of the solar air-conditioner. The first affects the temperature in the generator of the solar air-conditioner, while the second, the pressure in the power loop. The difficulty of Multi-Input/Multi-Output (MIMO) systems control is how to overcome the coupling effects among each degree of freedom. Initially, a traditional fuzzy-controller has been designed, its output being one of the components of the control signal for each DC motor driver. Subsequently, according to the characteristics of the system's dynamics coupling, an appropriate coupling fuzzy-controller (CFC) is incorporated into a traditional fuzzy-controller (TFC) to compensate for the dynamic coupling among each degree of freedom. This control strategy simplifies the implementation problem of fuzzy control, but can also improve the control performance. This mixed fuzzy controller (MFC) can effectively improve the coupling effects of the systems, and this control strategy is easy to design and implement. In Fig. 18, the flow chart implemented in the fuzzy-logic controller software is shown.

Sozen et al. [2004c] carried out a performance analysis of solar driven ejector-absorption refrigeration system (EARS) operated with aqua/ammonia. The performance of EARS was predicted using a fuzzy logic controller at different working conditions instead of complex rules and mathematical routines. Input data for the fuzzy logic controller are experimental results performed in the climatic conditions of Ankara, Turkey. Fuzzy input variables are generator temperature (T_g), evaporator temperature (T_e), condenser temperature (T_c), absorber temperature (T_a) and fuzzy output variables are the coefficient of performance (COP), exergetic coefficient of performance (ECOP) and circulation ratio (F). The results between analytic equations and by means of fuzzy logic controller were compared to evaluate the performance of the controller and found that the deviations of COP, ECOP, F for all working temperatures are less than 2, 5 and 0.2%, respectively. The statistical coefficient of multiple determination (R^2-value) equal to 1.0, 0.9996 and 1.0 for the COP, ECOP and F, respectively. Figure 19 shows the configuration of the fuzzy block.

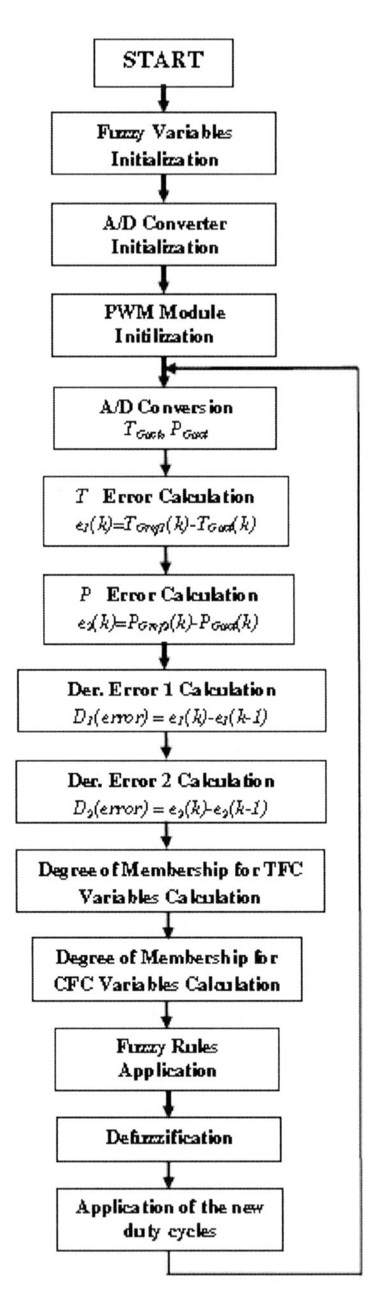

Fig. 18 Flow chart implemented in the TITO Fuzzy-controller software [Reprinted from *Applied Energy Journal*, Vol. 85, Lygouras et al., 2008, Variable structure TITO fuzzy-logic controller implementation for a solar air-conditioning system, pp. 190-203, Copyright 2008, with permission from Elsevier]

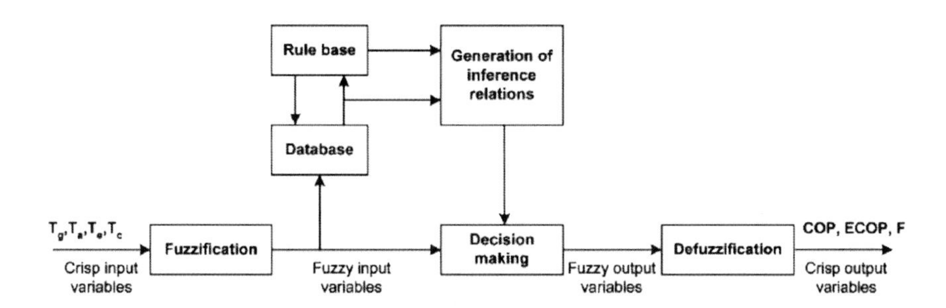

Fig. 19 The configuration of the fuzzy logic estimator [Reprinted from *Renewable Energy Journal*, Vol. 29, Sozen et al., 2004c, Performance prediction of a solar driven ejector-absorption cycle using fuzzy logic, pp. 53-71, Copyright 2004, with permission from Elsevier]

3.3 Applications of Genetic Algorithms

Table 6 summarizes various applications of genetic algorithms for absorption systems.

Table 6 Summary of absorption system applications of genetic algorithms

Authors	Year	Subject
Chow et al.	2002	Optimization of absorption systems

Only one application is currently available in literature on this subject carried out by Chow et al. [2002] who investigated the concept of integrating neural network and genetic algorithm in the optimal control of an absorption chiller system. Based on a commercial absorption unit, neural network was used to model the system characteristics and genetic algorithm as a global optimization tool. Figure 15, shown earlier, gives a brief outline of the optimization plan.

3.4 Applications of Data Mining

Table 7 summarizes various applications of data mining for absorption systems.

Table 7 Summary of absorption system applications of data mining

Authors	Year	Subject
Sencan et al.	2007	Modeling of absorption systems
Sencan	2007	Refrigerant-absorbent pairs

A theoretical modeling of an absorption heat transformer for the temperature range obtained from an experimental solar pond was presented by Şencan et al. [2007]. The working fluid pair in the absorption heat transformer is aqueous ternary hydroxide fluid consisting of sodium, potassium and caesium hydroxides in the proportions 40:36:24 (NaOH:KOH:CsOH). Different methods such as linear regression (LR), pace regression (PR), sequential minimal optimization (SMO), M5 model tree, M5 rules, decision table and back propagation neural network (BPNN) are used for modeling the absorption heat transformer. As seen before, the best results were obtained by the back propagation neural network model. However, the various data mining methods applied gave acceptable results for both the flow ratio (FR) and the coefficient of performance (COP) of the absorption heat transformer. The inputs to all methods tried were the evaporator temperature, absorber temperature, condenser temperature and generator temperature.

Şencan [2007] also used data mining process to determine the thermodynamic properties of two alternative refrigerant/absorbent couples (methanol/LiBr and methanol/LiCl). Linear regression (LR), pace regression (PR), sequential minimal optimization (SMO), M5 model tree, M5'Rules and back propagation neural network (BPNN) models are applied within the data mining process for determining the specific volume of the methanol/LiBr and methanol/LiCl fluid couples. The best result was obtained by using the back propagation model although the results obtained with the various data mining processes are very close. The R^2 value for the predicted specific volume of the methanol/LiBr fluid couple is 0.9840 and for the predicted specific volume of the methanol/LiCl fluid couple is 0.9966, which can be considered very satisfactory. The worsed results were obtained by the simple linear regression which gave respective values of 0.926 and 0.813 for the two fluid couples investigated. Finally, a new formulation is presented in the paper for the determination of the specific volumes of the two refrigerant/absorbent couples based on ANN results.

4 Conclusions

In this chapter, various soft computing techniques used for absorption systems modeling, prediction and control have been reviewed. A summary of available literature published in this area are presented. Soft computing techniques are becoming useful as alternate approaches to conventional techniques. Soft computing have been used and applied in different areas, such as engineering, economics, medicine, military, marine, etc. They have also been applied for modeling, identification, optimization, prediction and control of complex systems, such as absorption machines.

The importance of the using soft computing for applications in absorption systems can be seen from the many applications presented in this chapter. Soft computing techniques have been applied successfully in the wide range of absorption system applications.

References

Absorption Chillers (2010), `http://www.absorptionchillers.com` (accessed 12.12.2010)

Aly, A.A., Zeidan, E.-S.B., Hamed, A.M.: Performance evaluation of open-cycle solar regenerator using artificial neural network technique. Energy and Buildings 43, 454–457 (2011)

Chaturvedi, D.K.: Soft Computing: Techniques and Its Applications in Electrical Engineering. Springer, Heidelberg (2008) ISBN 978-3-540-77480-8

Chow, T.T., Zhang, G.Q., Lin, Z., Song, C.: Global optimization of absorption chiller system by genetic algorithm and neural network. Energy and Building 34, 103–109 (2002)

Cobaner, M., Unal, B., Kisi, O.: Suspended sediment concentration estimation by an adaptive neuro-fuzzy and neural network approaches using hydro-meteorological data. Journal of Hydrology 367, 52–61 (2009)

Colorado, D., Hernández, J.A., Rivera, W., Martínez, H., Juárez, D.: Optimal operation conditions for a single-stage heat transformer by means of an artificial neural network inverse. Applied Energy 88, 1281–1290 (2011)

Dipama, J., Teyssedou, A., Sorin, M.: Synthesis of heat exchanger networks using genetic algorithms. Applied Thermal Engineering 28, 1763–1773 (2008)

Efendigil, T., Onut, S., Kahraman, C.: A decision support system for demand forecasting with artificial neural networks and neuro-fuzzy models: A comparative analysis. Expert Systems with Applications 36, 6697–6707 (2009)

Ertunc, H.M., Hosoz, M.: Comparative analysis of an evaporative condenser using artificial neural network and adaptive neuro-fuzzy inference system. International Journal of Refrigeration 31, 1426–1436 (2008)

Foryuna, L., Rizzotto, G., Lavorgna, M., Nunnari, G., Xibilia, M.G., Caponetto, R.: Soft computing: new trends and applications. Springer-Verlag London Limited, Great Britain (2001) ISBN 1-85233-308-1

Goldberg, D.E.: Genetic Algorithms in Search, Optimization and Machine Learning. Addison-Wesley, Harlow (1989)

Gosselin, L., Tye-Gingras, M., Mathieu-Potvin, F.: Review of utilization of genetic algorithms in heat transfer problems. International Journal of Heat and Mass Transfer 52, 2169–2188 (2009)

Haykin, S.: Neural networks: a comprehensive foundation. Macmillan, New York (1994)

Hernandez, J.A., Bassam, A., Siqueiros, J., Juarez-Romero, D.: Optimum operating conditions for a water purification process integrated to a heat transformer with energy recycling using neural network inverse. Renewable Energy 34, 1084–1091 (2009)

Herold, K.E., Radermacher, R., Klein, S.A.: Absorption chilles and heat pumps. CRC Press, New York (1996)

Hoffmann, D., Apostolakis, J.: Crystal Structure Prediction by Data Mining. Journal of Molecular Structure 647, 17–39 (2003)

Holland, J.H.: Adaptations in Natural Artificial Systems. University of Michigan Press, Michigan (1975)

Jang, J.: ANFIS: adaptive-network-based fuzzy inference system. IEEE Transactions on Systems, Man and Cybernetics 23(3), 665–685 (1993)

Kalogirou, S.A.: Applications of artifcial neural-networks for energy systems. Applied Energy 67, 17–35 (2000)

Kalogirou, S.A.: Artificial neural networks in renewable energy systems applications: a review. Renewable and Sustainable Energy Reviews 5, 373–401 (2001)

Kalogirou, S.A.: Optimization of solar systems using artificial neural-networks and genetic algorithms. Applied Energy 77, 383–405 (2004)

Kucukali, S., Baris, K.: Turkey's short-term gross annual electricity demand forecast by fuzzy logic approach. Energy Policy 38, 2438–2445 (2010)

Lau, H.C.W., Cheng, E.N.M., Lee, C.K.M., Ho, G.T.S.: A fuzzy logic approach to forecast energy consumption change in a manufacturing system. Expert Systems with Applications 34, 1813–1824 (2008)

Li, S.T., Shue, L.Y.: Data Mining To Aid Policy Making in Air Pollution Management. Expert System With Applications 27, 331–340 (2004)

Lygouras, J.N., Botsaris, P.N., Vourvoulakis, J., Kodogiannis, V.: Fuzzy logic controller implementation for a solar air-conditioning system. Applied Energy 84, 1305–1318 (2007)

Lygouras, J.N., Kodogiannis, V.S., Pachidis, T., Tarchanidis, K.N., Koukourlis, C.S.: Variable structure TITO fuzzy-logic controller implementation for a solar air-conditioning system. Applied Energy 85, 190–203 (2008)

Manohar, H.J., Saravanan, R., Renganarayanan, S.: Modelling of steam fired double effect vapour absorption chiller using neural network. Energy Conversion and Management 47, 2202–2210 (2006)

MatLab (2010) MATLAB Neural Network Toolbox v. 4.0.4, `http://www.mathworks.com/access/helpdesk/help/pdf_doc/nnet/nnet.pdf`

Matlab Fuzzy logic toolbox user's guide (2010) Natick: The Math Works Inc., `http://www.mathworks.com/`

Paulescu, M., Gravila, P., Tulcan-Paulescu, E.: Fuzzy logic algorithms for atmospheric transmittances of use in solar energy estimation. Energy Conversion and Management 49, 3691–3697 (2008)

Rosiek, S., Batlles, F.J.: Modelling a solar-assisted air-conditioning system installed in CIESOL building using an artificial neural network. Renewable Energy 35, 2894–2901 (2010)

Sen, Z., Oztopal, A., Sahin, A.D.: Application of genetic algorithm for determination of Angström equation coefficients. Energy Conversion & Management 42, 217–231 (2001)

Sencan, A., Kalogirou, S.A.: A new approach using artificial neural networks for determination of the thermodynamic properties of fluid couples. Energy Conversion and Management 46, 2405–2418 (2005)

Sencan, A., Yakut, K.A., Kalogirou, S.A.: Thermodynamic analysis of absorption systems using artificial neural network. Renewable Energy 31, 29–43 (2006)

Sencan, A., Kızılkan, O., Bezir, N.C., Kalogirou, S.A.: Different methods for modeling absorption heat transformer powered by solar pond. Energy Conversion and Management 48, 724–735 (2007)

Sencan, A.: Modeling of thermodynamic properties of refrigerant/absorbent couples using data mining process. Energy Conversion and Management 48, 470–480 (2007)

Sozen, A., Arcaklioglu, E., Ozalp, M.: A new approach to thermodynamic analysis of ejector–absorption cycle: artificial neural networks. Applied Thermal Engineering 23, 937–952 (2003)

Sozen, A., Akcayol, M.A.: Modelling (using artificial neural-networks) the performance parameters of a solar-driven ejector-absorption cycle. Applied Energy 79, 309–325 (2004)

Sozen, A., Arcaklioglu, E., Ozalp, M.: Performance analysis of ejector absorption heat pump using ozone safe fluid couple through artificial neural Networks. Energy Conversion and Management 45, 2233–2253 (2004a)

Sozen, A., Özalp, M., Arcaklioglu, E.: Investigation of thermodynamic properties of refrigerant/absorbent couples using artificial neural Networks. Chemical Engineering and Processing 43, 1253–1264 (2004b)

Sozen, A., Kurt, M., Akcayol, M.A., Ozalp, M.: Performance prediction of a solar driven ejector-absorption cycle using fuzzy logic. Renewable Energy 29, 53–71 (2004c)

Sozen, A., Arcaklioglu, E., Ozalp, M.: Formulation based on artificial neural network of thermodynamic properties of ozone friendly refrigerant/absorbent couples. Applied Thermal Engineering 25, 1808–1820 (2005)

Takagi, T., Sugeno, M.: Fuzzy identification of systems and its applications to modeling and control. IEEE Transactions on Systems Man and Cybernetics 15(1), 116–132 (1985)

Xie, H.B., Jiang, Z.Y., Liu, X.H., Wang, G.D., Tieu, A.K.: Prediction of coiling temperature on run-out table of hot strip mill using data mining. Journal of Materials Processing Technology 177, 121–125 (2006)

Yuzgec, U., Becerikli, Y., Turker, M.: Nonlinear predictive control of a drying process using genetic Algorithms. ISA Transactions 45(4), 589–602 (2006)

A Comprehensive Overview of Short Term Wind Forecasting Models Based on Time Series Analysis

Athanasios Sfetsos

Environmental Research Laboratory,
Institute of Nuclear Technology and Radiation Protection,
National Centre for Scientific Research Demokritos,
153 10, Ag. Paraskevi, Greece
ts@ipta.demokritos.gr

Abstract. This chapter presents a comprehensive overview of short term wind forecasting models based on time series analysis. Several different approaches, presently considered as mature, are re-examined with an eye towards setting automated procedures to clarify grey areas in their application. Additionally, some approaches recently proposed in the literature are examined that include the application of localized linear models, and clustering algorithms coupled with linear and nonlinear models. Additionally, the impact of changing synoptic weather characteristics is captured, through the utilization of global meteorological variables and the subsequent development of a customized regime model. The application of the developed approach on an annual hourly wind speed data set is presented.

1 Introduction

The current large-scale introduction of wind power in the energy mix of European countries has undermined the necessity of power system operators, at all levels to better understand the dynamic behavior and variability of the wind characteristics. This would significantly reduce uncertainties in key wind power generation decision for optimal scheduling and dispatching. The forecasting of wind behaviour (either in terms of speed or directly power) has been identified as an important element to the decision making process that would be used to effectively incorporate the variability in wind power in the operation of power systems [NERC 2009, EWEA 2007, 2008].

The problem of forecasting wind resource can be split into temporal categories depending primarily on the time scale of the analysis and subsequently the intended application:

Very Short Term (seconds to a few minutes). Flow is dominated by turbulence, and forecasts depend on the current conditions and trends that cause changes over short periods of time. These forecasts are used for the operational aspects.

K. Gopalakrishnan et al. (Eds.): Soft Comput. in Green & Renew. Ener. Sys., STUDFUZZ 269, pp. 97–116.
springerlink.com

Short Range (up to 72h). Wind flow is on a combination of large scale atmospheric motion and microclimatic effects. Most forecasts over this period of time are based on statistical models which utilize relationships between observed quantities. These forecasts are of interest for their potential use in economic dispatch and unit commitment.

Medium Range (up to 7 days). Numerical Weather Prediction models that predict general circulation characteristics of the atmosphere and associated conditions are mainly applied, but recently statistical models (either alone or as part of Model Output Statistics) are also tested. This time horizon is primarily used for resource planning, e.g. fuels and maintenance issues.

Monthly and Seasonal. These forecasts usually have a structure that attempts to identify whether the specific quantity will be above, below or near normal over a specific interval. Such forecasts will be of use in resource planning and allocation.

Special Category. They involve forecasts of specific events or interest, for example periods of extreme winds or winds above the operation range of a wind turbine. Often, these forecasts take the form of probabilities that an event like this will occur under certain circumstances.

Additionally, the increased participation of wind producers in the electricity market environment relies on the knowledge of wind power contribution to the generation mix to effectively design more beneficiary trading strategies. The determination of both daily and hourly prices and their variation will influence the clearing prices for both energy and operating reserves. Thus the increased knowledge of the wind behavior is crucial for a number of purposes, such as: generation and transmission maintenance planning, determination of operating reserve requirements, unit commitment, economic dispatch, energy storage optimization (e.g., pumped hydro storage), and energy trading.

The introduction of Numerical Weather Prediction models have attracted the interest of the scientific community in recent years with several R&D efforts, such as ANEMOS [Kariniotakis 1999], POW'WOW[1], WILMAR[2] (both from Risoe), and ANEMOS.plus[3]. The installation and deployment of highly sophisticated NWP models such as the WRF (Skamarock et al. 2005), MM5 (Dudhia 1994), has become fairly straight-forward for anyone without advanced knowledge on meteorology. Furthermore the increase of available computational power has increased the spatial resolution of the applied model, reaching the order of 1km^2 for regional applications thus allowing for accurate representation of topographic impacts on the wind patterns (Vlachogiannis 2008). Another advantage of NWP models is that output is provided at many different heights, specified by the user, which wind generators are located (typically 50-100m agl). Finally, the performance of NWP can be enhanced with the introduction of Model Output Statistics (MOS) [Glahn 1972] for post-processing wind speed/ power predictions (e.g., Zephyr model [Nielsen et al 2001]).

---

[1] http://powwow.risoe.dk/

[2] http://www.wilmar.risoe.dk/index.htm.

[3] http://anemosplus.cma.fr/.

The literature has numerous statistical models based on the time series analysis, such as the Kalman Filters, Auto-Regressive Moving Average (ARMA), Auto-Regressive with Exogenous Input (ARX), and Box-Jenkins forecasting methods. On the majority of the applications these are univariate models and in only a handful of cases they treat exogenous variables (e.g., wind direction, temperature), which can improve the forecast error. A Kalman filter [Bossanyi 1985] with the last six measured values as inputs was applied for minutes ahead forecasting. The results were good when compared with the persistence for time horizons below 10 min. of averaged data, but poorer in longer averages and nonexistent for 1-hr averages. Contaxis et al. [1991] employed an AR model to forecast the wind speed for time horizons ranging between 30 min. and 5 hr for controlling an isolated hybrid diesel/wind system. Poggi et al. [2003] proposed an monthly update for autoregressive model in order to forecast the wind speed for the following 3 hr , whereas Torres et al. [2005] used five ARMA models to forecast the hourly average wind speed for ten hours in advance, reporting a 20% error reduction as compared to persistence. Kavasseri et al. [2009] presented the fractional-ARIMA (f-ARIMA) model to forecast the daily wind speed, being able to capture long-range correlations. El-Fouly et al. [2006] presented a new technique based on the Grey predictor model reporting an improvement against persistence in the range of 12% for the hourly wind forecast.

Alexiadis et al. [1998] proposed a NN model to forecast hourly wind speed using spatial inputs to the models, showing an improvement of 32% over persistence in the forecast error for a 1-hr horizon. Sfetsos [2002] in comparing different models found that the non-linear models overcame the performance of linear models and that all the non-linear models presented comparable RMSE. Maqsood et al. [2005] used more than one model to forecast wind speed for 24-hr-ahead. Four different NNs were trained for each season of the year. The best result was found when an ensemble of models was used. Barbounis and Theocharis [2006] and [152] employed locally recurrent neural networks to forecast wind speed and power 72 hr ahead, based on meteorological information. Abdel-Aal et al. [2009] applied abductive networks based on the group method of data handling to forecast the mean hourly wind speed. The model achieved an improvement of 8.2% compared to persistence in a 1-hr-ahead forecast.

Damousis and Dokopoulos [2001] and Damousis et al. [2004] present a Takagi-Sugeno FIS (optimized by a genetic algorithm) based on onsite and nearby locations of wind for a time horizon of between 30 and 240 min. The improvement over persistence ranged between 9.5% and 28.4%, depending on the time horizon (it increases with the time horizon). Potter [2006] developed an ANFIS to forecast the wind speed for a 2.5-min. time horizon. Wind speed data adjusted through splines considerably decrease the forecast error relative to persistence. Ramírez-Rosado and Fernández-Jiménez [2003] employed fuzzy time series to forecast the wind generation for a time horizon of 24 hr. Fuzzy linguistic information about wind allowed the forecasting method to register an improvement of 14.3% over persistence. Frías et al. [2007] developed a wind power model based on ANFIS and using online generation data of wind farms jointly with forecasts for the daily market.

Costa et al. [2003] tested a purely and fuzzy autoregressive, as well as an MLP NN, in order to forecast 10 steps ahead with 10-min. wind power data. The authors report the NN as having the best overall performance. Kusiak et al. in [2009] tested many different data-mining models to forecast the wind power: SVM, MLP NN, the M5P tree algorithm, the Reduced Error Pruning tree, and the bagging tree. The SVM and MLP NN returned the best forecasting ability, SVM for forecasts between 10 min. to 1 hr, whereas the MLP NN up to 4 hr. Fugon et al. [2008] compared the performance of data-mining models (NN, SVM, regression trees with bagging, and random forests for regression) and two reference linear regression models. All models outperformed persistence, and a superiority of the nonlinear models was found using data from three wind farms in France. Jursa [2007] compared different models for wind power forecasts, including MLP NN, mixture of experts, SVM, and nearest neighbor search with a Particle Swarm Optimization algorithm for feature selection. The results for 10 wind farms located in Germany showed that the best model was the ensemble with three different models (i.e., mixture of experts, nearest neighbor, and SVM), with a 15% improvement over an NN.

Pinson et al. [136] applied regime-switching models: the self-exciting threshold autoregressive (SETAR), the smooth transition autoregressive (STAR), and the Markov-switching autoregressive (MSAR). The performance of the models was evaluated on a one-step-ahead forecast in two Danish wind farms. In all test cases, the MSAR models significantly outperformed the rest. Ramírez-Rosado and Fernández-Jiménez [2004] developed a three-phase model using an FFT transform of the last 24 values of mean wind speed is computed, 23 fuzzy inference systems (Takagi-Sugeno) to forecast the coefficients of the Fourier transform, which were then used to forecast wind speed for the following hour.The reader is directed to the wealth of information published in the literature for detailed reviews on wind forecasting methodologies and results [Giebel 2003, Costa et al 2008, Wu et al 2007, Landberg et al 2003, Leia et al 2009,].

The present study aims to cover an aspect that is currently overlooked in the area of wind forecasting: *the analysis of the impact of exogenous meteorological variables on the development of basic and advanced forecasting models*. The examined variables included simple meteorological parameters that would be easily measured by a SCADA system on site. Additionally, data from the Mediterranean island of Corsica have been used as case studies to support the implementation of the developed models.

2 Model Inventory

This section describes the developed models that have been employed for the purposes of the present study. The different forecasting approaches that are employed during the course of this study, can be described by the generalized equation (1), that in principle combines the forecatsed value of the series, y_t, with past observations, y_{t-k}, exogenous variables, x_{t-j}, and previous error terms, e_{t-k}. The function f can be of any type, either linear or nonlinear.

$$y_t = f\,(y_{t-k}, x_{t-j}, e_{t-k}) \tag{1}$$

2.1 Linear Regression

The LR approach uses a linear equation to determine whether a variable of interest (y_t) is linearly related to past observations of the series (y_{t-p}), and exogenous parameters ($\mathbf{x}_q$), which in this case are meteorological parameters. The expression that governs this model is the following:

$$y_t = a + \sum_p \beta_p y_{t-p} + \sum_q \gamma_q x_q \tag{2}$$

The coefficients α, β and γ are usually estimated from a least squares algorithm. As inputs to the linear models, variables (here p and q) significantly different from zero on the 95% confidence level are selected using a *backwards stepwise elimination procedure*. Therefore, the final forecast is made with only those variables judged as statistically significant from the Student t-test statistic.

2.2 Feed Forward Artificial Neural Networks

Feed-forward Neural Networks are nowadays a common forecasting tool mainly due to their non-linear capabilities and ability to deal with large data sets. The operating principles are presented with many details in many Artificial Intelligence textbooks e.g. [Lin and Lee 1996]. The response of a neuron in the output layer as a function of its inputs is given from:

$$y_i = f_l \left(\sum_{q=1}^{l} w_{iq} f_2 \left(\sum_{j=1}^{m} v_{qj} x_j + b_j \right) + b_q \right) \tag{3}$$

where f1 and f2 can be non-linear sigmoid, linear or threshold activation functions.

The strength of neural networks lies in their ability to simulate any given problem, which is achieved from the modification of the network parameters through learning algorithms. Thus an ANN with the same structure can model a variety of different processes either linear or non-linear. In this study, the Levenberg-Marquardt algorithm is employed [Hagan and Menhaj 1996] because of its speed and robustness against the conventional back-propagation.

2.2.1 Neural Network Model Considerations

The most important issue concerning the introduction of ANN in time series forecasting is "generalization", which refers to their ability to produce reasonable forecasts on data sets other than those used for the estimation of the model parameters. This issue has two important parameters that should be accounted for. The first is data preparation, which involves pre-processing and the selection of the most significant variables. The second embraces the determination of the optimum model structure that is closely related with the estimation of the model parameters. Some useful insight can be found using statistical methods such as the correlation coefficients.

The second aspect can jointly be tackled under the cross-validation training scheme. The data set is split into three smaller sets the training (TS), the evaluation or validation (ES) and the prediction or testing (PS) sets. The model is initialized with few parameters. The next step is to train the model using data from the training set and when the error of the evaluation set is minimized, the model parameters and configuration are stored. The number of parameters is then increased and a new network is trained from the beginning. If ES error is lower compared to the previously found minimum, then the parameters of this new model are stored. This iterative process is terminated when a predefined number of iterations are reached (Fig. 1). The advantage of this procedure is that the model architecture is not defined prior to the training phase, but the entire process becomes more time consuming. The performance and forecasting ability of each model is measured on the totally unknown prediction or "out-of-sample" set.

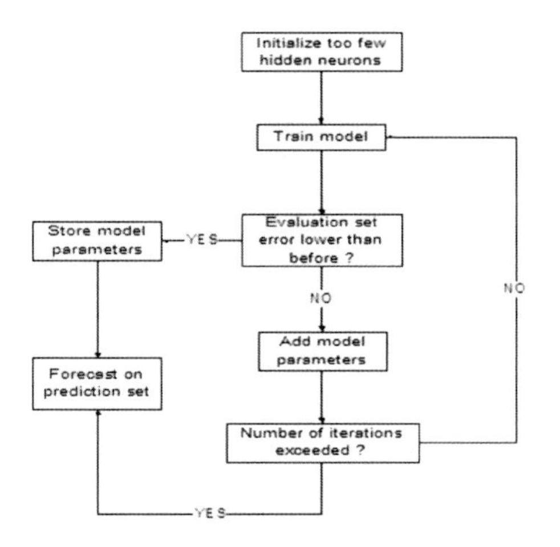

Fig. 1 Iterative Cross-validation Training Scheme

In this study, ES was formed using a Euclidean metric withholding a percent (here 25% is used) of the TS data that are located nearest to other data. The strength of this approach lies in the fact that TS covers more distinct characteristics of the process, thus, allowing for the development of a model with better generalization capabilities.

2.3 Nearest Neighbours

This class of hybrid models includes a local modeling and a function approximation to capture recent dynamics of the process. The idea behind these predictors is that segments of the series neighboring under some distance measure may correspond to similar future values. This claim was endorsed by the work of Farmer and

Sidorowich [1988] that showed that chaotic time-series prediction is several orders of magnitude better using local approximation techniques than universal approximators. The tricky part in these models is the selection of the embedding dimension, which effectively determines segments of the series, and the number of neighbors. For the purposes of this study the input to the NN model were the same as for the neural network:

$$y_t = [y_{t-p}, \ldots, x_q] \tag{4}$$

The number of neighbors was not pre-determined but was set to vary between predefined limits. A small number of neighbors increase the variance of the results, whereas a large number can compromise the local validity of a model and increase the bias of results. Once the nearest neighbors to y_t have been identified, an averaging procedure is followed in the present study to generate predictions.

2.4 ANFIS Basics

An Adaptive Neuro-Fuzzy Inference System [Jang 1993, 1995 can incorporate fuzzy if-then rules and also, provide fine-tuning of the membership function according to a desired input-output data pair. A first order Sugeno fuzzy model [1986] is used as a means of modeling fuzzy rules into desired outputs.

$$\text{if } x_1 = A_k; \ldots; \text{ and } x_n = B_j \text{ then } f_i = p_i x_1 + q_i x_n + r_i \tag{5}$$

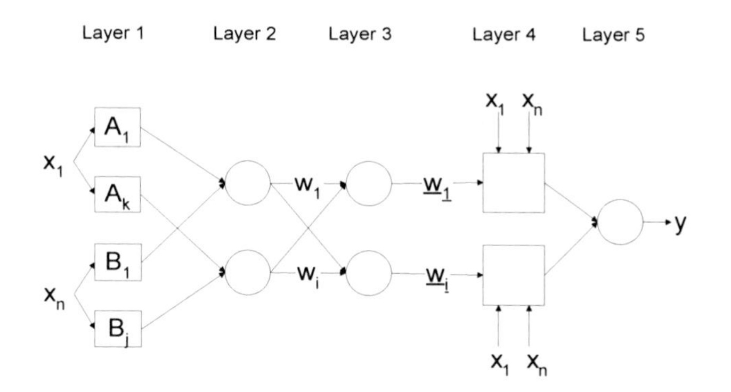

Fig. 2. ANFIS architecture

Each neuron in the first layer corresponds to a linguistic label and the output equals the membership grade of this linguistic label.

$$OL1_i = \mu_{Ak}(x_1) \tag{6}$$

Each neuron in layer 2 estimates the firing strength of a rule, wi, which is found from the multiplication of the incoming signals.

$$OL2_i = \mu_{Ak}(x_1) \cdot \mu_{Bj}(x_n) \tag{7}$$

Each neuron in layer 3 estimates the relative firing strength of a rule which is found as the ratio of the ith rule's firing strength to the sum of the firing strength of all rules, (j in total).

$$OL3_i = \underline{w_i} = \frac{w_i}{\sum_{i=1}^{j} w_i} \tag{8}$$

The output of layer 4 is the product of the previously found relative firing strength of the i-th rule and the following rule, f_i.

$$OL4_i = \underline{w_i} \cdot f_i = \underline{w_i} \cdot (p_i x_1 + q_i x_n + r_i) \tag{9}$$

The final layer computes the overall output as the summation of all incoming signals from layer 4.

$$OL5_i = \sum_{i=1}^{j} \underline{w_i f_i} = \frac{\sum_i w_i f_i}{\sum_i w_i} \tag{10}$$

The results are then defuzzified using a weighted average procedure. A back-propagation training method is employed to find the optimum value for the parameters of the membership functions and a least squares procedure for the linear parameters on the fuzzy rules, so that the error between the input and the output pairs is minimized. The total number of rules, j, equals the possible combinations of the number of memberships function of each variable.

2.5 Hybrid Clustering Algorithm (HCA)

The hybrid clustering algorithm is an iterative procedure that groups data, based on their distance from the hyper-plane that best describes their relationship. It is implemented through a series of steps, which are presented below:

(i) Determine the most important variables.
(ii) Form the set of patterns $H(t) = [y_t, y_{t-k}, \mathbf{x}_{t-k}]$.
(iii) Select the number of clusters n_h.
(iv) Initialize the clustering algorithm so that n_h clusters are generated and assign patterns.
(v) For each new cluster, apply a linear regression model to y_t using as explanatory variables the remaining of the set H_t.

(vi) Assign each pattern to a cluster based on their distance.
(vii) Go to **(v)** unless any of the termination procedures is reached.

The following termination procedures are considered: (a) the maximum pre-defined number of iterations is reached and (b) the process is terminated when all patterns are assigned to the same cluster as in the previous iteration in (vi). The selection of the most important lagged variables, (i), is based on the examination of the correlation coefficients of the data.

The proposed clustering algorithm is a complete time series analysis scheme with a dual output. The algorithm generates clusters of data, the identical characteristic of which is that they "belong" to the same hyper-plane, and synchronously, estimates a linear model that describes the relationship amongst the members of a cluster. Therefore, a set of n_h linear equations is derived.

$$y_{t,i} = a_{o,i} + \sum a_{i,k} y_{t-k} + \sum b_{i,j} X_{t-j} \quad , \quad i = 1 \ldots n_h \tag{11}$$

Like any other hybrid model that uses the target variables in the development stage, the model requires a secondary scheme to account for this lack of information in the forecasting phase. For HCA the only requirement is the determination of the cluster number, n_h and n_{cl} respectively, which is equivalent to the estimation of the final forecast.

The optimum number of HCA clusters is found from a modified cluster validity criterion. An estimate of under-partition (U_u) of the data was formed using the inverse of the average value of the coefficient of determination (R_i^2) on all regression models. U_o indicates the over-partitioning of the data set, and d_{min} is the minimum distance between linear models (eq 6). The optimum number is found from the minimization of a normalized combinatory expression of these two indices.

$$U_u = \frac{1}{\frac{1}{h}\sum_{i=1}^{h} R_i^2} \qquad U_o = \frac{c}{d_{\min}} \tag{12}$$

2.5.1 *Pattern Recognition*

A pattern recognition scheme with three alternative approaches was then applied to convert the LMCA and HCA output to the final predictions. Initially, a conventional clustering (k-means) algorithm was employed to identify similar historical patterns in the time series. The second was to determine n_{cl}/n_h at each time step, using information contained in the data of the respective cluster.

(p1) Select a second data vector : $P_t = [y_{t-k}, \mathbf{x}_{t-k}]$
(p2) Initialize a number of clusters n_k
(p3) Apply a k-means clustering algorithm on P_t.
(p4) Assign data vectors to each cluster, so that each of the n_k clusters should contain k_m, m = 1,..., n_k data.

To obtain the final forecasts the following three alternatives were examined:

(M1) From the members of the k-th cluster find the most frequent HCA cluster, i.e n_{cl} / n_h number.

(M2) From the members of the k-th cluster estimate the final forecast as a weighted average of the HCA clusters. Here p_i is the percentage of appearances of the LMCA / HCA cluster in the k-th cluster data.

$$y_t = \sum_i p_i y_{t,n} \qquad i = 1,...,k \;\; and \;\; n = 1,...,n_h \text{ or } n_{cl} \tag{13}$$

(M3) From the members of the k-th cluster estimate the final forecast as a distance weighted average of the HCA clusters.

$$\begin{aligned} y_t &= \sum_i t_i y_{t,n} \qquad i = 1,...,k \;\; and \;\; n = 1,...,n_h \text{ or } n_{cl} \\ d_i &= \left\| P_t - P_i \right\| \\ t_i &= \frac{d_i^{-a}}{\sum_i d_i^{-a}} \text{ and } a = 2 \end{aligned} \tag{14}$$

The optimal number of clusters for the pattern recognition stage was determined using the modified compactness and separation criterion for the k-means algorithm [Kim et al 2001].

2.6 Local Models Based on Clusters

The idea behind the application of clustering algorithms in time series analysis is to identify groups of data that share some common characteristics. On each of these groups, the relationships amongst the members are modelled through a single equation model. Consequently, each of the developed models has a different set of parameters. The process is described in the following steps:

(i) Selection of the input data for the clustering algorithm. This can contain lagged and/or future characteristics of the series, as well as other relevant information.

$$C(t) = [y_t, y_{t-k}, \mathbf{x}_{t-j}]. \tag{15}$$

Empirical evidence suggests that the use of the target variable y_t is very useful to discover unique relationships between input-output features. Additionally, higher quality modelling is ensured with the function approximation since the targets have similar properties and characteristics. However, this occurs to the expense of an additional process needed to account for this lack of information in the prediction stage.

(ii) Application of a clustering algorithm combined with a validity index or with user defined parameters, so that n_{cl} clusters will be estimated.

(iii) Assign all patterns from the training set to the n_{cl} clusters. For each of the clusters, apply a function approximation model,

$$y_t = f_i(y_{t-k}, \mathbf{x}_{t-j}) \qquad i = 1 \ldots n_{cl}, \tag{16}$$

so that n_{cl} forecasts are generated.

In this study, the k means clustering algorithm was selected [McQueen 1967]. It is a partitioning algorithm that attempts to directly decompose the data set into a set of groups through the iterative optimization of a certain criterion. More specifically, it re-estimates the cluster centres through the minimization of a distance-related function between the data and the cluster centres. The algorithm terminates when the cluster centres stop changing.

2.7 Error Metrics

In addition to previously described models, the ideal case of a perfect knowledge (PCF) of the ncl / nh parameter in the HCA is also presented. This indicates the predictive potential, or the least error that the respective methodology could achieve. Also, the base-case persistent approach ($y_t = y_{t-1}$) is shown as a relative criterion for model inter-comparison amongst different data sets. The ability of the models to produce accurate forecasts was judged against the following statistical performance metrics:

Root Mean Square Error

$$RMS = \sqrt{\frac{1}{k}\sum_{i=1}^{k}(r_i - y_i)^2} \tag{17}$$

Mean Absolute Percentage Error

$$MAPE = 100\frac{1}{k}\sum_{i=1}^{k}\frac{|r_i - y_i|}{r_i} \tag{18}$$

Index of Agreement

$$IA = 1 - \frac{\sum_{i=1}^{k}(r_i - y_i)^2}{\sum_{i=1}^{k}\left(|\bar{r} - y_i| + |\bar{r} - r_i|\right)^2} \tag{19}$$

Fractional Bias

$$FB = \frac{(\bar{r} - \bar{y})}{0.5 * (\bar{r} + \bar{y})} \tag{20}$$

3 Data Description

The examined data set contains hourly values of the following parameters: Wind Speed and Direction, Temperature, and Pressure made during the entire year of 1996. The data were acquired form the French Meteorological Office and are from the city of Ajaccio, at the island of Corsica. The exact geographical coordinates of this location are 41.550°N and 8.430°E.

The data set contains in total 8362 points, excluding missing values. The data set was split into the predictions set, which contained all available data for the months of February, May, August and November, totaling 2763 data points. This set was kept aside from any model during the development stage. The statistical properties of the two subsets are presented in Table 1 and the time series in Figure 3.

Table 1 Statistical properties of examined data sets

	Training Set	Prediction Set
Number of Points	5619	2763
Maximum (m/s)	17.70	15.7000
Minimum (m/s)	0	0.4000
Mean (m/s)	4.8395	5.5255
Stand. Dev. (m/s)	2.3952	2.5962

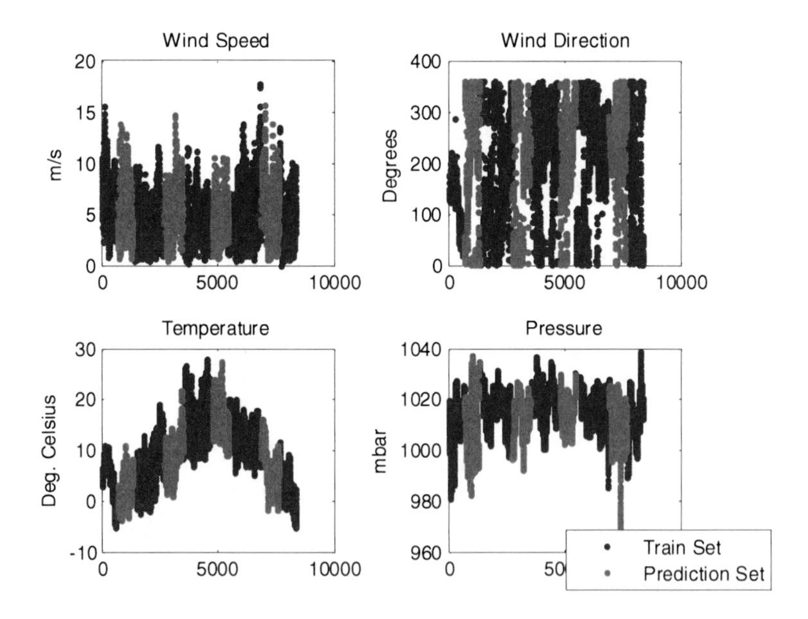

Fig. 3 Plot of Wind Speed and exogenous meteorological parameters

4 Modeling Results and Discussion

4.1 Univariate Models

Initially, the univariate approaches were examined. The application of the step-wise LR approach on the Training Set returned the WS(t-1), WS(t-2) and WS(t-24) as statistically significant variables. The model that is finally reached is introduced in Table 2. The most important variable is WS(t-1) followed by WS(t-2), which is an indication of the strong, short term correlation properties of the examined data series, evident in many wind speed studies.

Table 2 Model Development details

	Coef	**Std**	**t-stat**
Constant	0.2358	0.0274	8.6121
WS(t-1)	1.0298	0.0133	77.4915
WS(t-2)	-0.098	0.0133	-7.3615
WS(t-24)	0.02	0.0048	4.2067

The optimal model settings of the different models are:

- ANN: 4 layers in the hidden neuron with a sigmoid activation function
- ANFIS : 2 generalized Bell function for each input and a total of 8 rules
- NN: 5 Nearest neighbors
- HCA: ncl = 2 and nh = 15.

Table 3 Univariate model performance

	RMS	**MAPE**	**IA**	**FB**
Persistent	0.8335	14.0163	0.9736	-0.0004
LR	0.8189	14.1812	0.9731	0.0053
ANN	0.8168	14.3077	0.9735	0.0032
ANFIS	0.822	14.3227	0.973	0.0037
NN	0.879	15.5454	0.9683	0.0088
HCA-PCF	0.5387	9.2878	0.9888	0.0022
HCA-M3	0.8237	14.3011	0.9727	0.006

The results show that not all models are able to predict the wind speed on the Prediction set considerably better than the persistent approach. This pattern frequently occurs in wind forecasting analysis (e.g. Sfetsos 2002) and is attributed to the inability of hourly averages to represent structure in the time series on the high frequency side of the '*spectral gap*', lying at a period of typically around 1 hour.

4.2 Multivariate Models

The following step was concerned with the development of a multivariate series of models using exogenous meteorological parameters in addition to historical wind speed data. The finally derived model using the linear stepwise regression model is presented in Table 3.

The final variables of the models include historical wind speed parameters with WS(t-1), being the most significant of all, temperature and pressure differences of the forecast during the past 1h, 6h and 12h. The pressure at forecasted time, used to compute the differences has been estimated using a basic univariate model, with fairly accurate results. The latter set of variables can be considered as an indication of large scale forcings in the atmosphere.

The optimal model settings of the different models are:

- ANN: 8 layers in the hidden neuron with a sigmoid activation function
- ANFIS : 2 Gaussian function for each input
- NN: 5 Nearest neighbors
- HCA: ncl = 3 and nh = 12.

Table 4 Model Development details

	Coef	Std	t-stat
Constant	0.2321	0.0319	7.2756
WS(t-1)	0.9928	0.0133	74.67
WS(t-2)	-0.0863	0.0142	-6.0675
WS(t-6)	0.032	0.0071	4.4845
WS(t-24)	0.0187	0.005	3.7544
T(t-1)	0.1489	0.0159	9.3607
T(t-2)	-0.1514	0.0159	-9.519
P(t)-P(t-1)	-0.4098	0.0428	-9.565
P(t)-P(t-6)	0.0556	0.0137	4.0687
P(t)-P(t-12)	-0.0142	0.0063	-2.2725

Table 5 Univariate model performance

	RMS	MAPE	IA	FB
LR	0.7982	14.0761	0.9746	0.0046
ANN	0.7924	14.1094	0.9752	0.001
ANFIS	0.8026	14.0895	0.9744	0.0055
NN	1.1552	20.4315	0.9404	0.0267
HCA-PCF	0.3812	6.2724	0.9945	0.0019
HCA-M3	0.8087	14.1544	0.9741	0.0104

The analysis of the results (Table 5) shows that the error metrics of all models, with the exception of the NN, are better. This is an indication that the process of introducing exogenous variables in the forecasting phase could be beneficiary for the development of more accurate models. The overall best performance is found by the ANN (Fig 4), also being the better balanced one since FB is marginally different to 0.

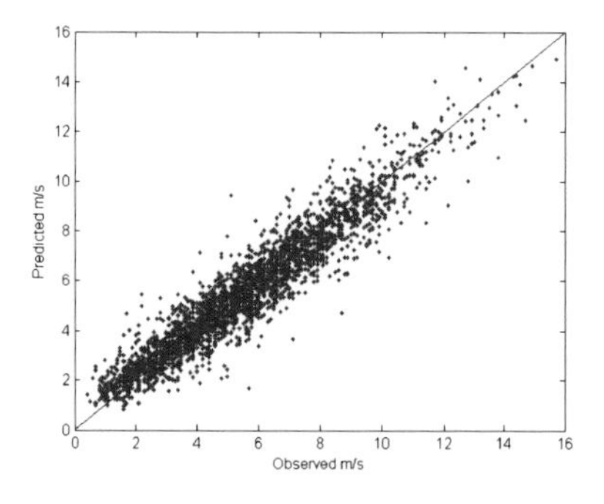

Fig. 3 ANN forecast (PS)

4.3 Cluster Models

The following analysis was based on the development of a set of clusters in an attempt to identify the finer properties of the wind speed properties. Using the k-means algorithm on the normalized set of variables found as most influential in the previous section, 8 clusters were identified. In each cluster, a different stepwise regression model was applied so that the parameters that are best describing the underlying wind behavior are determined.

Table 6 Cluster centers details

Cluster	WS(t-1)	WS(t-2)	WS(t-6)	WS(t-24)	T(t-1)	T(t-2)	P(t)-P(t-1)	P(t)-P(t-6)	P(t)-P(t-12)
1	10.0814	10.0912	9.2696	6.4321	9.7532	9.6873	-0.4666	-3.5929	-7.1864
2	3.9764	3.9548	3.6621	3.9162	19.2678	19.2507	-0.0853	-0.4723	-0.6877
3	7.4839	7.5712	7.8876	9.1625	7.2448	7.2744	0.6314	3.7131	6.288
4	2.693	2.6513	2.8165	3.5413	2.4098	2.4304	0.0051	0.1417	0.4474
5	6.5227	6.5464	6.556	5.944	0.5868	0.5968	0.0729	0.5025	1.023
6	6.4056	6.4556	6.4444	6.0067	13.7154	13.7387	0.0242	0.1452	0.0774
7	3.1916	3.1623	3.3928	3.7599	11.86	11.8512	0.0297	0.2049	0.3822
8	4.9931	4.9772	4.9367	4.9355	6.6671	6.6647	-0.1163	-0.5514	-0.7719

Each cluster can be described as representing one "wind regime", showing different characteristics

Cluster 1: Corresponds to high winds persisting for at least a day. The pressure exhibits a significant drop indicating the rapid arrival of a low pressure system. The most important variables for this model are the last recorded wind speed and pressure differences at the last hour and 6 hours before.

Table 7 Cluster 1, LR Model details

	Coef	Std	t-stat
Constant	-0.0257	0.3023	-0.085
WS(t-1)	0.9725	0.0297	32.6907
P(t)-P(t-1)	-0.9791	0.1206	-8.1169
P(t)-P(t-6)	0.0839	0.0271	3.0987

Cluster 2: The wind speed is considered medium to low, connected with high temperature values and a pressure drop of about 0.7 mbar per 12h. This pattern occurs on the warmer period of the year and the most important variables appear to be the last two recordings of wind speed, indicative of a rather turbulent process.

Table 8 Cluster 2, LR Model details

	Coef	Std	t-stat
Constant	0.7465	0.0937	7.9669
WS(t-1)	0.9643	0.0377	25.5635
WS(t-2)	-0.1501	0.0382	-3.9311

Cluster 3: Exhibits medium to high winds and an increase in pressure over the last 12 hours. This type of weather is mainly associated the fast arrival of a high pressure system or passage of a weather front. The correlation with the last recorded value, demonstrates short memory of the process and a possible indication of a rather turbulent and gusty regime.

Table 9 Cluster 3, LR Model details

	Coef	Std	t-stat
Constant	0.3139	0.2482	1.2646
WS(t-1)	0.9379	0.0334	28.1153

Cluster 4: is related with low to medium winds persisting over a period of 24h. It is mainly found in cold winter days (low temperature), with a slow increase in pressure. The most important variables are the last wind speed value in addition to the pressure difference of the last 12h.

Table 10 Cluster 4, LR Model details

	Coef	Std	t-stat
Constant	0.2099	0.0531	3.9527
WS(t-1)	0.959	0.0179	53.4919
P(t)-P(t-12)	-0.0163	0.0063	-2.5944

Cluster 5: It is associated with medium wind speeds, very low temperatures and an increase of about 1mbar of pressure in 12h. In addition to the short term correlation of wind speed, the pressure difference 1h and 6h appear to have a measureable impact on the variation of wind speed in that regime.

Table 11 Cluster 5, LR Model details

	Coef	Std	t-stat
Constant	0.4345	0.146	2.9761
WS(t-1)	0.9274	0.021	44.1778
P(t)-P(t-1)	-0.6715	0.1364	-4.9245
P(t)-P(t-6)	0.097	0.0286	3.3938

Cluster 6: It is associated with medium wind speeds occurring on days warmer than average an almost constant pressure gradient. This regime is mostly associated with calm conditions, and possibly appearances of sea breezes, justified by the appearance of temperature as an important factor explaining in quantitative terms the wind speed variation.

Table 12 Cluster 6, LR Model details

	Coef	Std	t-stat
Constant	0.8478	0.2253	3.7636
WS(t-1)	0.9089	0.0222	41.0269
T(t-1)	0.2474	0.0441	5.6097
T(t-2)	-0.2747	0.043	-6.385
P(t)-P(t-1)	-0.2731	0.0783	-3.4853

Cluster 7: This regime is associated with constant, low wind conditions and a very small increase in pressure. The temperature corresponds to the average yearly value. Due to the low variation of the atmospheric conditions, the wind speed exhibits a persistent behaviour, with only the last value defined as an important one.

Table 13. Cluster 7, LR Model details

	Coef	Std	t-stat
Constant	0.2236	0.0833	2.6831
WS(t-1)	0.8861	0.0182	48.7207

Cluster 8: Corresponds to medium wind speeds occurring during the colder period of the year. the pressure pattern is of constant drop with a rate of approximately 0.7 mbars per 12h. A seasonal pattern is observed, evident from the introduction of the respective parameter in Table 14.

Table 14 Cluster 8, LR Model details

	Coef	Std	t-stat
Constant	0.0918	0.1135	0.8086
WS(t-1)	0.951	0.0205	46.2799
WS(t-24)	0.0245	0.0094	2.6096
P(t)-P(t-1)	-0.3459	0.0629	-5.5019

5 Summary and Conclusions

This chapter introduced the development of statistical and time series based models to identify and determine in a quantitative manner the impact of easily measured meteorological quantities on the variation of hourly wind speed data. The introduced variables were wind direction, temperature and pressure, which are presently easily measured even in amateur meteorological stations. Using a clustering methodology with those exogenous variables that had a statistically significant role in determining the variation of hourly wind speed series, a number of different "wind regimes" could be identified. The finer analysis of each cluster together with the development of a customized linear model on each one, were able to identify the underlying characteristics of the hourly wind speed.

References

Abdel-Aal, R.E., Elhadidy, M.A., Shaahid, S.M.: Modeling and forecasting the mean hourly wind speed time series using GMDH-based abductive networks. Renewable Energy 34(7), 1686–1699 (2009)

Alexiadis, M.C., Dokopoulos, P.S., Sahsamanoglou, H.S., Manousaridis, I.M.: Short term forecasting of wind speed and related electric power. Solar Energy 63(1), 61–68 (1998)

Argonne National Laboratory (ANL 2009), Wind Power Forecasting: State-of-the-Art, ANL/DIS-10-1

Barbounis, T.G., Theocharis, J.B.: Long-term wind speed and power forecasting using local recurrent neural network models. IEEE Transactions on Energy Conversion 21(1), 273–284 (2006)

Bossanyi, E.: Short-Term Wind Prediction Using Kalman Filters. Wind Engineering 9(1), 1–8 (1985)

Contaxis, G.A., Kabouris, J.: Short term scheduling in a wind/diesel autonomous energy syste. IEEE Tr. on Power Systems 6(3), 1161–1167 (1991)

Costa, A., Crespo, A., Navarro, J., Lizcano, G., Madsen, H., Feitosa, E.: A review on the young history of the wind power short-term prediction. Renewable and Sustainable Energy 12, 1725–1744 (2008)

Costa, A., Crespo, A., Migoya, E.: First results from a prediction project. In: Proceedings of the European Wind Energy Conference, EWEC 2003, Madrid, Spain (2003)

Damousis, I.G., Dokopoulos, P.: A fuzzy model expert system for the fore-casting of wind speed and power generation in wind farms. In: Proceedings of the IEEE Int. Conf. on Power Industry Computer Applications, PICA 2001, pp. 63–69 (2001)

Damousis, I.G., Alexiadis, M.C., Theocharis, J.B., Dokopoulos, P.: A fuzzy model for wind speed prediction and power generation in wind farms using spatial correlation. IEEE Transactions on Energy Conversion 19(2), 352–361 (2004)

Dudhia, J., Grell, G., Stauffer, D.R.: A description of the fifth-generation Penn System/NCAR Mesoscale Model (MM5), NCAR Tech. Note NCAR/TN-39811A, p. 107 (1994)

El-Fouly, T.H.M., El-Saadany, E.F., Salama, M.M.A.: Grey Predictor for Wind Energy Conversion Systems Output Power Prediction. IEEE Transactions on Power System 21(3), 1450–1452 (2006)

EWEA, Making 180 GW a reality by 2020, The European Wind Energy Association Position Paper (October 2007)

EWEA. Pure Power: Wind Energy Scenarios up to 2030, The European Wind Energy Association (March 2008)

Farmer, J.D., Sidorowich, J.J.: Predicting chaotic dynamics. In: Kelso, J.A.S., Mandell, A.J., Shlesinger, M.F. (eds.) Dynamic Patterns in Complex Systems, pp. 265–292. World Scientific, Singapore (1988)

Frías, L., Gastón, M., Martí, I.: A new model for wind energy forecasting focused in the intra-daily markets. In: Proceedings of European Wind Energy Conference, EWEC 2007, Milan, Italy (2007)

Fugon, F., Juban, J., Kariniotakis, G.: Data mining for Wind Power Forecasting. In: Proceedings of the European Wind Energy Conference, EWEC 2008, Brussels, Belgium (2008)

Giebel, G., Kariniotakis, G., Brownsword, R.: State of the Art on Short-term Wind Power Prediction, ANEMOS Deliverable Report D1.1 (2003)

Glahn, H.R., Lowry, D.A.: The use of Model Output Statistics (MOS) in objective weather forecasting. Journal of Applied Meteorology 11, 1202–1211 (1972)

Hagan, M., Menhaj, M.: Training feedforward networks with the Marquardt algorithm. IEEE Transactions on Neural Networks 5, 989–993 (1996)

Jursa, R.: Wind power prediction with different artificial intelligence models. In: Proceedings of the European Wind Energy Conference, EWEC 2007, Milan, Italy (2007)

Kariniotakis, G., et al.: ANEMOS: Development of a Next Generation Wind Power Forecasting System for the Large-Scale Integration of Onshore & Offshore Wind Farms. In: Proceedings of the European Wind Energy Conference & Exhibition, EWEC 2003, Madrid, Spain, June 16-19 (2003)

Kusiak, A., Zheng, H.-Y., Song, Z.: Short-Term Prediction of Wind Farm Power: A Data-Mining Approach. IEEE Transactions on Energy Conversion 24(1), 125–136 (2009)

Kavasseri, R.G., Seetharaman, K.: Day-ahead wind speed forecasting using f-ARIMA models. Renewable Energy 34(5), 1388–1393 (2009)

Kim, D.J., Park, Y.W., Park, D.J.: A Novel Validity Index for Determination of the Optimal Number of Clusters. IEICE Tr. on Information and Systems E84(2), 281–285 (2001)

Landberg, L., Giebel, G., Nielsen, H., Nielsen, T., Madsen, H.: Short-term Pre-diction – An Overview. Wind Energy 6(3), 273–280 (2003)

Leia, M., Shiyana, L., Chuanwen, J., Honglinga, L., Yana, Z.: A review on the forecasting of wind speed and generated power. Renewable and Sustainable Energy Reviews 13, 915–920 (2009)

Lin, C.T., Lee, C.S.: Neural fuzzy systems. A neuro-fuzzy synergism to intelligent systems. Prentice Hall, Englewood Cliffs (1996)

Maqsood, I., Khan, M., Huang, G., Abdalla, R.: Application of soft comput-ing models to hourly weather analysis in southern Saskatchewan, Canada. Engineering Applications of Artificial Intelligence 18(1), 115–125 (2005)

McQueen, J.B.: Some Methods for Classification and Analysis of Multivari-ate Observations. In: Proc. of 5th Berkley Symposium on Mathematical Statistics and Probability, pp. 281–297 (1967)

NERC (North American Electric Reliability Corporation) Accommodating High Levels of Variable Generation, Special Report (April 2009), http://www.nerc.com/news_pr.php?npr=283

Nielsen, T.S., Madsen, H., Nielsen, H., Landberg, L., Giebel, G.: Zephyr – The Prediction Models. In: Proceedings of the European Wind Energy Con-ference, EWEC 2001, Copenhagen, Denmark, July 2-6, pp. 868–871 (2001)

Pinson, P., Christensen, L.E.A., Madsen, H., Sørensen, P., Donovan, M.H., Jensen, L.E.: Regime-switching modelling of the fluctuations of offshore wind generation. Journal of Wind Engineering & Industrial Aerodynamics 96(12), 2327–2347 (2008)

Poggi, P., Muselli, M., Notton, G., Cristofi, C., Louche, A.: Forecasting and simulating wind speed in Corsica by using an autoregressive model. Energy Conversion and Management 14(20), 3177–3196 (2003)

Potter, C.W., Negnevistky, M.: Very short-term wind forecasting for Tasmanian power generation. IEEE Transactions on Power Systems 21(2), 965–972 (2006)

Ramírez-Rosado, I.J., Fernández-Jiménez, L.A.: Next-day wind farm electric energy generation forecasting using fuzzy time-series. In: Proceedings Int. Conf. on Modeling, Identification and Control, Innsbruck, Austria, pp. 237–240 (2003)

Ramírez-Rosado, I.J., Fernández-Jiménez, L.A.: An advanced model for short-term forecasting of mean wind speed and wind electric power. Control and Intelligent Systems 31(1), 21–26 (2004)

Sfetsos, A.: A comparison of various forecasting techniques applied to mean hourly wind speed time series. Renewable Energy 21(1), 23–35 (2000)

Sfetsos, A.: A novel approach for the forecasting of mean hourly wind speed time series. Renewable Energy 27(2), 163–174 (2002)

Skamarock, W.C., Klemp, J.B., Dudhia, J., Gill, D.O., Barker, D.M., Wang, W., Pow-ers, J.G.: A Description of the Advanced Research WRF Version 2, NCAR/TN–468 (2005)

Torres, J.L., García, A., de Blas, M., de, F.: A Forecast of hourly av-erages wind speed with ARMA models in Navarre. Solar Energy 79(1), 65–77 (2005)

Vlachogiannis, D., Sfetsos, A., Andronopoulos, S., Gounaris, N., Yiotis, A., Stubos, A.K.: The Demokritos air quality web based system for the Greater Athens Area. In: iEMSs 2008, International Congress on Environmental Modelling and Software, Barcelona, Catalonia, July 7-10 (2008)

Wu, Y.K., Hong, J.S.: A literature review of wind forecasting technology in the world. In: Proceedings of IEEE Power Tech Conference, Lausanne, Switzerland, July 1-5, pp. 504–509 (2007)

Load Flow with Uncertain Loading and Generation in Future Smart Grids

Olav Krause[1] and Sebastian Lehnhoff[2]

[1] The University of Queensland, School of Information Technology and Electrical Engineering, Brisbane, Queensland 4072, Australia
o.krause@uq.edu.au
[2] Carl von Ossietzky University, Department of Computing Science, Escherweg 2, 26121 Oldenburg, Germany
sebastian.lehnhoff@uni-oldenburg.de

Abstract. The growing amount of renewable and fluctuating energy sources for the production of electrical energy increases the volatility and level of uncertainty in the operation of power systems. Whether it is the growing number of photovoltaic installations harnessing solar energy or large-scale wind farms, these new class of environmentally dependent appliances increase the unpredictability of load situations hitherto known only from consumer behavior. One of the mayor concerns in grid operation under increasing feed-in from unpredictable generation and consumption is the detection of peaks in network strain. In order to limit investments into grid infrastructure to a reasonable level node-specific limitations for power injections are introduced to reduce the probability of such peaks that may pose a threat to a stable operation of the power system. In order to support the ongoing integration of renewable generation into the grid, a trade-off has to be found between investment costs and imposed operational constraints. In order to determine the probability of congestions under these unpredictable conditions, mathematical algorithms are employed that are able to estimate the probability of certain line loading levels from the probabilistic data derived from the appliances' behavior.

This chapter will cover a variety of approaches to solve (probabilistic) load flow problems, ranging from currently deployed state-of-the-art procedures to the newest advances in probabilistic load flow calculation and determination. Advantages and drawbacks of those methods are discussed in detail.

1 Introduction

In general, electric network states or congestions are determined and calculated using power flow calculations. The most popular method of solving the non-linear system of power flow equations is the Newton-Raphson (NR) method. The NR method starts from initial and likely guesses of all unknown variables (voltage angles, voltage magnitudes at load and generator buses). Next, a Taylor Series is

K. Gopalakrishnan et al. (Eds.): Soft Comput. in Green & Renew. Ener. Sys., STUDFUZZ 269, pp. 117–156.
springerlink.com

formulated for each of the power balance equations included in the equation system. The resulting linearized system of power flow equations is solved to determine the next iteration (a refined guess) of the voltage angles and magnitudes for which the procedure is repeated. This process continues until a stopping criterion is met, e.g. the difference of two subsequent results for voltage angles and magnitudes being beyond a specified threshold.

With a high amount of renewable – volatile and unpredictable – generation, standard power flow calculations reach their limits when applied to power system planning due to the continuous fluctuation of necessary data from the grid's feed-in nodes making it difficult to guess likely starting points for the initial step of the NR method. Choosing starting points for the NR method that deviate too far from the sought solution may cause the iterative method to diverge [Kornerup and Muller 2006]. Even when choosing appropriate starting values, the iterative method may be insufficient for timely detection of congestions in highly dynamic scenarios due to the number of (possibly computational complex and time consuming) iterations until the algorithm converges. In order to determine the probability – and thus mean time of occurrence – of congestions under these conditions, mathematical algorithms are employed that are able to estimate the probability of certain line loading levels from the probabilistic data derived from the generators' behavior. Hence, the stochastic behavior of (non-deterministic) renewable generation as well as loads is no longer described through clearly defined values, but given as a range of possible states together with their corresponding probability. This representation allows the prediction of the behavior for any given generator or load with a certain amount of probability or "softness". This is a sharp contrast to conventional power flow calculations, which precisely determine the state of the network on the basis of correct data for the (deterministic) behavior of every conventional generator and load connected to the network. Conventional power flow calculation by design is not able to cope with fuzzy input data and is very sensitive to misguesses, in the sense that a poorly chosen value in a scenario-based congestion analysis may lead to false results and ultimately to congestions not being detected [Kornerup and Muller 2006]. Probabilistic load flow calculation can tolerate this up to a certain extend without the result becoming useless. The concept of probabilistic load flow calculation is known for almost 40 years and appropriate research has been conducted in [Borkowska 1974][Dopazo et al. 1975][Allan and Alshakarchi 1976][Allan and Alshakarchi 1977] [Aboytes 1978][Allan et al. 1981][Silva et al. 1985][Silva and Arienti 1990].

In this chapter the authors will introduce the mathematical basis of probabilistic load flow calculations, current reference approaches and algorithms and give an outlook on further developments in this field.

1.1 Structure of Public Power Systems

In order to understand the differences in the mathematical formulation and problem-solving strategies, it is important to first get a basic understanding of the structure and operation of a public electric power system. In general, large power systems are composed of multiple voltage levels that can be distinguished into the *transmission system* and the *distribution system* (see Fig. 1).

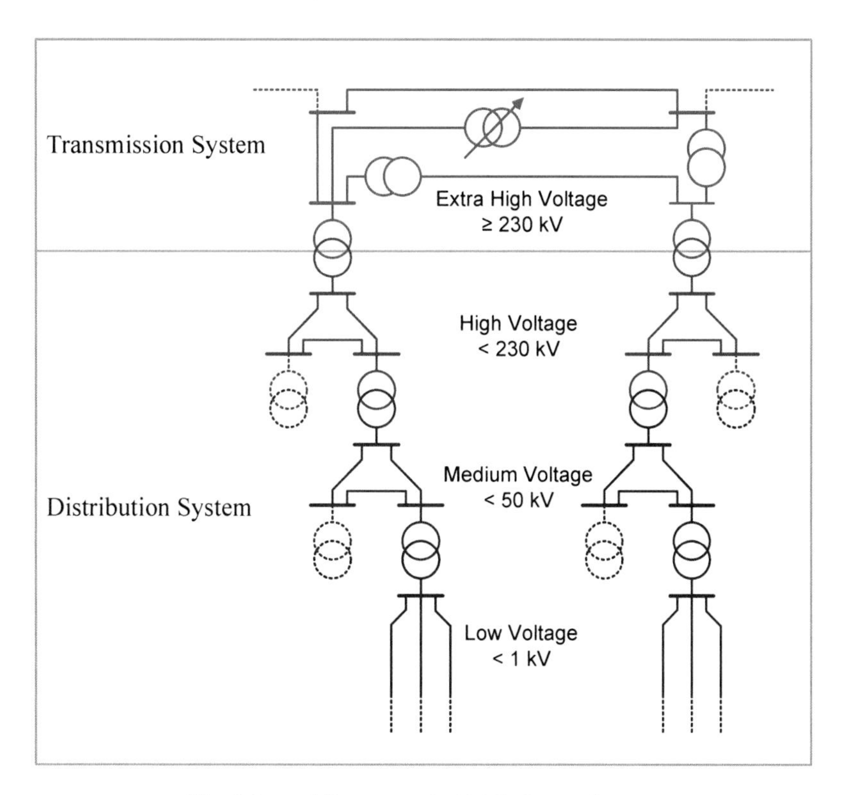

Fig. 1 Overall Structure of a Public Power System

The purpose of the transmission system is the transmission of large amounts of power across long distances at high voltages and thus with reduced losses. Most of the large (fossil) power plants are connected to the transmission system due to their high nominal power. Furthermore, the transmission system's power grid is usually highly interconnected (meshed). It is the network level at which active power is balanced. The latter is an important fact for the calculation of probabilistic load flows to which we will come back later within this chapter.

1.2 Differences between Deterministic and Probabilistic Calculation

A model of a complex technical system can be described as black box having a set of input parameters x_1 to x_n that may influence a system's state. In addition to the system's input parameters there is a set of state variables y_1 to y_m characterizing the operational state of the system (see Fig. 2).

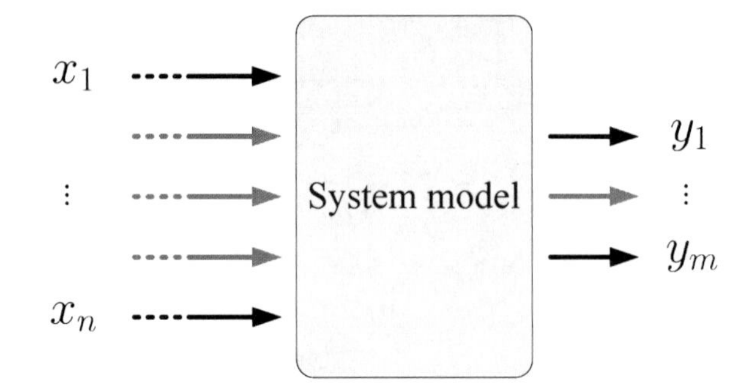

Fig. 2 Input/Output Characteristics of a System Model

In deterministic calculations of a system's operational state the set of input parameters are exactly determined. Hence, the goal of deterministic calculations is to derive the exact operational state of the system determined by the input parameters. For a power system this may not be possible (as we will demonstrate later on in this chapter).

In probabilistic calculations based on real numbers input parameters and state variables are described by probability functions (PF) and probability density functions (PDF). In Fig. 3 an exemplary PDF for the input parameter x_1 is sketched.

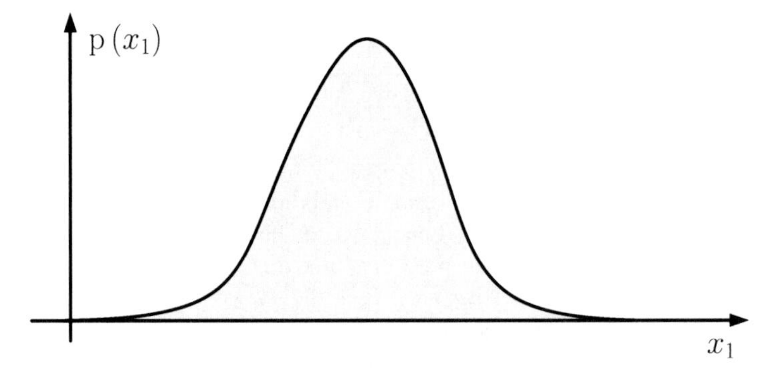

Fig. 3 Example of a Probability Density Function

The probability density function p(x) is a function that describes the relative likelihood for the variable x to occur at a given point. The probability for the x to fall within a particular region is given by the integral of this variable's density over the region. The probability density function is nonnegative everywhere, and its integral over the entire space is equal to one because the parameter or variable x has to take a value from its possible range (1).

$$\int_{-\infty}^{\infty} \mathrm{p}(x_1)\mathrm{d}x_i = 1 \tag{1}$$

On continuous numbers the probability of a single, exactly determined value is zero. Instead a non-zero probability of occurrence can only exist for a given interval $[a, b]$ of values (see (2)).

$$\int_a^b \mathrm{p}(x_1)\mathrm{d}x_i = \mathrm{P}_{a,b} \quad ; a < b \tag{2}$$

The main goal of probabilistic calculus in this context is to determine all possible operational states of the system that may result from possible input parameters and to assign a certain probability to them. The most popular approaches for Probabilistic Load Flow Calculation currently applied and often referred to as reference algorithms are based on experiments that are performed on the basis of deterministic load flow calculation. Thus, in the following section we will briefly introduce the deterministic load flow calculation in before addressing Probabilistic Load Flow Calculation.

2 The Deterministic Load Flow Problem

In probabilistic load flow calculation the solution of a deterministic load flow is the basis or at least an integral part of the solution strategy. Thus, in the following this technique is briefly explained. As its name already indicates the deterministic load flow calculates an exactly determined system state from exactly determined input parameters. More precisely, load flow calculation is a sub-problem of *state estimation* and may be calculated based on different kinds of input parameters. In this chapter we want to restrict ourselves to the simplest and most frequently used approach to deterministic load flow calculation. The intention of the load flow calculation is to calculate all complex-valued nodal voltages, uniquely describing the network's state and thus every other value within the network may be easily calculated from it.

Before going into details some basic mathematical relationships and model principles will be recapped in the following section.

2.1 Relation between AC Voltage and Current

In today's electric power systems alternating current (AC) is used for power transmission (with the exception of dedicated High Voltage Direct Current (HVDC) links). This is mainly due to the possibility of easily transforming power between different voltage levels using regular inductive transformers. This allows for the transmission of electric power at much higher voltage levels than initially being generated at and thus significantly reducing transmission losses. Furthermore, it is possible to provide different voltages to different kinds of loads at an acceptable level of technical effort. Under normal conditions all voltages and currents in AC systems are periodic and sinusoidal. Thus, they may be described by their amplitude and a time delay given in angular displacement against each other.

This is called *phasor representation* and will be discussed in detail in the following subsection.

2.1.1 Phasor Representation

A sinusoidal voltage can be uniquely described by the three parameters: *frequency* ω, *amplitude* $\hat{v}$ and *phase angle* δ against a reference oscillation of the same frequency. Although a formulation based on the harmonic sine function would be possible, traditionally the cosine function is used to describe AC voltages in the time domain (see (3)).

$$v(t) = \hat{v} \cdot \cos(\omega \cdot t + \delta) \tag{3}$$

Under steady state conditions (to which we want to limit ourselves here) the system exhibits a unique frequency ω. Thus, it is sufficient to specify the voltage magnitude $\hat{v}$ and phase angle δ against the reference oscillation of ω to uniquely describe the voltage $v(t)$ by just two parameters. The operational frequency may differ for various power systems (e.g. 50 Hz in the European ENTSO-E grid, or 60 Hz in the North-American grids).

In power engineering it is common to represent the oscillating voltages, currents, etc. using complex numbers. In general, a complex number is a two-dimensional number consisting of a real and an imaginary part, which are independent of each other. It can be written in the form $a + \mathrm{i} \cdot b$, where a and b are real numbers and i is the standard imaginary unit $\mathrm{i} = \sqrt{-1}$. Although the imaginary unit is often represented by i, in power engineering j is used more commonly. The latter will also be used within this chapter.

Based on complex numbers the current value of a complex voltage $V(t)$ is given in (4) where two possible representations are used. On the left hand side the voltage is expressed by real and imaginary parts and on the right hand side the polar representation is used.

$$\begin{aligned} V(t) &= \hat{v}\left(\underbrace{\cos(\omega \cdot t + \delta)}_{Real\ part} + \mathrm{j} \cdot \underbrace{\sin(\omega \cdot t + \delta)}_{Imaginary\ Part}\right) \\ &= \hat{v} \cdot e^{j \cdot \omega \cdot t + \delta} \end{aligned} \tag{4}$$

The current complex voltage $V(t)$ can be interpreted as a rotating pointer in the complex plane (see Fig. 4). Formula (4) shows that the real part of $V(t)$ is the current real voltage $v(t)$. The polar form of $V(t)$ can also be split into a constant complex part $\hat{u} \cdot e^{\mathrm{j} \cdot \delta}$ and a complex part $e^{\mathrm{j} \cdot \omega \cdot t}$, rotating with the operational frequency ω (see (5)).

$$U(t) = \hat{u} \cdot e^{\mathrm{j} \cdot \omega \cdot t + \delta} = \hat{u} \cdot e^{\mathrm{j} \cdot \delta} \cdot e^{\mathrm{j} \cdot \omega \cdot t} \tag{5}$$

Fig. 4 illustrates this distinction. The pointer in black is the complex part rotating counter-clockwise with frequency ω while the red one is following the black pointer with an angular distance of δ (with $\delta < 0$ in this case).

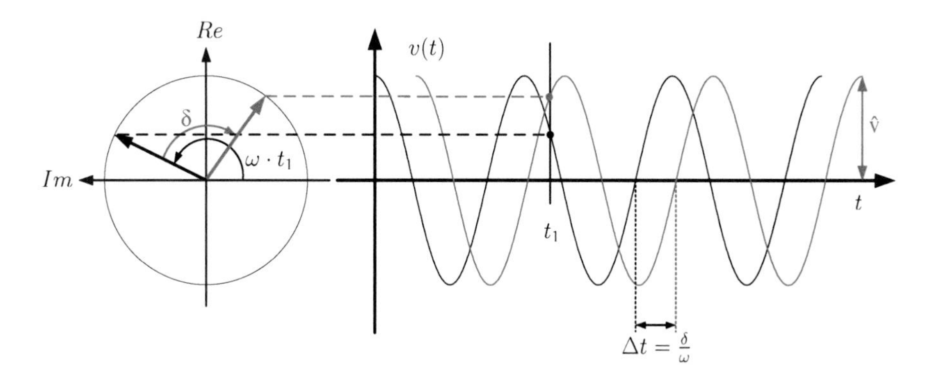

Fig. 4 Time Domain Signal and Phasor Representation

Since the rotating component is the same for every voltage, current, etc. it may be omitted and $v(t)$ only be represented by the complex phasor U (6) consisting of the so called absolute value $\hat{v}$ and the so called argument δ.

$$U = \hat{u} \cdot e^{\mathrm{j} \cdot \delta} \tag{6}$$

The real valued current voltage value $\hat{v}(t)$ can be reformulates using (7)

$$u(t) = U \cdot e^{\mathrm{j} \cdot \omega \cdot t} \tag{7}$$

Under normal conditions the currents in an AC-network are sinusoidal, thus this representation is also used to describe alternating currents.

2.1.3 Modeling of Linear Circuit Elements

Since most elements of a power grid exhibit a linear behavior in the sense of current caused by an applied voltage, they can be described with acceptable precision by linear models. As the calculations will be conducted based on complex-valued phasor representations, the linear elements have to be complex-valued as well in order to model time shifts between current and voltage caused by inductivities or capacities.

In general, these complex elements have two possible representations, *impedance* and *admittance*. The impedance Z of an element determines the applied voltage U in order to induce a certain current I (8).

$$U = Z \cdot I \tag{8}$$

The real part of the complex-valued impedance Z is known as the *resistance* R, while the complex part is known the *reactance* X (see (9)).

$$\underbrace{Z}_{Impedance} = \underbrace{R}_{Resistance} + \mathrm{j} \cdot \underbrace{X}_{Reactance} \tag{9}$$

The admittance Y of an element is a measure of how much current will flow through it if a certain voltage is applied to the element (10).

$$I = Y \cdot U \tag{10}$$

The real part of the admittance Y is known as the *conductance* G, while the imaginary part is known as the *susceptance* B (see (11)).

$$\underbrace{Y}_{Admittance} = \underbrace{G}_{Conductance} + \mathrm{j} \cdot \underbrace{B}_{Susceptance} \tag{11}$$

In case of a non-zero Impedance the corresponding admittance can be calculated as the inverse of the impedance as stated in (12).

$$Y = \frac{1}{Z} = \frac{1}{(R + \mathrm{j} \cdot X)} = \frac{(R - \mathrm{j} \cdot X)}{(R + \mathrm{j} \cdot X)(R - \mathrm{j} \cdot X)} = \frac{(R - \mathrm{j} \cdot X)}{R^2 + X^2} \tag{12}$$

In grid modeling this admittance representation is particularly important since it offers the possibility to model open links. The model of an open link has an admittance equal to zero and infinite impedance. If an element has a non-zero admittance the corresponding impedance can be calculated as stated in (13).

$$Z = \frac{1}{Y} = \frac{1}{(G + \mathrm{j} \cdot B)} = \frac{(G - \mathrm{j} \cdot B)}{(G + \mathrm{j} \cdot B)(G - \mathrm{j} \cdot B)} = \frac{(G - \mathrm{j} \cdot B)}{G^2 + B^2} \tag{13}$$

2.2 *Network Representation*

From an electrical point of view, the most relevant parts a power grid are the overhead lines, cables and transformers. Although they are quite different in functionality, shape and physical features, they can be modeled in a similar fashion. Most models base on the so called π-equivalent circuit, named after the arrangement of the model's elements. In the following subsections the basis of this modeling and as well as a complete exemplary network model will be explained and demonstrated.

2.2.1 π-Equivalent Line Model

The most frequently used model for cables and overhead lines in power system analysis is the so called π-equivalent circuit (as its structure closely resembles the Greek letter). We will give a brief introduction on this circuit model. For more

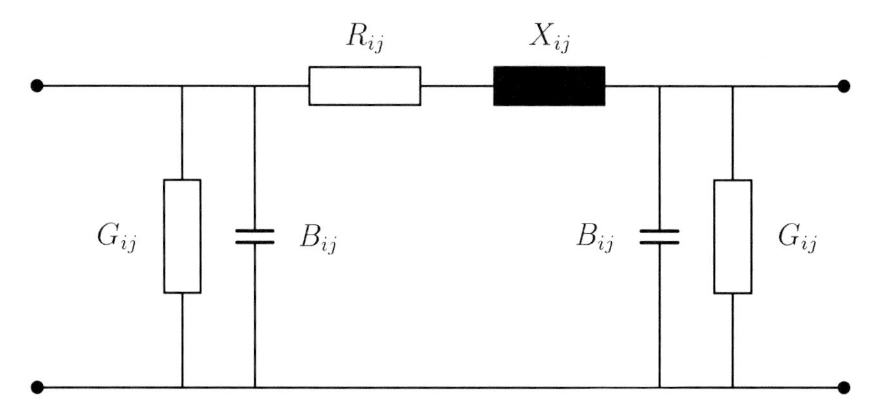

Fig. 5 ***π***-equivalent circuit

details about the model and especially the modeling of electrically long lines we refer the reader to [Grainger and Stevenson 1994].

The π-equivalent circuit of a line connecting node i with node j consists of four elements of which two occur twice. These elements represent the following phenomena:

- **Serial resistance** $\boldsymbol{R_{ij}}$**:** Every regular conductor (except super conductors) has a specific resistance. It is dependent on the conductor's material, temperature, length and cross-section area. In general, the specific resistance of a material is specified as a function of its temperature. This temperature dependence is usually neglected in power systems studies, resulting in the specific resistance to be a constant value. For conductors with a constant cross-section the resistance R_{ij} can be calculated as the product of the specific resistance and length divided by the cross-section area.
- **Serial reactance** $\boldsymbol{X_{ij}}$**:** Every conductor conducting a current creates a magnetic field around itself that is proportional to the flowing current. If the current changes over time, the resulting magnetic field change induces voltages in neighboring circuits and along the conductor itself. This is because even a single, linear conductor can be interpreted as being an induction loop with the returning wire being infinitely far away. These phenomena can be modeled as mutual inductivity (induction due to changing currents in another conductor) and self-inductivity (induction due to changing currents in the conductor itself). In case of a multi-phase symmetrically operated power line (to which we want to limit ourselves here for the sake of conceptual clarity) both, self and mutual inductivity, can be represented by a fictitious operational inductivity L_{ij}. The voltage induced in the conductor of a power line works against the voltage

driving the current with a 90° shift ahead. In complex numbers this can be represented as an imaginary impedance X_{ij}. As the induced voltage depends on how fast the magnetic flux changes, it is not only a function of the current generating the magnetic field, but also a function of the change rate – the derivative of the current over time. As power systems are regularly operated with sinusoidal voltages and currents – and thus the amplitude of the derivative changes proportionally with the frequency – the reactance increases linearly with the frequency of operation. Thus, with constant operational frequency ω_0 the reactance is given by $X_{ij} = \omega_0 \cdot L_{ij}$. For further details please refer to [Grainger and Stevenson 1994] and [Kundur 1994]

- **Shunt Conductivity G_{ij}**
  - The shunt conductivity is a model for insulation leak currents, corona losses, etc. As there is much effort put into increasing insulation qualities the value of shunt conductivity is much lower than of all other elements of a line's π-model. This element is even omitted in many cases since it has no significant effect on the result.
- **Shunt Susceptance B_{ij}**
  - Every conductor being brought to a certain voltage has to be charged to reach it. This is because electrons have to be brought into it or withdrawn from it in order to build up the electric field of a strength corresponding to the voltage the conductor should be brought to. The amount of charge needed to reach a certain voltage is called capacity C. It depends on its shape, its size, the distance to the counter pole and also of the insulation material surrounding the conductor. The shunt susceptance is of crucial importance when it comes to the modeling of cables.

2.2.2 Network Model

Power grids are usually modeled as a composition of π-equivalent circuits for which appropriate admittance matrices are generated. In the center of the calculation is always the vector of complex-valued nodal voltages, since it unambiguously describes the operational state of the grid and every other value can be calculated from it (see (14)). The two most important admittance matrices are the line admittance matrix and the nodal admittance matrix. Their composition and features are discussed in the following.

$$\mathbf{V}_{\text{node}} = \begin{bmatrix} V_1 \\ \vdots \\ V_n \end{bmatrix} \tag{14}$$

Line admittance matrix

With the help of the line admittance matrix it is possible to calculate the lines' and transformers' complex-valued currents from the vector of complex-valued nodal voltages. In order to set up this matrix the following technique is used. The current through a line depends on the voltage difference between the two nodes connected by the line, since the voltage difference is the voltage being active along the line. For the current flowing through a line only the two serial elements resistance R_{ij} and reactance X_{ij} are relevant, since they are the only elements connecting both nodes. With the current being calculated from the voltage the lines admittance is needed which can be calculated as stated in (15).

$$Y_{ij} = \left(R_{ij} + \mathrm{j} \cdot X_{ij}\right) \tag{15}$$

In order to set up the line admittance matrix the subtraction within is executed implicitly by expanding the term in braces of formula (16) and assigning the negative sign to the second instance of the lines admittance.

$$I_{ij} = Y_{ij} \cdot \left(V_i - V_j\right) = Y_{ij} \cdot V_i - Y_{ij} \cdot V_j \tag{16}$$

For the four-node network depicted in Fig. 2 this results in the line admittance stated in (17).

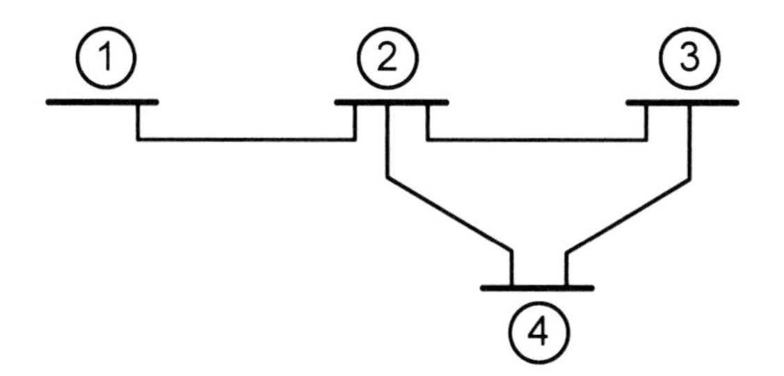

Fig. 6 Four Node Example Network

With this formulation it is possible to calculate all line and transformer currents I_{ij} with a single matrix vector multiplication in one step (see (17)).

$$\begin{bmatrix} I_{12} \\ I_{23} \\ I_{34} \\ I_{24} \end{bmatrix} = \underbrace{\begin{bmatrix} Y_{12} & -Y_{12} & & \\ & Y_{23} & -Y_{23} & \\ & & Y_{34} & -Y_{34} \\ & Y_{24} & & -Y_{24} \end{bmatrix}}_{\mathbf{Y}_{\text{line}}} \cdot \begin{bmatrix} V_1 \\ V_2 \\ V_3 \\ V_4 \end{bmatrix} \tag{17}$$

The line admittance matrix is essential for calculating line loadings after determining the network's operational state with the Load Flow Calculation.

Nodal admittance matrix

The second essential matrix when modeling electric power grids is the nodal admittance matrix. In contrast to the line admittance, the shunt elements have to be considered when stating the nodal admittance matrix. With the nodal admittance matrix the nodal currents of a network can be calculated from the vector of complex nodal voltages. Since all currents in a node have to sum up to zero (according to Kirchhoff's point rule), the current flowing from a node or into a node is the sum of all currents flowing away from or into the node within the network. Having the π-equivalent circuit in mind, these are the currents flowing through the shunt elements of all lines at the particular node and the currents flowing through the lines to other nodes (see Fig. 7). There are two different reference-arrow systems for this representation that differ in the orientation of positively counted nodal current I_i. The orientation affects whether a positive sign of the nodal power indicates power consumption or power feed-in. For the generator reference arrow system (GRAS) current flowing in the direction of a node (into the node) are counted positively (see Fig. 7).

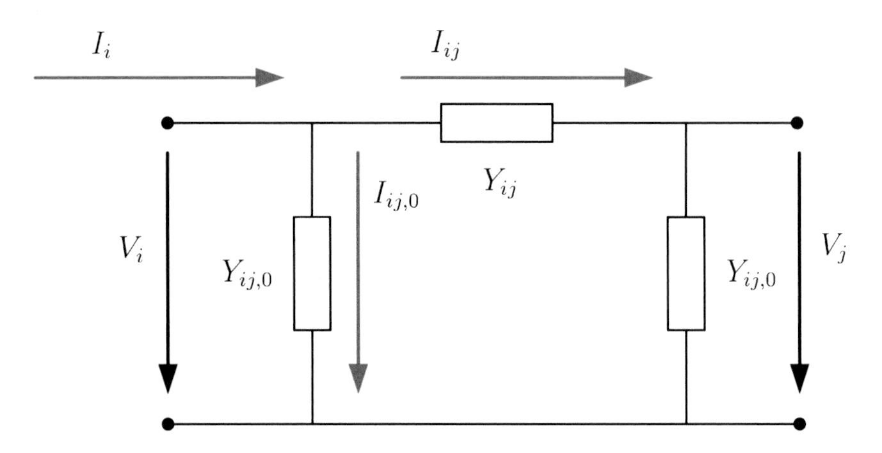

Fig. 7 Voltages and currents in a $\boldsymbol{\pi}$-equivalent circuit (GRAS)

The shunt elements of a line are represented as $Y_{ij,0}$ on both ends of the π-equivalent circuit each representing half of the complex shunt admittance of the respective line. Thus, $Y_{ij,0}$ is $\frac{1}{2}\left(G_{ij} + \mathrm{j} \cdot B_{ij}\right)$

Assuming zero admittance between nodes that are not directly connected with each other the aforementioned sum of currents within the generator reference-arrow system results in the form of (18) where the nodal current I_i is the positive sum of all other currents flowing from or towards the node.

$$\sum_{\substack{j=1 \\ j \neq i}}^{n} Y_{ij} \cdot \left(V_i - V_j\right) + \sum_{\substack{j=1 \\ j \neq i}}^{n} Y_{ij,0} \cdot V_i = I_i \tag{18}$$

The second reference-arrow system is the load reference-arrow system (LRAS). In LRAS currents flowing in the direction out and away from a node are counted positively (see Fig. 8). This inverses the sign of the nodal power in comparison to the generator reference-node system and lets power consumption appear with a positive sign of the nodal power.

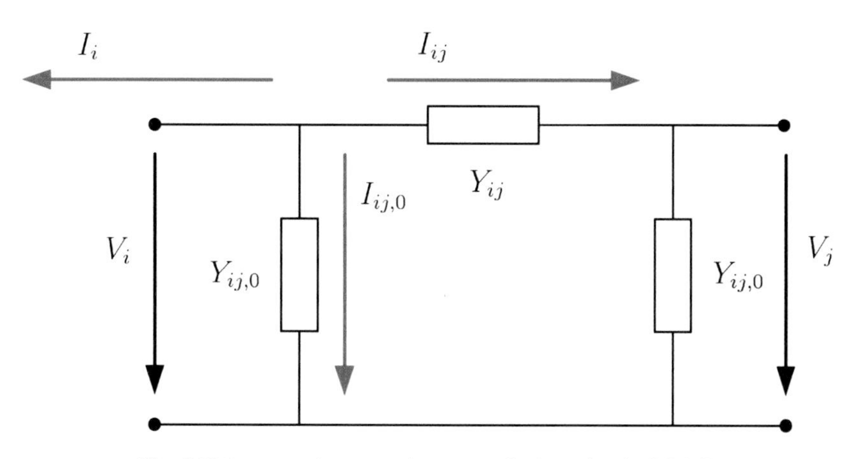

Fig. 8 Voltages and currents in a π-equivalent circuit (LRAS)

Within LRAS the sum of all positively signed currents has to be zero. This results in the nodal current I_i being the negative sum of all other currents from and to node i (19).

$$-\sum_{\substack{j=1\\j\neq i}}^{n} Y_{ij}\cdot\left(V_i - V_j\right) - \sum_{\substack{j=1\\j\neq i}}^{n} Y_{ij,0}\cdot V_i = I_i \tag{19}$$

Within this chapter we will use the generator reference-arrow system for the further explanations. It is important to know about the existence of these two different reference-arrow systems, since they affect the set-up of the nodal admittance matrix. For the study and further understanding of additional material on this topic beyond this chapter it is important to recognize the difference.

Next, we will transpose formula (18) into the term given in (20):

$$\sum_{\substack{j=1\\j\neq i}}^{n}\left(Y_{ij} + Y_{ij,0}\right)\cdot V_i - \sum_{\substack{j=1\\j\neq i}}^{n} Y_{ij}\cdot V_j = I_i \tag{20}$$

With the elements of the π-equivalent circuit being $Y_{ij} = (R_{ij} + \mathrm{j}\cdot X_{ij})^{-1}$ and $Y_{ij,0} = G_{ij} + \mathrm{j}\cdot B_{ij}$ (compare Fig. 5) and having in mind the structure of (20), the nodal admittance matrix can be stated as follows (21):

$$\begin{bmatrix} I_1\\ \vdots\\ I_n\end{bmatrix} = \underbrace{\begin{bmatrix} \sum_{j=1}^{n}(Y_{1j} + Y_{1j,0}) & \cdots & -Y_{1n}\\ \vdots & \ddots & \vdots\\ -Y_{n1} & \cdots & \sum_{j=1}^{n}(Y_{ij} + Y_{nj,0})\end{bmatrix}}_{\mathbf{Y}_{\text{node}}} \cdot \begin{bmatrix} V_1\\ \vdots\\ V_n\end{bmatrix} \tag{21}$$

It is common standard to index the elements of the nodal admittance matrix in lower-case y_{ij} (see (22))

$$\begin{bmatrix} I_1 \\ \vdots \\ I_n \end{bmatrix} = \underbrace{\begin{bmatrix} y_{11} & \cdots & y_{1n} \\ \vdots & \ddots & \vdots \\ y_{n1} & \cdots & y_{nn} \end{bmatrix}}_{\mathbf{Y}_{\text{node}}} \cdot \begin{bmatrix} V_1 \\ \vdots \\ V_n \end{bmatrix} \tag{22}$$

The elements y_{ij} of the nodal admittance matrix are then given by (23) and (24).

$$y_{ii} = \sum_{j=1}^{n} (Y_{ij} + Y_{ij,0}) \tag{23}$$

$$y_{ij} = -Y_{ij}\ ; i \neq j \tag{24}$$

In order to give a practical example, the nodal admittance matrix for the four-node example network depicted in Fig. 6 would look similar to (25).

$$\begin{bmatrix} Y_{12} + \cdots \\ \ldots Y_{12,0} & -Y_{12} & 0 & 0 \\ -Y_{12} & \begin{matrix} Y_{12} + Y_{23} + Y_{24} + \cdots \\ \ldots Y_{12,0} + Y_{23,0} + Y_{24,0} \end{matrix} & -Y_{23} & -Y_{24} \\ 0 & -Y_{23} & \begin{matrix} Y_{23} + Y_{24} + \cdots \\ \ldots Y_{23,0} + Y_{24,0} \end{matrix} & -Y_{34} \\ 0 & -Y_{24} & -Y_{34} & \begin{matrix} Y_{24} + Y_{34} + \cdots \\ \ldots Y_{24,0} + Y_{34,0} \end{matrix} \end{bmatrix} \tag{25}$$

2.4 Power Flow Equations and Their Features

The central element in Power Flow Calculation is the complex-valued nodal power S_i. It is computed from the complex nodal voltage V_i at the particular node i and the conjugate-complex nodal current I_i (see (26)).

$$S = U \cdot I^* \tag{26}$$

Setting up (26) in real numbers yields (27) with the both types of power, active power P and reactive power Q.

$$\underbrace{(P + \mathrm{j} \cdot Q)}_{S} = \underbrace{(e + \mathrm{j} \cdot f)}_{V} \cdot \underbrace{(a - \mathrm{j} \cdot b)}_{I^*} \tag{27}$$

By transposing (27) to (28) both types of power can be separated and calculated separately.

$$\underbrace{(P + \mathrm{j} \cdot Q)}_{S} = \underbrace{(ea + fb)}_{P} + \mathrm{j} \cdot \underbrace{(af - eb)}_{Q} \tag{28}$$

Both formulations assume that the current I is constant and does not change or more precisely, that I is not a function of V. However, in power systems this assumption does not hold, as a change in voltage at a certain node also changes the voltage differences between the particular node and all other nodes within the network. This has a direct effect on the currents flowing from or to the particular node.

In order to understand these interdependencies, the origin, structure and features of the Power Flow Equations (PFE) for computation of nodal power in a network are explained in the following section.

Nodal Power in a Network

The basic equations for calculating the power flow into or out of a node in case of an interconnected network are similar to the ones given in (26), but set up separately for every node i of a network (see (29)).

$$S_i = V_i \cdot I_i^* \tag{29}$$

As mentioned before, in an interconnected network the nodal current I_i is a function of all nodal voltages within the network. Using the elements y_{ij} of the nodal admittance matrix, the nodal current I_i can set up as (30).

$$I_i = \sum_{j=1}^{n} y_{ij} \cdot V_j \tag{30}$$

Inserting this expression into formula (29) yields the Power Flow Equations in their complex-valued form (31). Please note that all elements within the sum have to be conjugate-complex values.

$$S_i = U_i \cdot \sum_{j=1}^{n} y_{ij}^* \cdot V_j^* \tag{31}$$

Setting up (31) based on real numbers yields (32), where g_{ij} and b_{ij} are real and imaginary parts of the element y_{ij} of the nodal admittance matrix, respectively.

$$(P_i + \mathrm{j} \cdot Q_i) = (e_i + \mathrm{j} \cdot f_i) \cdot \sum_{j=1}^{n} (g_{ij} - \mathrm{j} \cdot b_{ij}) \cdot (e_j - \mathrm{j} \cdot f_j) \tag{32}$$

With the intermediate step (33) the Power Flow Equations can be separated into active power P_i (34) and reactive power Q_i (35).

$$(P_i + \mathrm{j} \cdot Q_i) = (e_i + \mathrm{j} \cdot f_i) \cdot \sum_{j=1}^{n} (g_{ij}e_j - b_{ij}f_j) - \mathrm{j} \cdot (b_{ij}e_j + g_{ij}f_j) \tag{33}$$

$$P_i = e_i \cdot \sum_{j=1}^{n} \left(g_{ij} e_j - b_{ij} f_j \right) + f_i \cdot \sum_{j=1}^{n} \left(b_{ij} e_j + g_{ij} f_j \right) \tag{34}$$

$$Q_i = f_i \cdot \sum_{j=1}^{n} \left(g_{ij} e_j - b_{ij} f_j \right) - e_i \cdot \sum_{j=1}^{n} \left(b_{ij} e_j + g_{ij} f_j \right) \tag{35}$$

With these Power Flow Equations it is then possible to calculate all nodal powers from the vector of complex-valued nodal voltages. However, in general the vector of complex-valued nodal voltages is not known, but has to be determined from the given (metered) vector of complex-valued nodal powers. In order to do so and to understand the Power Flow Equations and their behavior in detail, their derivative will be set up and analyzed. It will be shown, that the Power Flow Equations are not complex differentiable and the resulting effects on power flow calculations are discussed.

Partial Differential Equations

In order to calculate the complex derivative of a complex-valued function – as are our Power Flow Equations – the partial derivatives of the real-valued version of the particular function have to be set up first. Due to the structure of the nodal admittance matrix – and thus the Power Flow Equations – two cases have to be distinguished. The first one is a nodal power derived against a component of the voltage at the same node (case $j = i$)). The second one is the general case of $j \neq i$. The derivatives of the Power Flow Equations are given in the following formulas (see (36) through (39)).

$$\frac{\partial P_i}{\partial e_j} = \begin{cases} 2e_i g_{ii} + \sum\limits_{\substack{k=1 \\ k \neq i}}^{n} (g_{ik} e_k - b_{ik} f_k) & ; j = i \\ e_i g_{ij} + f_i b_{ij} & ; j \neq i \end{cases} \tag{36}$$

$$\frac{\partial P_i}{\partial f_j} = \begin{cases} 2f_i g_{ii} + \sum\limits_{\substack{k=1 \\ k \neq i}}^{n} (b_{ik} e_k + g_{ik} f_k) & ; j = i \\ -e_i b_{ij} + f_i g_{ij} & ; j \neq i \end{cases} \tag{37}$$

$$\frac{\partial Q_i}{\partial e_j} = \begin{cases} -2e_i b_{ii} - \sum_{\substack{k=1 \\ k \neq i}}^{n} (b_{ik} e_k + g_{ik} f_k) & ; j = i \\ f_i g_{ij} - e_i b_{ij} & ; j \neq i \end{cases} \tag{38}$$

$$\frac{\partial Q_i}{\partial f_j} = \begin{cases} -2f_i b_{ii} + \sum_{\substack{k=1 \\ k \neq i}}^{n} (g_{ij} e_j - b_{ij} f_j) & ; j = i \\ -f_i b_{ij} - e_i g_{ij} & ; j \neq i \end{cases} \tag{39}$$

For a complex-valued function to be complex differentiable, the so called *Cauchy-Riemann partial differential equations* have to be fulfilled. In the case of the Power Flow Equations they have the following form (40).

$$\frac{\partial P_i}{\partial f_j} = -\frac{\partial Q_i}{\partial e_j} \quad ; \quad \frac{\partial P_i}{\partial e_j} = \frac{\partial Q_i}{\partial f_j} \tag{40}$$

It is obvious that they are only fulfilled in case of all nodal voltages being equal to zero. Thus, the Power Flow Equations are not complex differentiable and do not have a derivative. But it is possible to state the so called Jacobian matrix of the Power Flow Equation that in most cases behaves similar to what would be expected from a derivative. Some effects of the aforementioned deficit will be discussed in the following section together with the set-up of the Jacobian matrix.

Jacobian matrix

The Jacobian matrix describes the behavior of the Power Flow Equations in the vicinity of an operational point. The operational point is given by the vector of complex nodal voltages. The dependency of the Jacobian matrix on the current vector of nodal voltages becomes clear when having a closer look at the partial derivatives (36) to (39). Although the derivatives are a function of the vector of nodal voltages, they will be stated here without argument for the sake of readability. The overall structure of the Jacobian matrix $\mathbf{J}(\mathbf{V})$ is depicted in (41).

$$\underbrace{\begin{bmatrix} \Delta P_1 \\ \Delta Q_1 \\ \vdots \\ \Delta P_n \\ \Delta Q_n \end{bmatrix}}_{\Delta \mathbf{S}} = \underbrace{\begin{bmatrix} \frac{\partial P_1}{\partial e_1} & \frac{\partial P_1}{\partial f_1} & \cdots & \frac{\partial P_1}{\partial e_n} & \frac{\partial Q_1}{\partial f_n} \\ \frac{\partial Q_1}{\partial e_1} & \frac{\partial Q_1}{\partial f_1} & \cdots & \frac{\partial Q_1}{\partial e_n} & \frac{\partial Q_1}{\partial f_n} \\ \vdots & \vdots & \ddots & \vdots & \vdots \\ \frac{\partial P_n}{\partial e_1} & \frac{\partial P_n}{\partial f_1} & \cdots & \frac{\partial P_n}{\partial e_n} & \frac{\partial P_n}{\partial f_n} \\ \frac{\partial Q_n}{\partial e_1} & \frac{\partial Q_n}{\partial f_1} & \cdots & \frac{\partial Q_n}{\partial e_n} & \frac{\partial Q_n}{\partial f_n} \end{bmatrix}}_{\mathbf{J}(\mathbf{V})} \cdot \underbrace{\begin{bmatrix} \Delta e_1 \\ \Delta f_1 \\ \vdots \\ \Delta e_n \\ \Delta f_n \end{bmatrix}}_{\Delta \mathbf{V}} \tag{41}$$

The Jacobian matrix of the Power Flow Equations translates a change in nodal voltage $\mathbf{\Delta V}$ into a change in nodal power $\mathbf{\Delta S}$ (see (41)). As already discussed for the Power Flow Equations themselves, the inverse direction is of higher practical importance. It is more important to translate a change in the network's input parameters – the nodal power – to a change of the network state – the nodal voltages. This can be done using the inverse $\mathbf{J}^{-1}(\mathbf{V})$ of the Jacobian matrix $\mathbf{J}(\mathbf{V})$ (see (42)).

$$\Delta\mathbf{V} = \mathbf{J}^{-1}(\mathbf{V}) \cdot \Delta\mathbf{S} \tag{42}$$

The Jacobian matrix of the Power Flow Equations suffers from close-to-singular conditions and rank deficiency conditions for certain operational states of the network. This is closely related to certain instability phenomena like angle-instabilities or voltage-instabilities. For more details about these instability phenomena, please refer to [Kundur 1994]. These close-to-singular and rank-deficiency issues have a significant impact on the results of some modern Probabilistic Load Flow approaches and will be discussed in this context later throughout this chapter.

Under moderate and regular operational conditions of the network the Jacobian matrix is of full rank and thus has an inverse $\mathbf{J}^{-1}(\mathbf{V})$.

2.5 Newton-Raphson Load Flow Calculation

As already mentioned before, the main goal of Power Flow Calculations is the determination of a network's operational state by determining the vector of complex nodal voltages from a given vector of nodal powers. This can be interpreted as finding a root or null of the equations given in (43). Since the PFE do not have an inverse function, the principle strategy is to tune a vector of complex nodal voltages to make the result of the PFE match given (metered) values for the nodal powers with an acceptable precision.

$$\mathbf{S}_{\text{given}} - \text{PFE}(\mathbf{V}) = 0 \tag{43}$$

The approach most frequently used is the Newton-Raphson method. The Newton-Raphson method bases on a shortened Taylor-Series, but suffers from convergence issues under certain conditions. To point out the source of these convergence issues the Newton-Raphson iterative loop will be derived from a Taylor-Series and it will by pointed out, why some of the preconditions for the application of the Taylor-Series technique are not fulfilled.

Taylor-Series

The Taylor-Series representation tries to match, or at least to approximate, a function by the Taylor-polynomial that bases on the derivatives of a function at a certain expansion point. The Taylor-Series of a scalar function with one argument and one value is stated in (44). A Taylor-Series always refers to a certain expansion point a and is developed up to the degree m. In case the derivatives of degree larger than m are not equal to zero, there is a non-zero residual R_m.

$$f(x) = f(a) + \sum_{i=1}^{m} \left(\frac{1}{i!} f^{(i)}(a) \cdot (x-a)^i \right) + R_m \tag{44}$$

Prerequisite for a Taylor-Series is the differentiability of a function $f(x)$ up to degree m at least [Bronstein et al. 2000]. In case of a single-value, multi-argument function there is no total derivative of function $f(x_1, \ldots, x_n)$. Thus the Taylor-Series is developed using the partial derivatives (see (45)).

$$\begin{aligned} f(x_1 + \Delta x_1, \cdots, x_n + \Delta x_n) &= f(x_1, \cdots, x_n) \\ &+ \sum_{i=1}^{m} \frac{1}{i!} \left(\frac{\partial}{\partial x_1} \Delta x_1 + \cdots + \frac{\partial}{\partial x_n} \Delta x_n \right)^i f(x_1, \cdots, x_n) + R_m \end{aligned} \tag{45}$$

For multi-value, multi-argument functions (45) changes to (46).

$$\begin{aligned} f_i(x_1 + \Delta x_1, \cdots, x_n + \Delta x_n) &= f_i(x_1, \cdots, x_n) \\ &+ \sum_{j=1}^{m} \frac{1}{j!} \left(\frac{\partial}{\partial x_1} \Delta x_1 + \cdots + \frac{\partial}{\partial x_n} \Delta x_n \right)^j f_i(x_1, \cdots, x_n) + R_m \end{aligned} \tag{46}$$

In this case the single components $f_i(x_1, \ldots, x_n)$ are developed independently. The prerequisite for differentiability remains.

Newton-Technique

The Newton-technique bases on the development of a Taylor-Series of the respective function in order to approximate the behavior of the function in the vicinity of a given expansion point. With the Newton-technique the Taylor-Series is only developed up to the degree $m = 1$ and the residual R_m is omitted. Thus, (46) may be simplified according to (47).

$$\begin{aligned} f_i(x_1 + \Delta x_1, \cdots, x_n + \Delta x_n) &\approx f_i(x_1, \cdots, x_n) \\ &+ \left(\frac{\partial}{\partial x_1} \Delta x_1 + \cdots + \frac{\partial}{\partial x_n} \Delta x_n \right) \cdot f_i(x_1, \cdots, x_n) \end{aligned} \tag{47}$$

The error introduced by omitting the residual R_m depends on the values of the derivatives of the function with a degree higher than 1. The partial derivatives of degree 1 may then be assembled to form the Jacobian matrix of the multi-value, multi-argument function $\mathbf{f}(\mathbf{x})$ at expansion point $\mathbf{x}$ (see (48)).

$$\mathbf{J}_{\mathbf{f}}(\mathbf{x}) = \begin{bmatrix} \frac{\partial f_1(\mathbf{x})}{\partial x_1} & \cdots & \frac{\partial f_1(\mathbf{x})}{\partial x_n} \\ \vdots & \ddots & \vdots \\ \frac{\partial f_n(\mathbf{x})}{\partial x_1} & \cdots & \frac{\partial f_n(\mathbf{x})}{\partial x_n} \end{bmatrix} \tag{48}$$

Transposing (47) with the help of (48) to its matrix representation yields (49).

$$\mathbf{f}(\mathbf{x} + \Delta\mathbf{x}) \approx \mathbf{f}(\mathbf{x}) + \mathbf{J}_{\mathbf{f}}(\mathbf{x}) \cdot \Delta\mathbf{x} \tag{49}$$

If the Jacobian matrix is full-rank and thus has an inverse, (49) can be transformed to (50) by expansion with the inverse $\mathbf{J_f^{-1}(x)}$ of Jacobian matrix $\mathbf{J_f(x)}$.

$$\mathbf{J_f^{-1}(x)} \cdot \left(\mathbf{f(x+\Delta x)} - \mathbf{f(x)}\right) \approx \mathbf{J_f^{-1}(x)} \cdot \mathbf{J_f(x)} \cdot \mathbf{\Delta x} \tag{50}$$

Since multiplying a full-rank matrix with its inverse yields the identity matrix, (50) can be simplified to (51).

$$\mathbf{J_f(x)} \cdot \left(\mathbf{f(x+\Delta x)} - \mathbf{f(x)}\right) \approx \mathbf{\Delta x} \tag{51}$$

With the Newton-technique (51) is used as the corrected argument $\mathbf{x}$ in order to reach a certain given value of $\mathbf{f(x)}$. To achieve this, an iterative loop is used in which the value $\mathbf{f(x+\Delta x)}$ is used for the subsequent iteration. This approach starts with an initial argument $\mathbf{x}^k$ and the given value $\mathbf{f_{given}}$. Formula (51) is utilized to find an argument improvement $\mathbf{\Delta x}^k$.

$$\mathbf{J_f(x}^k) \cdot \left(\mathbf{f}_{\text{given}} - \mathbf{f(x}^k)\right) \approx \mathbf{\Delta x}^k \tag{52}$$

Fig. 9 illustrates this approach for a single-value, single-argument function $f(x)$ iteratively approaching the root (or null) of $f(x)$ in two cycles.

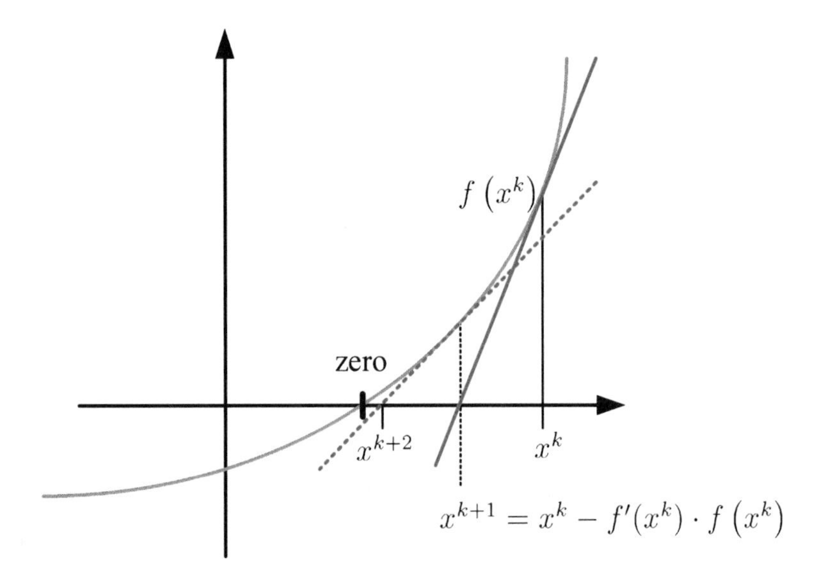

Fig. 9 Iterative cycle of the Newton-technique

With the improvement of the argument Δx^k the improved argument x^{k+1} can be calculated using (52)

$$\mathbf{x^{k+1}} = \mathrm{x^k} + \mathbf{\Delta x}^k \tag{52}$$

For functions that do not have high-order derivatives of significant value, the Newton-technique converges quite fast and reliably. In case of the Newton-Raphson Load Flow Calculation this is not always true due to the non-differentiability of the Power Flow Equations.

Newton-Raphson Load Flow Calculation

The input parameters for a Newton-Raphson Load Flow vary in type according to the behavior of the power equipment connected to the nodes of the network. In general, three types of nodes are distinguished; Slack-node, PQ-node and PU-node. First, the Newton-Raphson Load Flow Calculation will be explained only with Slack- and PQ-nodes. PU-nodes will be explained later in this chapter.

In every Load Flow Problem there has to be one node being assigned to be the Slack-node. The reason is that the sum of all nodal power balances (active as well as reactive) and the network's power losses or gains (only reactive power) has to be equal to zero. But since the network's active power losses and reactive losses and gains are a function of the nodal power balances and the voltages within the network, they are a result of the Power Flow Calculation and previously unknown. Therefore, the Power Flow Calculation cannot find a solution with only PQ-nodes being modeled in the network. Furthermore, the transmission of power can be done at multiple voltages. Thus, there is a need for at least one node defining and enforcing a certain voltage at its node. To solve this, a Slack-node is introduced that has a defined voltage as the input argument for the Load Flow Calculation and its active and reactive power balance as a return value. By this, the Slack-node balances the overall active and reactive power balance (including possible losses) throughout the network. A technical interpretation of the Slack-node in the context of transmission and distribution systems will be given at the end of this section.

A PQ-node models a generator or load that is controlled for its active and reactive power to be held constant. This is only an approximation of the real behavior, as active power consumption of generation is a function of the voltage and active power consumption or generation at the particular node. However, the model provides a good estimate which can be iteratively improved by recalculation after adjusting the respective power balance according to a more complex load or generator model and the results of a previously solved Load Flow.

The third standard-type of nodes being modeled is the PU-node. A PU-node models a generator (mathematically also a load, although technically rarely done) that has an automatic voltage regulator. Such generators adjust their reactive power balance in order to manipulate the absolute value of the voltage at controlled node. Under regular conditions, an increase in reactive power generation generally increases the absolute value of the voltage at the node it is injected. A decrease or consumption lowers the absolute value of the voltage. This effect is used for automatic voltage regulation. In case the generator reaches its individual limit of active power generation or consumption, which is in general a function of the generator's active power balance, the active power balance will remain at its maximum and the PU-node model has to be replaced by a PQ-node model during the Power Flow Calculation.

In this chapter only the Power Flow Calculation with one Slack-node and multiple exclusive PQ-model nodes will be explained. For more details about modeling PU-nodes, please refer to [Grainger 1994].

Any of the nodes of the network model may be assigned to be the Slack-node. Throughout the following explanations node number 1 will be assigned to be the slack node without loss of generality. In order to set up a solvable problem the active and reactive power of the slack node will be omitted within the iteration loop as they are no argument but the value of the Load Flow Calculation. Thus the reduced Power Flow Equations used in the iterative loop have the following form (see (53))

$$\begin{bmatrix} \cancel{P_1} \\ \cancel{Q_1} \\ P_2 \\ Q_2 \\ \vdots \\ P_n \\ Q_n \end{bmatrix} = \begin{bmatrix} \cancel{f_{P_1}(V_1, \dots, V_n)} \\ \cancel{f_{Q_1}(V_1, \dots, V_n)} \\ f_{P_2}(V_1, \dots, V_n) \\ f_{Q_2}(V_1, \dots, V_n) \\ \vdots \\ f_{P_n}(V_1, \dots, V_n) \\ f_{Q_n}(V_1, \dots, V_n) \end{bmatrix} \tag{53}$$

Since the Jacobian matrix of the Power Flow Equations are used to derive an improvement for the vector of nodal voltages from the given active and reactive power balances at all nodes, the aforementioned reduction requires that the two rows of the Jacobian matrix, corresponding to the active and reactive power at the slack node, are also omitted. Furthermore, the complex-valued voltage at the Slack-node is fixed and not to be changed. Thus, the two columns of the Jacobian matrix, which correspond to the voltage at the slack node, also have to be omitted (see (54)). Otherwise, the inverse of the Jacobian matrix would also yield changes in the Slack-node's voltage.

$$\underbrace{\begin{bmatrix} \cancel{\Delta P_1} \\ \cancel{\Delta Q_1} \\ \Delta P_2 \\ \Delta Q_2 \\ \vdots \\ \Delta P_n \\ \Delta Q_n \end{bmatrix}}_{\Delta \mathbf{S}_{\mathrm{red}}} = \underbrace{\begin{bmatrix} \cancel{\frac{\partial P_1}{\partial e_1}} & \cancel{\frac{\partial P_1}{\partial f_1}} & \cancel{\frac{\partial P_1}{\partial e_2}} & \cancel{\frac{\partial P_1}{\partial f_2}} & \cdots & \cancel{\frac{\partial P_1}{\partial e_n}} & \cancel{\frac{\partial Q_1}{\partial f_n}} \\ \cancel{\frac{\partial Q_1}{\partial e_1}} & \cancel{\frac{\partial Q_1}{\partial f_1}} & \cancel{\frac{\partial Q_1}{\partial e_2}} & \cancel{\frac{\partial Q_1}{\partial f_2}} & \cdots & \cancel{\frac{\partial Q_1}{\partial e_n}} & \cancel{\frac{\partial Q_1}{\partial f_n}} \\ \cancel{\frac{\partial P_2}{\partial e_1}} & \cancel{\frac{\partial P_2}{\partial f_1}} & \frac{\partial P_2}{\partial e_2} & \frac{\partial P_1}{\partial f_2} & \cdots & \frac{\partial P_2}{\partial e_n} & \frac{\partial P_2}{\partial f_n} \\ \cancel{\frac{\partial Q_2}{\partial e_1}} & \cancel{\frac{\partial Q_2}{\partial f_1}} & \frac{\partial Q_2}{\partial e_2} & \frac{\partial Q_2}{\partial f_2} & \cdots & \frac{\partial Q_2}{\partial e_n} & \frac{\partial Q_2}{\partial f_n} \\ \vdots & \vdots & \vdots & \vdots & \ddots & \vdots & \vdots \\ \cancel{\frac{\partial P_n}{\partial e_1}} & \cancel{\frac{\partial P_n}{\partial f_1}} & \frac{\partial P_n}{\partial e_2} & \frac{\partial P_n}{\partial f_n} & \cdots & \frac{\partial P_n}{\partial e_n} & \frac{\partial P_n}{\partial f_n} \\ \cancel{\frac{\partial Q_n}{\partial e_1}} & \cancel{\frac{\partial Q_n}{\partial f_1}} & \frac{\partial Q_n}{\partial e_2} & \frac{\partial Q_n}{\partial f_2} & \cdots & \frac{\partial Q_n}{\partial e_n} & \frac{\partial Q_n}{\partial f_n} \end{bmatrix}}_{\mathbf{J}_{\mathrm{red}}(\mathbf{V})} \cdot \underbrace{\begin{bmatrix} \cancel{\Delta e_1} \\ \cancel{\Delta f_1} \\ \Delta e_2 \\ \Delta f_2 \\ \vdots \\ \Delta e_n \\ \Delta f_n \end{bmatrix}}_{\Delta \mathbf{V}_{\mathrm{red}}} \tag{54}$$

For the start of a Newton-Raphson Load Flow Calculation an initial vector of nodal voltages is needed. To avoid convergence issues, it is common to set all nodal voltages to the value of the slack node (in general also the nominal voltage throughout the network; see (55)).

$$\mathbf{V}^0 = \begin{bmatrix} V_1 \\ V_2^k = V_1 \\ \vdots \\ V_n^k = V_1 \end{bmatrix} \quad (55)$$

With this vector (and all successive vectors $\mathbf{V}^k$ during the iteration) the resulting nodal powers for all nodes except the slack node are calculated. Based on this result and the given nodal powers within the Load Flow Calculation the difference between the presently assumed network state and the targeted result is calculated using (56).

$$\begin{bmatrix} \Delta P_2^k \\ \Delta Q_2^k \\ \vdots \\ \Delta P_n^k \\ \Delta Q_n^k \end{bmatrix} = \begin{bmatrix} P_{2,\mathrm{given}} \\ Q_{2,\mathrm{given}} \\ \vdots \\ P_{n,\mathrm{given}} \\ Q_{n,\mathrm{given}} \end{bmatrix} - \begin{bmatrix} f_{P_2}(V_1, V_2^k, \dots, V_n^k) \\ f_{Q_2}(V_1, V_2^k, \dots, V_n^k) \\ \vdots \\ f_{P_n}(V_1, V_2^k, \dots, V_n^k) \\ f_{Q_n}(V_1, V_2^k, \dots, V_n^k) \end{bmatrix} \quad (56)$$

Knowing the difference between the given (metered) nodal powers and the nodal powers corresponding to the assumed network state, an improvement of the voltage vector can be calculated using (57).

$$\begin{bmatrix} \Delta e_2^k \\ \Delta f_2^k \\ \vdots \\ \Delta e_n^k \\ \Delta f_n^k \end{bmatrix} = \mathbf{J}_{\mathrm{red}}^{-1}(\mathbf{V}^k) \cdot \begin{bmatrix} \Delta P_2^k \\ \Delta Q_2^k \\ \vdots \\ \Delta P_n^k \\ \Delta Q_n^k \end{bmatrix} \quad (57)$$

With the voltage improvement determined using (57) the improved vector of nodal voltage $\mathbf{V}^{k+1}$ for the next iteration can be calculated using (58).

$$\mathbf{V}^{k+1} = \begin{bmatrix} V_1 \\ V_2^{k+1} = V_2^k + (\Delta e_2^k + j \cdot \Delta f_2^k) \\ \vdots \\ V_n^{k+1} = V_n^k + (\Delta e_n^k + j \cdot \Delta f_n^k) \end{bmatrix} \quad (58)$$

Based on the improved vector of nodal voltages the nodal powers and the remaining difference to the given values is calculated and compared against a pre-defined convergence limit ϵ (59), (60).

$$\left|\Delta P_i^k\right| < \epsilon\,; \forall i \in \{2, \dots, n\} \quad (59)$$

$$\left|\Delta Q_i^k\right| < \epsilon\,; \forall i \in \{2, \dots, n\} \quad (60)$$

Once all absolute values of the remaining differences at all nodes are below the convergence limit, the iteration is stopped and the found voltage vectors may be interpreted as the desired result.

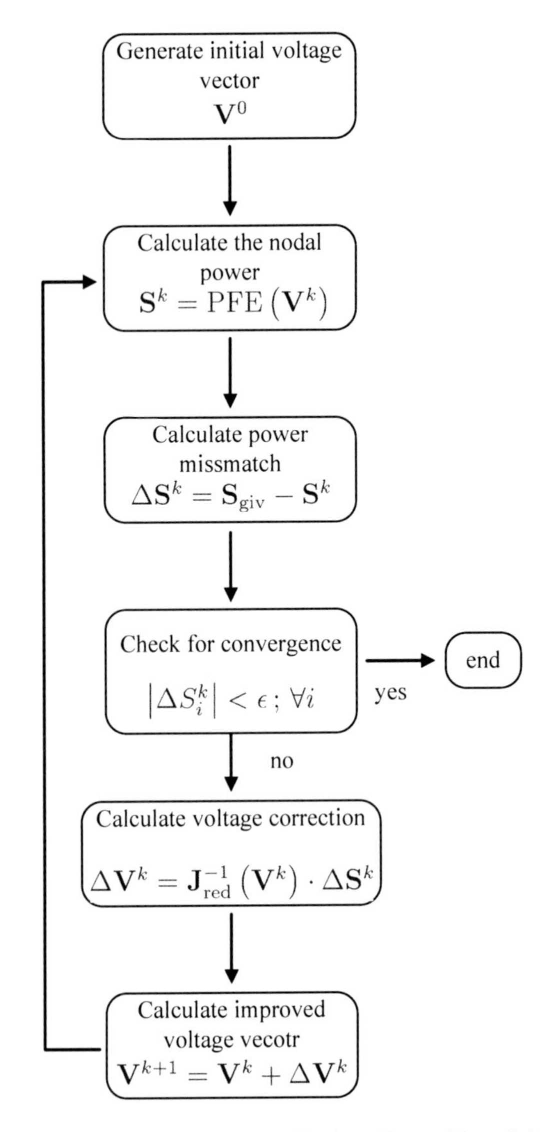

Fig. 10 Iterative loop of the Newton-Raphson Power Flow Calculation

3 Probabilistic Load Flow

There are two main kinds of Probabilistic Load Flow Calculation. The first kind is based on huge number of experiments, while the other tries to estimate the behavior of the power flow equations in the vicinity of a given operational point. Probabilistic Load Flow calculation is still a field of intense research activities. In this chapter we will introduce three approaches that are widely accepted in the scientific community. Two are experiment-based and the last one is an example of how probabilistic calculus is applied to reduce computation time.

3.1 Sampling with Newton-Raphson

The first approach for Probabilistic Load Flow Calculation, presented in this chapter, is an experiment-based one. It is frequently used as a reference algorithm. However, its applicability to problems of a reasonable size and practical complexity is limited due to the vast number of experiments needed. The first step of this approach is to determine a number of discrete probabilities from a continuous Probability Density Function (PDF; see Fig. 11).

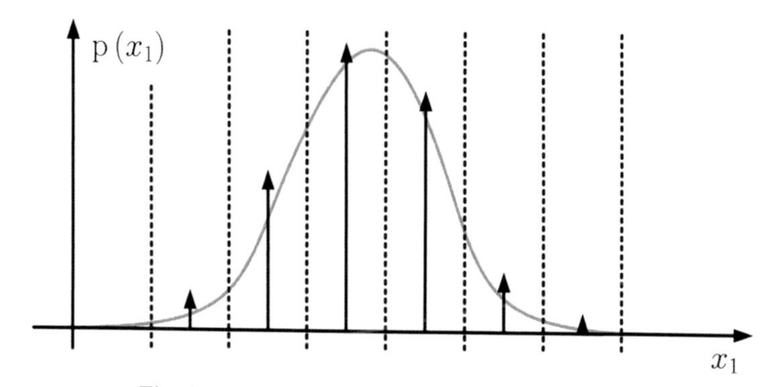

Fig. 11 Transition to one-dimensional discrete PDF

Assuming a given PDF $\mathrm{p}(x_1)$ and an interval T over variable x_1 the probability $\mathrm{P}(n)$ that x_1 is going to take a value of interval n can be determined using (61). Thus, it is possible to get a set of discrete probabilities that are ing $\mathrm{p}(x_1)$. Obviously, the precision of this approximation depends on T, while the range in which $\mathrm{p}(x_1)$ is approximated depends on T and the number of intervals.

$$\mathrm{P}(n) = \int_{\left(n-\frac{1}{2}\right)\cdot T}^{\left(n+\frac{1}{2}\right)\cdot T} \mathrm{p}(x_1) \cdot \mathrm{d}x_1 \tag{61}$$

In case of the Probabilistic Load Flow Calculations explained in this chapter, the input data for each node is not a one-dimensional PDF, but a two-dimensional PDF reflecting the probability of certain combinations of active and reactive power at each node. Fig. 12 illustrates this in the complex plane of a node's complex power. The probability density is indicated by the coloration intensity.

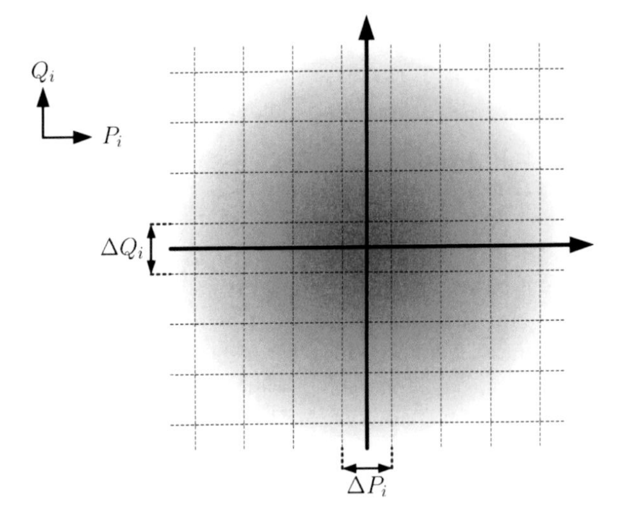

Fig. 12 Transition to two-dimensional discrete PDF

In order to derive a set or discrete probabilities from this two-dimensional PDF, intervals for both dimensions have to be defined. In case of the complex nodal power at node i, these are ΔP_i and ΔQ_i.

Similar to the one-dimensional case, the probability for the complex nodal power S to take a value in the interval, specified by P_i and Q_i, is given by (62).

$$\mathrm{P}_i(P_i, Q_i) = \int\limits_{P_i-\frac{1}{2}\Delta P_i}^{P_i+\frac{1}{2}\Delta P_i} \int\limits_{Q_i-\frac{1}{2}\Delta Q_i}^{Q_i+\frac{1}{2}\Delta Q_i} \mathrm{p}_i(p,q) \cdot \mathrm{d}p \cdot \mathrm{d}q \tag{62}$$

With the intervals defined and the discrete probabilities determined, the procedure of experiment-sampling consists of solving the Load Flow Problems for all possible discrete combinations of nodal powers, weighting the results with the combined probability of the single probabilities associated with the respective nodal powers and collecting them. In the following explanation the experiment index $k \in \{1, \dots, o\}$ will be used.

For every experiment k the input argument for the Load Flow Calculation is a certain combination $\mathbf{S}_k$ of nodal powers for all nodes, except the Slack-node (63).

$$\mathbf{S}_k = \begin{bmatrix} P_{2,k} \\ Q_{2,k} \\ \vdots \\ P_{n,k} \\ Q_{n,k} \end{bmatrix} \tag{63}$$

The probability $\mathbf{P}_k(\mathbf{S}_k)$ of occurrence of $\mathbf{S}_k$ is the combined probability $P_i(P_i, Q_i)$ of the occurrence of all single nodal values of S_k (64).

$$\mathrm{P}_{\boldsymbol{k}}(\mathbf{S}_k) = \prod_{i=2}^{n} \mathrm{P}_i\left(P_{i,k}, Q_{i,k}\right) \tag{64}$$

The experiment itself is the Load Flow Calculation (LFC) for $\mathbf{S}_k$. The result is, as described in the precious section, the vector of complex nodal voltages $\mathbf{V}_k$ for experiment k (see (65)).

$$\mathbf{V}_k = \mathrm{LFC}(\mathbf{S}_k) \tag{65}$$

The probability of occurrence $\mathrm{P}(\mathbf{V}_k)$ of the determined vector of nodal voltages $\mathbf{V}_k$ is equal to the probability of occurrence $\mathrm{P}(\mathbf{S}_k)$ of the input argument $\mathbf{S}_k$ (see (66)).

$$\mathrm{P}(\mathbf{V}_k) = \mathrm{P}(\mathbf{S}_k) \tag{66}$$

Since the main question of Probabilistic Load Flow Calculation is with which probability nodal voltages occur within a certain interval of values, the results of the experiments have to be collected and analyzed in a second step. Fig. 13 illustrates a typical distribution of a complex nodal voltage within the complex plane.

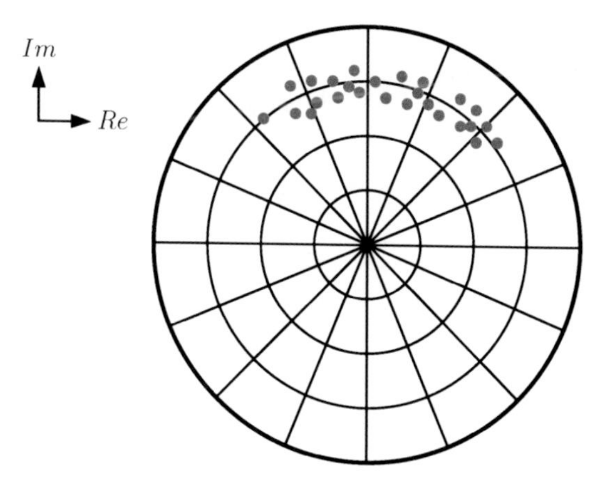

Fig. 13 Typical distribution of nodal voltages

The typical question to be answered with the results of the Probabilistic Load Flow Calculation is the probability of an absolute voltage value at a node i being within an allowable voltage band. In order to calculate this probability the auxiliary function $\sigma_{i,k}(V_{i,\min}, V_{i,\max})$ is defined according to (67).

$$\sigma_{i,k}(V_{i,\min}, V_{i,\max}) = \begin{cases} 1 & V_{i,\min} \leq |V_{i,k}| \leq V_{i,\max} \\ 0 & else \end{cases} \tag{67}$$

With this auxiliary function the probability $\mathrm{P}(V_{i,\min} \leq |V_i| \leq V_{i,\max})$ of the absolute value $|V_i|$ of the complex nodal voltage V_i being within a defined voltage band can be calculated using (68).

$$\mathrm{P}(V_{i,\min} \leq |V_i| \leq V_{i,\max}) = \sum_{k=1}^{o} \sigma_{i,k}(V_{i,\min}, V_{i,\max}) \cdot \mathrm{P}(\mathbf{V}_k) \tag{68}$$

The second important question to be answered from the results of the Probabilistic Load Flow Calculation is the probability of the absolute value $|I_j|$ of the complex-valued line current I_j flowing through line $j \in \{1, \dots, m\}$ not exceeding the line's individual maximum value $I_{j,\max}$. In order to answer this question, the vector of complex-valued line currents $\mathbf{I}_{\text{line},k}$ has to be calculated from the previously determined vectors of complex-valued nodal voltages $\mathbf{V}_k$ for each experiment k. As described in the previous sections, the vector $\mathbf{I}_{\text{line},k}$ can be determined by multiplying the line admittance matrix $\mathbf{Y}_{\text{line}}$ with the vector of nodal voltages $\mathbf{V}_k$ for each experiment k (see (69)).

$$\mathbf{I}_{\text{line},k} = \mathbf{Y}_{\text{line}} \cdot \mathbf{V}_k \tag{69}$$

The probability of occurrence $\mathrm{P}(\mathbf{I}_{\text{line},k})$ of this result is equal to the probability of occurrence $\mathrm{P}(\mathbf{V}_k)$ of the corresponding network state, represented by $\mathbf{V}_k$ (see (70)).

$$\mathrm{P}(\mathbf{I}_{\text{line},k}) = \mathrm{P}(\mathbf{V}_k) \tag{70}$$

Similar to the nodal voltages, these results have to be further analyzed in order to answer the initial question about the probability of the absolute value of the line's current being below or equal to the line's individual limit $I_{j,\max}$. Fig. 14 illustrates a typical distribution of values of $I_{j,k}$ in the complex plane of the line's complex-valued current.

As already used before, an auxiliary function $\sigma_{j,k}(I_{j,\max})$ will be utilized to calculate the probability of the absolute value $|I_j|$ of line j being smaller or equal to the line's limit $I_{j,\max}$ (see (71)).

$$\sigma_{j,k}(I_{j,\max}) = \begin{cases} 1 & |I_{j,k}| \leq I_{j,\max} \\ 0 & else \end{cases} \tag{71}$$

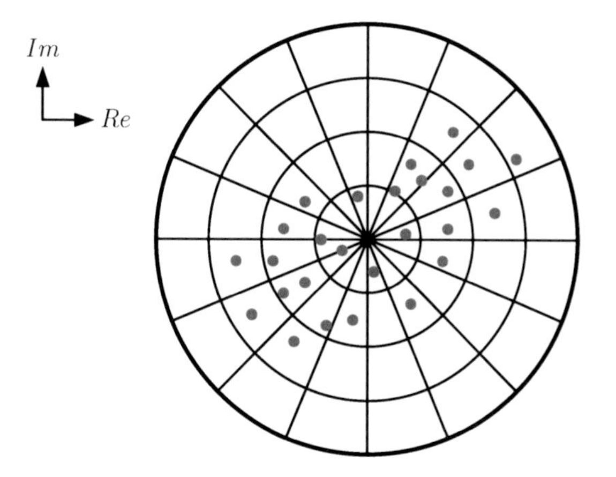

Fig. 14 Typical distribution of line currents

With $\sigma_{j,k}(I_{j,\max})$ defined according to (71) the aforementioned probability $\mathrm{P}(|I_j| \leq I_{j,\max})$ can be calculated according to (72).

$$\mathrm{P}(|I_j| \leq I_{j,\max}) = \sum_{k=1}^{o} \sigma_{j,k}(I_{j,\max}) \cdot \mathrm{P}(\mathbf{I}_{\mathrm{line},k}) \quad (72)$$

The main drawback of this approach is the enormous number of experiments needed for sampling the distribution. If, for example, active and reactive power on a 20-node network is sampled with 10 intervals each, the resulting number of experiments is $10^{2 \cdot 20}$. Obviously, this approach is only feasible for very small problems and serves as the means of verification of simplified approaches.

3.2 Sampling with Newton-Raphson and Monte-Carlo

One possibility to reduce the computational burden in comparison to the previously explained approach is to randomly select experiments while skipping most of them. This approach is known as the Monte-Carlo approach. By calculating only the results for randomly chosen experiments, the computational burden may be eased. With an increasing number of experiments the results of the Monte-Carlo approach converge to the results of the reference algorithm.

However, the main benefit of the Monte-Carlo approach is the approximation of the results through repeated random sampling for large-scale problems, not solvable with the reference algorithm.

Monte-Carlo methods are often used for the calculation and simulation of real-world physical systems. Due to the repeated similar computation of random or pseudo-random numbers, these methods are qualified for calculation by high-performance computer architectures and tend to be used when it is unfeasible or even impossible to compute exact solutions with deterministic (reference) algorithms.

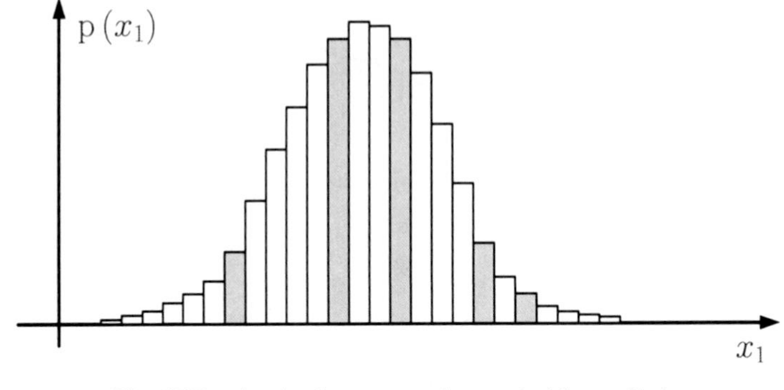

Fig. 15 Randomly chosen experiments in Monte-Carlo

Monte-Carlo simulation methods are particularly useful when studying systems with a large number of coupled degrees of freedom, such as in Load Flow Calculations in electrical power systems with a wide range of phenomena with significant uncertainties in their inputs [Chen et al. 2008], as have been discussed so far.

3.3 Convolution Based on Jacobian Matrix

Another promising approach for easing the computational burden is the utilization of probabilistic calculus instead of statistical analysis of a huge number of experiments. The principle idea is to avoid having to perform a complete and time consuming Load Flow Calculation for every experiment. There are different approaches that make use of convolution-based methods out of which we will examine the one most often found in the literature. It is based on the approximation of the behavior of the Power Flow Equations in the vicinity of an operational point with the help of the Jacobian matrix. First, we will cover some basics on probabilistic calculus and explain the convolution in multi-dimensional cases.

The first example is a simple addition of two values x and y that are not clearly determined, but may vary within a certain range of values (see Fig. 16). The general question is which combination of values yields a certain result z.

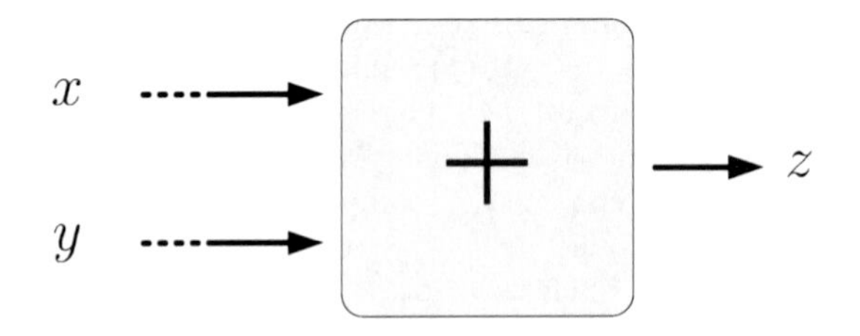

Fig. 16 Two argument one value summing unit

The trivial relation between x, y and z is (73).

$$z = y + x \tag{73}$$

In order to determine which combinations of x and y yield a certain value for z (73) has to be transposed to (74). With z given, x may be varied with (74) always providing the corresponding value of y that yields the result of (73) being equal to the specified value z.

$$y = z - x \tag{74}$$

Provided with the PDF of x and y being $\mathrm{p}_x(x)$ and $\mathrm{p}_y(y)$ equation (74) can be used to find all combinations of x and y that give a certain z. Based on this the probability $\mathrm{p}_z(z)$ of a certain z can be determined by (75) since the integral runs over all combinations that have z as a result in (73).

$$\mathrm{p}_z(z) = \int_{-\infty}^{\infty} \mathrm{p}_x(x) \cdot \mathrm{p}_y(z - x) \cdot \mathrm{d}x \tag{75}$$

This procedure can also be extended to a multi-argument single-value summing unit (see Fig. 17) where the value y depends on n arguments x_1 to x_n.

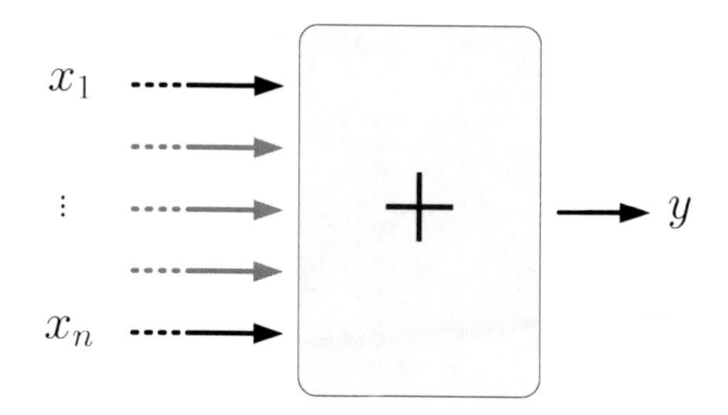

Fig. 17 Multi-argument single-value summing unit

The mathematical operation of the multi-input single-value summing unit can be expressed as the sum of all its arguments (76).

$$y = \sum_{i=1}^{n} x_i \tag{76}$$

In order to be able to determine the probability $\mathrm{p}_y(y)$ of a certain value y, again all combinations that yield y through (76) have to be determined first. Similar to the two-argument single-value summing unit, this leads to one argument (in this example x_n) being expressed as a function of the given value y and all other arguments x_i of the summing unit (see (77)).

$$x_n = y - \sum_{i=1}^{n-1} x_i \tag{77}$$

With (77) always yielding the right value for x_n so that the result of (76) is a given z, the probability of a certain value z can be calculated with the nested integral (78) as it considers all combined probabilities of all combinations of $x_1, \dots, x_n$ that result in z.

$$\mathrm{p}_y(y) = \int_{-\infty}^{\infty} \cdots \int_{-\infty}^{\infty} \mathrm{p}_{x_1}(x_1) \dots \mathrm{p}_{x_n}\left(y - \sum_{i=1}^{n-1} x_i\right) \mathrm{d}x_1 \dots \mathrm{d}x_{(n-1)} \tag{78}$$

Extending the multi-argument single-value summing unit to a weighted multi-argument single-value summing unit (see Fig. 18) does not significantly increase the complexity of the solution.

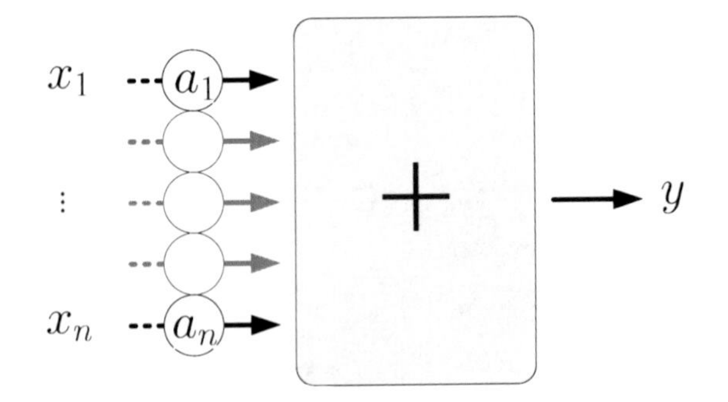

Fig. 18 Weighted multi-argument single-value summing unit

The key to the calculation of the PDF $\mathrm{p}_y(y)$ again is the mathematical operation (79).

$$y = \sum_{i=1}^{n} a_i \cdot x_i \tag{79}$$

As already demonstrated equation (79) has to be transposed to (80) in order to express one of the arguments (here x_n) as a function of y and the other arguments $x_1, \dots, x_n$. In this case it has to be assured, that the weighting factor a_n of the argument x_n is not equal to zero. For practical application it is important to choose the argument to be expressed as a function of the value and the other arguments that has the highest absolute weighting factor.

$$x_n = \frac{1}{a_n}\left(y - \sum_{i=1}^{n-1} a_i \cdot x_i\right) \tag{80}$$

Similar to above simpler examples, with (80) the PDF $\mathrm{p}_y(y)$ of y can be calculated using (81).

$$\mathrm{p}_y(y) = \int_{-\infty}^{\infty} \cdots \int_{-\infty}^{\infty} p_{x_1}(x_1) \dots p_{x_n}\left(\frac{y - \sum_{i=1}^{n-1} a_i \cdot x_i}{a_n}\right) dx_1 \dots dx_{(n-1)} \tag{81}$$

This weighted multi-argument single-value summing unit is the blueprint for the convolution based on the Jacobian matrix since the convolutions corresponding to a single row of the Jacobian matrix can be calculated separately. Its mathematical behavior is exactly the same as for the weighted multi-argument single-value summing unit.

The Probabilistic Load Flow Calculation based on convolution techniques on the Jacobian matrix starts with the approximation of the behavior of the Power Flow Equations in the vicinity of an expansion point $\mathbf{V}_0$. Two issues have to be addressed by the Probabilistic Load Flow: first, the PDF of nodal voltages and the second, the PDF of line loadings. In the remainder of this section we will focus on determining the PDF of nodal voltages after which we determine the PDF of the line loadings.

In practical application, a reasonable choice for the expansion point is the expected vector of the nodal powers. Since the nodal powers are complex-valued the expected value is complex-valued too. With continuous numbers it has the form of (82). The expected values $\mathrm{E}(S_i)$ will serve as the expansion point for the Jacobian matrix later.

$$\mathrm{E}(S_i) = \iint_{-\infty}^{\infty} (P_i + \mathrm{j} \cdot Q_i) \cdot \mathrm{p}_i(p, q) \cdot \mathrm{d}p \cdot \mathrm{d}q \tag{82}$$

Assuming that $\mathrm{p}_i(P_i, Q_i)$ is sampled within the interval $(-s, \dots, s)$ with an interval size ΔP_i and ΔQ_i respectively, in discrete numbers the expected values can be calculated for each node using (83).

$$\mathrm{E}(S_i) = \sum_{p=-s\cdot\Delta P_i}^{s\cdot\Delta P_i} \sum_{q=-s\cdot\Delta Q_i}^{s\cdot\Delta Q_i} (p + \mathrm{j} \cdot q) \cdot \mathrm{P}_i(p, q) \tag{83}$$

In either cases the complex-valued expected values of nodal powers have to be separated into their real and imaginary part, $\mathrm{E}(P_i)$ and $\mathrm{E}(Q_i)$ respectively, in order to calculate the voltage vector $\mathbf{V}_0$ serving as the expansion point (see (84)).

$$\big(\mathrm{E}(P_i) + \mathrm{j} \cdot \mathrm{E}(Q_i)\big) = \mathrm{E}(S_i) \tag{84}$$

With (84) the voltage vector V_0 of the expansion point is then determined using a regular deterministic Load Flow Calculation (85).

$$\underbrace{\begin{bmatrix} e_{1,0} \\ f_{1,0} \\ \vdots \\ e_{n,0} \\ f_{n,0} \end{bmatrix}}_{\mathbf{V}_0} = \mathrm{LFC}\left(\underbrace{\begin{bmatrix} \mathrm{E}(P_2) \\ \mathrm{E}(Q_2) \\ \vdots \\ \mathrm{E}(P_n) \\ \mathrm{E}(Q_n) \end{bmatrix}}_{\mathbf{E}(\mathbf{S})} \right) \tag{85}$$

With the inverse $\mathbf{J}_{\mathrm{red}}^{-1}(\mathbf{V})$ of the reduced Jacobian matrix $\mathbf{J}_{\mathrm{red}}(\mathbf{V})$ as stated in (54) and the particular deviation ΔS_i from the expected value $\mathrm{E}(S_i)$ for each nodal power, the impact of a change in nodal power on the nodal voltages can be estimated in the vicinity of $\mathbf{V}_0, \mathbf{E}(\mathbf{S})$ using (86).

$$\begin{bmatrix} e_2 \\ f_2 \\ \vdots \\ e_n \\ f_n \end{bmatrix} \approx \underbrace{\begin{bmatrix} e_{2,0} \\ f_{2,0} \\ \vdots \\ e_{n,0} \\ f_{n,0} \end{bmatrix}}_{\mathbf{V}_0} + \mathbf{J}_{\mathrm{red}}^{-1}(\mathbf{V}_o) \cdot \left(\underbrace{\begin{bmatrix} P_2 \\ Q_2 \\ \vdots \\ P_n \\ Q_n \end{bmatrix} - \underbrace{\begin{bmatrix} \mathrm{E}(P_2) \\ \mathrm{E}(Q_2) \\ \vdots \\ \mathrm{E}(P_n) \\ \mathrm{E}(Q_n) \end{bmatrix}}_{\mathbf{E}(\mathbf{S})}}_{\Delta S} \right) \tag{86}$$

When indexing the elements of $\mathbf{J}_{red}^{-1}(\mathbf{V}_o)$ with $J_{\mathrm{red},ij}$ the single approximated voltage components e_i and f_i can be calculated as (87) and (88) respectively.

$$\begin{aligned} e_i = e_{i,0} &+ \sum_{j=2}^{n} J_{\mathrm{red},(2i-1)(2j-3)} \cdot \left(P_j - \mathrm{E}(P_j)\right) \\ &+ \sum_{j=2}^{n} J_{\mathrm{red},(2i-1)(2j-2)} \cdot \left(Q_j - \mathrm{E}(Q_j)\right) \end{aligned} \tag{87}$$

$$\begin{aligned} f_i = f_{i,0} &+ \sum_{j=2}^{n} J_{\mathrm{red},2i(2j-3)} \cdot \left(P_j - \mathrm{E}(P_j)\right) \\ &+ \sum_{j=2}^{n} J_{\mathrm{red},2i(2j-2)} \cdot \left(Q_j - \mathrm{E}(Q_j)\right) \end{aligned} \tag{88}$$

Although they can be calculated separately they are interconnected closely through corresponding input parameters, since they are real and imaginary parts of the same complex number. Thus, equation (87) and (88) are real and imaginary parts of the same function (89).

$$(e_i + \mathrm{j} \cdot f_i) = f(P_2, \dots, P_n, Q_2, \dots, Q_n) \tag{89}$$

In order to determine the PDF of the nodal voltage V_i (89) has to be transposed in a fashion so that V_i is an input argument and the two real-valued nodal powers are its values. As an example we assume that P_k and Q_l were assigned to be the values for the function (90).

$$(P_k + \mathrm{j} \cdot Q_l) = f(e_i, f_i, P_2, \dots, \cancel{P_k}, \dots, P_n, Q_2, \dots, \cancel{Q_l}, \dots, Q_n) \tag{90}$$

With (90) any combination of nodal powers can be found that would result in (86) yielding the given value for V_i. When $f_{P_k}(\dots)$ and $f_{Q_l}(\dots)$ are the real and imaginary part of (90), this can be utilized to determine the searched value of the PDF $\mathrm{P}_{V_i}(e_i + \mathrm{j} \cdot f_i)$ of V_i (91).

$$\begin{aligned} &P_{V_i}(e_i + \mathrm{j} \cdot f_i) \\ &= \int_{-\infty}^{\infty} \cdots \int_{-\infty}^{\infty} P_1(P_1, Q_1) \dots P_k\left(f_{P_k}(\dots), Q_k\right) \dots P_l\left(P_l, f_{Q_l}(\dots)\right) \dots \end{aligned} \tag{91}$$

As mentioned before, the second important issue to be addressed by the Probabilistic Load Flow Calculation is the PDF of line loadings. The procedure is very similar to the one described before, determining the PDFs of nodal powers.

As stated in the previous section, the vector of complex-valued line currents can be determined by multiplying the line admittance matrix $\mathbf{Y}_{\text{line}}$ with the voltage vector representing the network state of interest. In this case, in which not only one state, but the PDF of line loadings should be estimated by using probabilistic calculus and a simplified network model based on the Jacobian matrix, (86) has to be modified. Since (86) does not include values for the voltage at the reference node, the line admittance matrix has to be stated in a modified way. Using the example of section 2.2.2 and the corresponding line admittance matrix as stated in (17), the modified matrix that separates the influence of the reference node's voltage is (92).

$$\begin{bmatrix} I_{12} \\ I_{23} \\ I_{34} \\ I_{24} \end{bmatrix} = \underbrace{\begin{bmatrix} Y_{12} \\ 0 \\ 0 \\ 0 \end{bmatrix}}_{\mathbf{Y}_{\text{line,ref}}} \cdot V_1 + \underbrace{\begin{bmatrix} -Y_{12} & & \\ Y_{23} & -Y_{23} & \\ & Y_{34} & -Y_{34} \\ Y_{24} & & -Y_{24} \end{bmatrix}}_{\mathbf{Y}_{\text{line,red}}} \cdot \begin{bmatrix} V_2 \\ V_3 \\ V_4 \end{bmatrix} \tag{92}$$

Based on this separation of the influence of the reference node's voltage that is assumed to be constant, the influence of certain combinations of nodal powers in the vicinity of $\mathbf{V}_0, \mathbf{E}(\mathbf{S})$ can be estimated using (93). Note that $\mathbf{Y}_{\text{line,ref}}$ is the column of $\mathbf{Y}_{\text{line}}$ that corresponds to the reference node's voltage. The example of $\mathbf{Y}_{\text{line,ref}}$ stated in (93) is only valid for the exemplary network described in section 2.2.2.

$$\underbrace{\begin{bmatrix} a_1 \\ b_1 \\ \vdots \\ a_m \\ b_m \end{bmatrix}}_{\mathbf{I}_{\text{line}}} \approx \underbrace{\begin{bmatrix} Y_{12} \\ 0 \\ 0 \\ 0 \end{bmatrix}}_{\mathbf{Y}_{\text{line,ref}}} \cdot V_1 + \mathbf{Y}_{\text{line,red}} \cdot \left(\underbrace{\begin{bmatrix} e_{2,0} \\ f_{2,0} \\ \vdots \\ e_{n,0} \\ f_{n,0} \end{bmatrix}}_{\mathbf{V}_0} + \mathbf{J}_{\text{red}}^{-1}(\mathbf{V}_o) \cdot \left(\underbrace{\begin{bmatrix} P_2 \\ Q_2 \\ \vdots \\ P_n \\ Q_n \end{bmatrix} - \underbrace{\begin{bmatrix} \mathrm{E}(P_2) \\ \mathrm{E}(Q_2) \\ \vdots \\ E(P_n) \\ E(Q_n) \end{bmatrix}}_{\mathbf{E}(\mathbf{S})}}_{\Delta S} \right) \right) \tag{93}$$

Separating the constant and variable part of (93) gives the basic line loading $\mathbf{I}_{\text{line,0}}$ corresponding to the expansion point $\mathbf{V}_0, \mathbf{E}(\mathbf{S})$ (see (94))

$$\underbrace{\begin{bmatrix} a_{1,0} \\ b_{1,0} \\ \vdots \\ a_{m,0} \\ b_{m,0} \end{bmatrix}}_{\mathbf{I}_{\text{line,0}}} = \underbrace{\begin{bmatrix} Y_{12} \\ 0 \\ 0 \\ 0 \end{bmatrix}}_{\mathbf{Y}_{\text{line,ref}}} \cdot V_1 + \mathbf{Y}_{\text{line,red}} \cdot \underbrace{\begin{bmatrix} e_{2,0} \\ f_{2,0} \\ \vdots \\ e_{n,0} \\ f_{n,0} \end{bmatrix}}_{\mathbf{V}_0} \tag{94}$$

For the variable part of (93) the two matrices $\mathbf{Y}_{\text{line,red}}$ and $\mathbf{J}_{\text{line,red}}^{-1}(\mathbf{V}_o)$ can be summarized with $\mathbf{J}_{\text{line,red}}^{-1}(\mathbf{V}_0)$ according to (95).

$$\underbrace{\begin{bmatrix} \Delta a_1 \\ \Delta b_1 \\ \vdots \\ \Delta a_m \\ \Delta b_m \end{bmatrix}}_{\Delta\mathbf{I}_{\text{line}}} = \underbrace{\mathbf{Y}_{\text{line,red}} \cdot \mathbf{J}_{\text{red}}^{-1}(\mathbf{V}_o)}_{\mathbf{J}_{\text{line,red}}^{-1}(\mathbf{V}_0)} \cdot \left(\underbrace{\begin{bmatrix} P_2 \\ Q_2 \\ \vdots \\ P_n \\ Q_n \end{bmatrix} - \underbrace{\begin{bmatrix} E(P_2) \\ E(Q_2) \\ \vdots \\ E(P_n) \\ E(Q_n) \end{bmatrix}}_{\mathbf{E}(\mathbf{S})}}_{\Delta S} \right) \tag{95}$$

In order to be able to calculate the PDF of the complex-valued line currents, (95) has to be transformed to take Δa_j and Δb_j of the respective line current as an input argument and yields two active and/or reactive powers for one or two nodes that correspond to a certain complex-valued line current. When indexing the elements of $\mathbf{J}_{\text{line,red}}^{-1}(\mathbf{V}_0)$ as $J_{line,ij}$, a_j and b_j of line j can be calculated using (96) and (97), respectively.

$$\begin{aligned} a_j = a_{j,0} &+ \sum_{i=2}^{n} J_{\text{line,red},(2j-1)(2i-3)} \cdot \left(P_j - \mathrm{E}(P_j)\right) \\ &+ \sum_{i=2}^{n} J_{\text{line,red},(2j-1)(2i-2)} \cdot \left(Q_j - \mathrm{E}(Q_j)\right) \end{aligned} \tag{96}$$

$$\begin{aligned} b_j = b_{j,0} &+ \sum_{j=2}^{n} J_{\text{line,red},2j(2i-3)} \cdot \left(P_j - \mathrm{E}(P_j)\right) \\ &+ \sum_{j=2}^{n} J_{\text{line,red},2j(2i-2)} \cdot \left(Q_j - \mathrm{E}(Q_j)\right) \end{aligned} \tag{97}$$

Although they can be calculated separately they are interconnected closely through corresponding input parameters, since they are real and imaginary parts of the same complex number. Thus, equation (96) and (97) are real and imaginary parts of the same function (98).

$$\left(a_j + \mathrm{j} \cdot b_j\right) = f(P_2, \dots, P_n, Q_2, \dots, Q_n) \tag{98}$$

In order to determine the PDF of the line current I_j (98) has to be transposed in a fashion so that I_j serves an input argument and two real-valued nodal powers as its values. As an example we assume that P_k and Q_l were assigned to be the values for the function (99).

$$(P_k + \mathrm{j} \cdot Q_l) = f(a_j, b_j, P_2, \dots, \cancel{P_k}, \dots, P_n, Q_2, \dots, \cancel{Q_l}, \dots, Q_n) \tag{99}$$

With (99) any combination of nodal powers can be found that would result in (93) taking the given value of I_j. If $f_{P_k}(\dots)$ and $f_{Q_k}(\dots)$ are real and imaginary parts of (99), respectively, this can then be used to determine the searched value of the PDF $\mathrm{P}_{I_j}\left(a_j + \mathrm{j} \cdot b_j\right)$ of I_j (100).

$$\begin{aligned} &P_{I_j}\left(a_j + \mathrm{j} \cdot b_j\right) \\ &= \int_{-\infty}^{\infty} \cdots \int_{-\infty}^{\infty} P_1(P_1, Q_1) \dots P_k\left(f_{P_k}(\dots), Q_k\right) \dots P_l\left(P_l, f_{Q_l}(\dots)\right) \dots \end{aligned} \tag{100}$$

The approach presented here is just one popular example for reducing the computational complexity of Probabilistic Power Flow Calculation. There are a number of approaches, nearly all of which are based on network models, computational techniques and approximation of the network's behavior, techniques which we have presented in this chapter. It is meant to be a starting point for the diverse spectrum of different methods and ongoing research in Probabilistic Power Flow Calculation [Dondera et al. 2007][dong et al. 2010].

3.5 Conclusions and Final Remarks

In this chapter we covered the basic principles behind the determination of an electrical grid's operational state through load flow calculation. The major challenge for grid operators in future Smart Grids will be the translation of metered and derived probabilistic values of nodal power into nodal voltages and line currents. These two are subject to operational restrictions and constraints for a stable and reliable electrical power supply. The non-linearity of the complex-valued power flow equations require the utilization of the so called Newton-Raphson method, an iterative algorithm starting from likely guesses of all unknown variables and formulating a Taylor Series for each of the power balance equations included in the equation system. The resulting linearized system of power flow equations is solved to determine a refined iteration of the voltage angles and magnitudes for which the procedure is repeated until a stop (precision) criterion is met.

Under increased power feed-in from renewable resources and technological innovation with more intelligent applications mainly in the form of dispatchable loads and smaller, more geographically distributed Generation units such as combined heat and power cogeneration, photovoltaics, windcraft and in the mid-term, plug-in electric or plug-in hybrid electric vehicles (PEV, PHEV), respectively, the grid's complexity is increasing drastically [Williams and Crawford 2010]. First attempts on applying these Newton-Raphson-based methods to such highly dynamic Smart Grid Scenarios and dealing with related robustness and real-time issues of the algorithm are presented in [Krause and Lehnhoff 2008], [Krause and Lehnhoff 2009] and [Krause and Lehnhoff 2010].

Under conventional centralized organization and (thus limited individual) control, these distributed and highly stochastical power sources could adversely affect power standards and quality, generation efficiency or violate capacity limits or reliability constraints of the existing infrastructure (feeders, transformers and lines especially in low-voltage distribution grids).

Traditionally, grid utilization through load flows can be relatively well predicted on the basis of past loads generation schedules but distributed generation from renewable sources and high-capacity demand (e.g. from heat pumps, PEV or PHEV) is disruptive and stochastic in nature.

The major weakness of deterministic transmission planning and load flow calculation is their inability to take account of such probabilistic characteristics in power systems including uncertainties in load forecasting, generation schedules and random system faults. These situations are very difficult to identify using classical deterministic power flow mechanisms.

With the methods of probabilistic load flow calculation covered in this chapter, we have demonstrated state-of-the-art approaches to tackle these issues. The downside of experiment-based and even Monte-Carlo-based methods is the computational overhead when simulating large-scale networks (or the lack of precision for the latter one). In order to apply probabilistic load flow calculation to realistic (trans-national) large-scale power grids or even envision on-line algorithms for this task probabilistic calculus-based approaches are a major part of current research on this topic. In this chapter we have covered a convolution-based method.

Another promising approach is the propagation of characteristic measures through the Jacobian matrix or other linearized models [Zhang and Lee 2004][Patra and Misra 1993]. With this contribution we have covered the basics that allow prospective readers to dive deeper into related further publications mentioned throughout this chapter.

References

Aboytes, F.: Stochastic Contingency Analysis. IEEE Transactions on Power Apparatus and Systems 97, 335–341 (1978)

Allan, R.N., Alshakarchi, M.R.G.: Probabilistic AC Load Flow. Proceedings of the Institution of Electrical Engineers 123, 531–536 (1976)

Allan, R.N., Alshakarchi, M.R.G.: Probabilistic Techniques in AC Load-Flow Analysis. Proceedings of the Institution of Electrical Engineers 124, 154–160 (1977)

Allan, R.N., da Silva, A.M.L., Burchett, R.C.: Evaluation Methods and Accuracy in Probabilistic Load Flow Solutions. IEEE Transactions on Power Apparatus and Systems 100, 2539–2546 (1981)

Borkowska, B.: Probabilistic Load Flow. IEEE Transactions on Power Apparatus and Systems PAS93, 752–759 (1974)

Bronstein, I.N., Semendjajev, K.A., Musiol, G., Mühlig, H.: Taschenbuch der Mathematik. Verlag Harri Deutsch, Frankfurt am Main (2000) ISBN 3-8171-2004-4

Chen, P., Chen, Z., Bak-Jensen, B.: Probabilistic Load Flow: A Review. In: Proceedings of the 3rd IEEE International Conference on Electric Utility Deregulation and Restructuring and Power Technologies 2008, pp. 1586–1591 (2008)

da Silva, A.M.L., Allan, R.N., Soares, S.M., Arienti, V.L.: Probabilistic Load Flow Considering Network Outages. IEEE Proceedings on Generation, Transmission and Distribution 132, 139–145 (1985)

da Silva, A.M.L., Arienti, V.L.: Probabilistic Load Flow by a Multilinear Simulation Algorithm. IEEE Proceedings on Generation, Transmission and Distribution 137, 276–282 (1990)

Dopazo, J.F., Klitin, O.A., Sasson, A.M.: Stochastic Load Flows. IEEE Transactions on Power Apparatus and Systems 94, 299–309 (1975)

Dondera, D., Popa, R., Velicescu, C.: The Multi-Area Systems Reliability Estimation Using Probabilistic Load Flow by Gramm-Charlier Expansion. In: Proceedings of the IEEE International Conference on Computer as a Tool (EUROCON), pp. 1470–1474 (2007)

Dong, L., Cheng, W., Bao, H., Yang, Y.: Probabilistic Load Flow Analysis for Power System Containing Wind Farms. In: Proceedings oft the IEEE Power and Energy Engineering Conference (APPEEC), pp. 1–4 (2010)

Grainger, J.J., Steveson, W.D.: Power System Analysis. Mcgraw-Hill Education, New York (1994)

Kornerup, P., Muller, J.M.: Choosing Starting Values for Certain Newton–Raphson Iterations. In: Theoretical Computer Science - Real Numbers and Computers Archive, vol. 351(1). Elsevier, Amsterdam (2006)

Krause, O., Lehnhoff, S., Rehtanz, C., Handschin, E., Wedde, H.F.: On-line Stable State Determination in Decentralized Power Grid Management. In: Proceedings of the 16th Power Sys-tems Computation Conference (PSCC 2008). IEEE Press, Los Alamitos (2008)

Krause, O., Lehnhoff, S., Rehtanz, C., Handschin, E., Wedde, H.F.: On Feasibility Boundaries of Electrical Power Grids in Steady State. International Journal of Electric Power & Energy Systems (IJEPES) 31(9), 437–444 (2009)

Krause, O., Lehnhoff, S., Rehtanz, C.: Linear Constraints for Remaining Transfer Capability Allocation. In: Proceedings of the IEEE International Conference on Innovative Smart Grid Technologies Europe 2010. IEEE Press, Los Alamitos (2010)

Kundur, P.: Power System Stability and Control. Mcgraw-Hill Professional, New York (1994)

Patra, S., Misra, R.B.: Probabilistic Load Flow Solution us-ing Method of Moments (1993)

Williams, T.J., Crawford, C.: Probabilistic Power Flow Modeling: Renewable Energy and PEV Grid interactions. In: Proceedings of The Canadian Society for Mechanical Engineering Forum 2010 (2010); In Proceedings of the IEEE International Conference on Advances in Power System Control, Operation and Management (APSCOM 1993), p. 922 (1993)

Zhang, P., Lee, S.T.: Probabilistic Load Flow Computation using the Method of Combined Cumulants and Gram-Charlier Expansion. IEEE Transactions on Power Systems 19(1), 676–682 (2004)

Evaluation of Green and Renewable Energy System Alternatives Using a Multiple Attribute Utility Model: The Case of Turkey

İhsan Kaya[1] and Cengiz Kahraman[2]

[1] Yıldız Technical University, Department of Industrial Engineering, 34349, Yıldız, Istanbul, Turkey
[2] Istanbul Technical University, Department of Industrial Engineering, 34367, Maçka, Istanbul, Turkey

Abstract. Energy is a critical foundation for economic growth and social progress. It is estimated that 70% of the world energy consumption could be provided from renewable resources by the year 2050 so that renewable energy which is the inevitable choice for sustainable economic growth, for the harmonious coexistence of human and environment as well as for the sustainable development is very important for the humanity. The aim of this chapter is to evaluate the renewable energy alternatives as a key way for resolving the Turkey's energy-related challenges because of the fact that Turkey's energy consumption has risen dramatically over the past three decades as a consequence of economic and social development. In order to realize this aim, the Multi-Attribute Utility Theory (MAUT) is used for the evaluation of renewable energy alternatives. According to MAUT, the overall evaluation U(x) of an object x is defined as a weighted addition of its evaluation with respect to its relevant value dimensions. In the evaluation phase, 4 main attributes and 17 sub-attributes are used to determine the most appropriate renewable energy alternative among Solar, Wind, Hydropower, Biomass, and Geothermal.

1 Introduction

Energy is essential for economic and social development and improved quality of life in all countries. For that reason, energy constitutes one of the main inputs for economic and social development. In line with the increasing population, urbanization, industrialization, spreading of technology and rising of wealth, energy consumption is increasing. Energy consumption and consequently energy supply at minimum amount and cost is the main objective, within the approach of a sustainable development that supports economic and social development. Much of the world's energy, however, is currently produced and consumed in ways that could

K. Gopalakrishnan et al. (Eds.): Soft Comput. in Green & Renew. Ener. Sys., STUDFUZZ 269, pp. 157–182.
springerlink.com

not be sustained if technology were to remain constant and if overall quantities were to increase substantially. Energy has an important role in our daily life. Moreover, energy sources affect the strategies of a country directly. In the world, two kinds of energy are available; non-renewable and renewable. Renewable energy is the energy derived from natural sources. Clean, domestic and renewable energy is commonly accepted as the key for future life. This is primarily because renewable energy resources have some advantages when compared to fossil fuels. Renewable energy sources are also often called alternative sources of energy. Renewable energy resources that use domestic resources have the potential to provide energy services with zero or almost zero emissions of both air pollutants and greenhouse gases. Main renewable energy resources are biomass energy, hydro energy, geothermal energy, solar energy, and wind energy. When we try to select any alternative using some attributes, we have to take into account conflicting issues among the considered attributes. For example, two attributes that could be used in selecting a renewable energy alternative might be reliability and implementation cost. These are two conflicting attributes since an attempt to increase reliability possibly causes an increase in implementation cost. The selection among renewable energy alternatives is a multiattribute problem with many conflicting attributes. We have to evaluate some alternatives by taking into account their advantages and disadvantages based on selection attributes. Hence, this problem should be solved by a multiattribute method.

To assess the environmental impacts of the renewable energy alternatives, life-cycle assessment (LCA) is an important tool. Selection of product design, materials, processes, reuse or recycle strategies, and final disposal options requires careful examination of energy and resource consumption as well as environmental discharges associated with each prevention or design alternative. To accomplish this task, LCA models have been developed and software products are available. Because of the difficulty in estimating resource consumption and environmental discharges produced by processes associated with the life cycle of a renewable energy alternative, the scope of a LCA analysis is limited (simplified) by drawing an ad hoc system boundary that excludes all but a few upstream and downstream processes (Hendrickson et al., 1998). Therefore, the multiattribute utility theory used in our chapter evaluates Green and Renewable Energy System Alternatives by taking into account their benefits and costs along their life cycles through the selection attributes.

Utility functions give us a way to measure investor's preferences for wealth and the amount of risk they are willing to undertake in the hope of attaining greater wealth. The risk aversion property states that the utility function is concave or, in other words, that the marginal utility of wealth decreases as wealth increases. Different investors can and will have different utility functions. Theoretically, decision makers comprise three types: risk averse, risk neutral, and risk taker (risk prone or risk seeking).

The principle of expected utility maximization states that a rational investor, when faced with a choice among a set of competing feasible investment alternatives, acts to select an investment which maximizes his expected utility of wealth.

For the construction of utility functions, the decision-maker's preferences for gambles are often analyzed by the method suggested by Bell et al. (1978) and Keeney and Raiffa (1993).

Multiattribute utility models can take into consideration the decision maker's preferences in the form of utility function which is defined over a set of attributes. Utility is a measure of desirability or satisfaction and provides a uniform scale to compare and / or combine tangible and intangible attributes. A utility function is a devise which quantifies the preferences of a decision maker by assigning a numerical index to varying levels of satisfaction of an attribute. A utility value is an abstract equivalent of the abstract being considered from natural units such as years, or $, into a series of commensurable units on an interval scale of zero to one. Such transformation of value, may, or may not, be a linear function. This is primarily dependent upon the decision maker(s) or expert(s) from which such functions were derived. The utility models may be multiplicative or additive. The decision maker must first check which model is suitable for the considered problem. Turkish energy consumption has risen dramatically over the past 20 years due to the combined demands of industrialization and urbanization. Turkey's primary energy consumption has increased from 32 mtoe (million tons of oil equivalents) in 1980 to 74 mtoe in 1998. According to the planning studies, Turkey's final consumption of primary energy is estimated to be 171 mtoe in 2010 and 298 mtoe in 2020. In other words, in 1999, domestic energy production met 36% of the total primary energy demand and will probably meet 24% in 2020. The level of Turkey's energy consumption is still low relative to similar sized countries, such as France and Germany, with gross inland consumptions of 235 and 339 mtoe in 1995 and with estimated values of 290 and 350 mtoe in 2020, respectively (Hepbaslı and Ozalp, 2003). When the case of Turkey is considered, it can be said that Turkey is heavily dependent on imported energy resources placing a big burden on the economy. Air pollution is also becoming a great environmental concern in the country. In this situation, renewable energy resources appear to be the one of the most efficient and effective solutions for clean and sustainable energy development in Turkey. Turkey's geographical location has several advantages for extensive use of most of these renewable energy sources. As Turkey's economy has expanded in recent years, the consumption of primary energy has increased. Presently in order to increase the energy production from domestic energy resources, to decrease the use of fossil fuels as well as to reduce of green house gas emissions, different renewable energy sources are used for energy production in Turkey. Among these renewable energy resources, hydropower, biomass, biogas, bio-fuels, wind power, solar energy and geothermal energy are the most favorite ones in the future. The selection of the best alternative for Turkey takes an important role for energy investment decisions. There are various decision-making methodologies developed by researches in the literature. Among the most used multi-criteria decision making methods for renewable energy investments, it can be counted Analytic Hierarchy Process (AHP), Analytic Network Process (ANP), Multi Attribute Utility Models, Preference Ranking Organization Method for Enrichment Evaluation

(PROMETHEE), the elimination and choice translating reality (ELECTRE), a hybrid of ELECTRE III, and PROMETHEE II.

In this chapter, to the best of our knowledge, the MAUT is first time used for the evaluation of green and renewable energy alternatives. The rest of this chapter is organized as follows: Multiple attribute utility models and recent studies on decision making by using MAUT are explained in Section 2. Renewable energy and alternatives are briefly summarized in Section 3. The application is detailed in Section 4 and the obtained results are discussed in Section 5.

2 Multiple Attribute Utility Models

Multiple attribute utility theory (MAUT) which is an analytical method for decision-making based on multiple criteria originated in the eighteenth century was developed by Keeney and Raiffa (1976).One important class of methods in multicriteria decision making (MCDM) is based on constructing a utility or value function $U(x)$, which represents the overall strength of support in favor of the alternative x. This approach is known as the multiple attribute utility theory (MAUT) (Beliakov and Warren, 2001). MAUT is one of the major analytical tools associated with the field of decision analysis. The MAUT analysis of alternatives explicitly identifies the measures that are used to evaluate the alternatives, and helps to identify those alternatives that perform well on a majority of these measures, with a special emphasis on the measures that are considered to be relatively more important. MAUT can be used instead of a costing approach when good cost data are not available or when cost is not suitable as a measure of performance. Alternatively, MAUT can be used to embellish costing information that is considered to be incomplete (e.g., to account for the intangibles) (Butler et al., 2001).

In general, the utility $U(x)=U(x_1, x_2, x_3, ..., x_n)$, of any combination of outcomes $(x_1, x_2, x_3, ...x_n)$ for n attributes $(X_1, X_2, X_3, ..., X_n)$ can be expressed as either (i) an additive or (ii) a multiplicative function of the individual attribute utility functions $U_1(x_1), U_2(x_2), U_3(x_3), ..., U_n(x_n)$ provided that each pair of attributes is preferentially independent of its complement and utility independent of its complement (Canada and Sullivian, 1989).

2.1 Additive Utility Model

In this case, the attributes should be additively independent. This will be true if $\sum_{i=1}^{n} k_i = 1$ in the model as given in Eq. 1. The utility $U(x)$ of any combination of outcomes for n attributes $(X_1, X_2, X_3, ...X_n)$ can be expressed as follows (Canada and Sullivian, 1989):

$$U(x)=\sum_{i=1}^{n}U\left(x_i,\overline{x}_i^0\right)=\sum_{i=1}^{n}k_iU_i\left(x_i\right) \tag{1}$$

where $U\left(x_i,\overline{x}_i^0\right)$ is the utility of the outcome for the ith criterion, x_i, and the worst possible outcome for the complement of the ith attribute, $\overline{x}_i^0$, k_i is the weight (scaling factor) for the ith attribute, and $U_i\left(x_i\right)$ is the utility of the outcome x_i for the ith attribute. Further conditions and explanations for the additive utility model are as follows (Canada and Sullivian, 1989):

- U is normalized by $U\left(x_1^0,x_2^0,x_3^0,...,x_n^0\right)=0$ and $U\left(x_1^*,x_2^*,x_3^*,...,x_n^*\right)=1.00$ (Note: x_i^0 means the worst outcome of x_i and x_i^* means the best outcome of x_i).
- U_i is a conditional utility function of x_i normalized by $U_i\left(x_i^0\right)=0$ and $U_i\left(x_i^*\right)=1$, for $i=1,2,...,n$ attributes.
- $k_i=U\left(x_i^*,\overline{x}_i^0\right)$, for $i=1,2,...,n$ attributes.

2.2 *Multiplicative Utility Model*

The utility $U(x)$ of any combination of outcomes of n attributes can be obtained from the solution to the following equation (Canada and Sullivian, 1989):

$$KU(x)+1=\prod_{i=1}^{n}\left[Kk_iU_i\left(x_i\right)+1\right] \tag{2}$$

Solving for $U(x)$ gives

$$U(x)=\frac{\prod_{i=1}^{n}\left[Kk_iU_i\left(x_i\right)+1\right]-1}{K} \tag{3}$$

where

- $U(x)$ is normalized by $U\left(x_1^0,x_2^0,x_3^0,...,x_n^0\right)=0$ and $U\left(x_1^*,x_2^*,x_3^*,...,x_n^*\right)=1$.
- $U_i\left(x_i\right)$ is a conditional utility function of X_i normalized by $U_i\left(x_i^0\right)=0$ and $U_i\left(x_i^*\right)=1$, for $i=1,2,...,n$.
- $k_i=U\left(x_i^*,\overline{x}_i^0\right)$.

- K is a scaling constant that is a solution to

$$1+K=\prod_{i=1}^{n}\left(1+Kk_i\right) \tag{4}$$

and must be found iteratively. When utility independence applies, as assumed by the model, $-1<K<0$.

2.3 Determination of Utility Functions for Individual Attributes

To use either of the models above, a utility function must be specified for each attribute, X_i, where $U_i\left(x_i^0\right)=0$ and $U_i\left(x_i^*\right)=1.0$. The shape of each utility function depends on the decision maker's subjective judgments on the relative desirability of various outcomes. This can be done by obtaining answers to a series of questions such as the following: For attribute X_i, what certain outcome, x_i, would be equally as desirable as a *P%* chance of the highest outcome and *(1-P)%* chance of the lowest outcome? This can be expressed in utility terms, using the extreme values, x_i^* and x_i^0, as (Canada and Sullivian, 1989):

$$U\left(x_i=?\right)=PU\left(x_i^*\right)+\left(1-P\right)U\left(x_i^0\right) \tag{5}$$

To obtain plotting plots for the utility function, one can vary *P* as desired. Alternatively, one could specify the certain outcome, x_i over a range of values and ask questions such as: At what *P* is the certain outcome x equally desirable as

$$PU\left(x_i^*\right)+\left(1-P\right)U\left(x_i^0\right)?$$

Note that the utility of the certain outcome=the probability of the best outcome, which saves calculations. The points on possible curve can be determined and plotted until one is satisfied with the "accuracy" of his or her utility representation.

2.4 Determination of Weighting or Scaling Factors

Once utility functions for all criteria have been determined, the next step is to determine the weighting for each attribute, k_i. From the explanations for both the additive and multiplicative utility models, $k_i=U\left(x_i^*,\overline{x}_i^0\right)$, where $0\le k_i\le 1$. In words, k_i is the utility if the outcome for attribute *i* is at its best value, x_i^*, and the outcome for all attributes except *i* are at their respective worst values, $\overline{x}_i^0$. Two types of questions often helpful in assessing the k_i's are given below (Canada and Sullivian, 1989):

Question A

For what probability *P* are you indifferent between:

- The lottery giving a *P* chance at $x^* = \left(x_1^*, x_2^*, x_3^*, \ldots, x_n^*\right)$ and *(1-P)* chance at $x^0 = \left(x_1^0, x_2^0, x_3^0, \ldots, x_n^0\right)$
- The consequence $\left(x_i^0, \ldots, x_{i-1}^0, x_i^*, x_{i+1}^0, \ldots, x_n^0\right)$

The above is shown diagrammatically and in words in Figure 1. The result of such an assessment is that $P = k_i$.

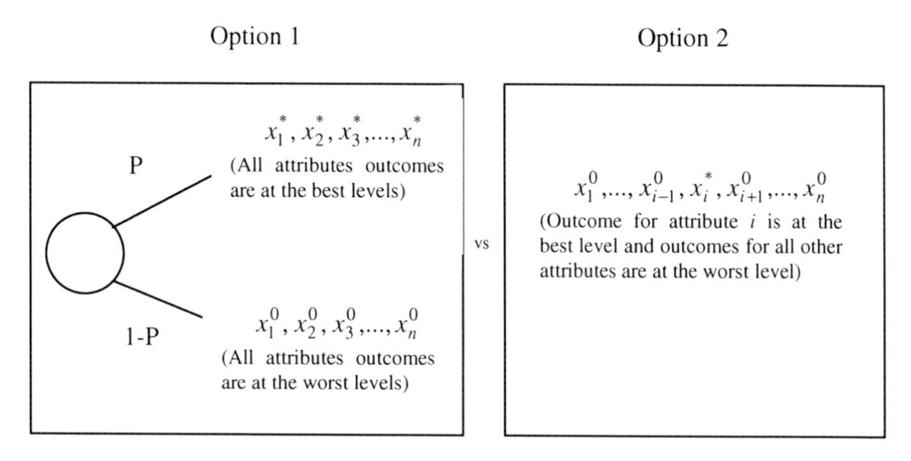

Fig. 1 Illustration of Question A for finding weighting or scaling factor, k_i, for the *i*th attribute (Canada and Sullivian, 1989)

Question B

Select a level of $X_i\,(\text{e.g. } x_i')$ for attribute *i* and a level of $X_j\,(\text{e.g. } x_j')$ for attribute j so that you are indifferent between:

- An outcome yielding x_i' and x_j^0 together, and
- An outcome yielding x_j' and x_i^0 together.

Thus one can use the relation

$$k_i U_i(x_i') = k_j U_j(x_j')$$

To solve for either k_i or k_j, depending on which is unknown.

Suggested good practice in assessing the k 's would be first to rank them, then to use question A to evaluate the largest k_i, and then to use question B successively to evaluate the magnitude of the other k_i's relative to the largest k_i, or relative to any other k_i which has already been determined The rank ordering of the k_i's can be done by intuitive judgment or by asking the decision maker, for instance, whether the prefers $\left(x_1^*, \bar{x}_1^0\right)$ or $\left(x_2^*, \bar{x}_2^0\right)$. If the former is preferred, $k_1 > k_2$, and if the latter is preferred, $k_2 > k_1$. This can be done for as many combinations of weighting factors as deemed needed to check for consistency. (Note: $\bar{x}_1^0$ means all x attributes other than x_1 are at their respective worst values.) (Canada and Sullivian, 1989).

The usage of MAUT in the literature has been briefly summarized in the following: De Melo Brito et al. (2010) developed several decision models by using different multi-criteria methods. They integrated utility functions with the variable interdependent parameters method to evaluate alternatives through an additive value function regarding mean time to repair, contract cost, the geographical spread of the candidate's service network, the candidate's reputation and the compatibility of company cultures. Zhang and Xing (2010) presented a fuzzy-multi-objective particle swarm optimization (PSO) to solve the fuzzy time–cost–quality tradeoff (TCQT) problem. They described the time, cost and quality as fuzzy numbers and a fuzzy multiattribute utility methodology incorporated with constrained fuzzy arithmetic operations was adopted to evaluate the selected construction methods. They applied PSO to search for the TCQT solutions by incorporating the fuzzy multi-attribute utility methodology. Nishizaki et al. (2010) proposed a method for the sensitivity analysis of multiattribute utility functions in multiplicative form, taking into account the imprecision of the decision maker's judgment in the procedures for determining attribute weights. Streicher-Porte et al. (2009) applied MAUT to the supply of computers to schools in Colombia by evaluating three different supply scenarios. Wang et al. (2009) introduced net promoter score technology to help firms target satisfied or passive consumers, and allow them to highlight the additional value to consumers of environmentally-friendly products. To achieve the above goals, MAUT was used to develop an aggregated fulfillment level in relation to obtaining such products. Yang et al. (2009) developed a new hybrid methodology to explain the role of Bayesian Networks in MAUT. They proposed a novel utility function, which can appropriately represent the risk results produced and avoid the arguments resulting from exclusive states expressed by linguistic variables with fuzzy nature and the ignorance/incomplete representation of context dependency between decision attributes. Cirtita and Ilieş (2009) proposed a tool to define the best network alternative in downstream supply chain, based on MAUT, creating a value function with scalable importance criteria coefficients. Abouelnaga et al. (2009) used MAUT to optimize the selection process for energy alternatives which are nuclear, hydroelectric, gas/oil, and solar in Egypt. Jimenez et al. (2009) considered the situation where there was the least knowledge of the alternative consequences or performances, i.e. when there is no knowledge whatsoever of the performance of several alternatives for some attributes, i.e. neither a precise performance nor a probability distribution can be specified in MAUT. Zhang (2008) proposed a framework of multi-objective simulation optimization for optimizing equipment-configurations of earthmoving operations by integrating an activity object-oriented simulation, MAUT, a statistical approach like the two-stage ranking and selection procedure and particle swarm optimization algorithm. The MAUT was applied to evaluate the performances generated through simulation by considering multiple criteria and the preference of decision-makers. Kainuma and Tawara (2006) extended the range of the supply chain to include re-use and recycling throughout the life cycle of products and services and proposed the MAUT method for assessing a supply chain. Xu and Huang (2006) proposed a quantitative setup plan evaluation system driven by MAUT coupled with manufacturing error simulation to serve three purposes: (*i*) to clarify what is optimality of setup plans, (*ii*) to provide a systematic method of evaluating setup plan alternatives quantitatively, and (*iii*) to incorporate in existing automatic setup planning systems a human interface to fulfill their potential values. Jimenez et al. (2003) described a decision support system based on an additive or multiplicative multiattribute utility model for identifying the optimal strategy. Butler et al. (2001) described the application and detailed of how they used the simulator, MAUT, and statistical ranking and selection to select the best project configuration of possible configurations. Sohn et al. (2001) proposed a method to aggregate multi-stakeholder opinions and assimilate the public opinions during the course of the decision making process. The analytic hierarchy process (AHP) and MAUT were employed, and for uncertainty analysis, a fuzzy set based approach was adopted in the aggregation phase. Malakooti and Subramanian (1999) developed a generalized decomposable multiattribute utility function for representing the decision maker's preferential behavior. Malakooti (1989) introduced a quasi-concave nonlinear multiple attribute utility function to rank multiple criteria alternatives.

3 Renewable Energy

Renewable energy sources have been important for humans since the beginning of civilization. In the following, each renewable energy alternative is briefly explained (Kaygusuz, 2002; 2003; Ulutaş, 2005; Demirbaş, 2008; Kahraman et al., 2009):

Biomass

Biomass refers to living and recently dead biological material that can be used as fuel or for industrial production. Most commonly, biomass refers to plant matter grown for use as biofuel, but it also includes plant or animal matter used for production of fibres, chemicals or heat. Biomass may also include biodegradable wastes that can be burnt as fuel. It excludes organic material which has been transformed by geological processes into substances such as coal or petroleum. Unlike other renewable energy sources, biomass can be converted directly into liquid fuels for transportation needs. It can be used as a diesel additive to reduce vehicle emissions or in its pure form to fuel a vehicle.

Hydropower

Hydropower or hydraulic power is the power derived from the force or energy of moving water, which may be harnessed for useful purposes. Hydropower is obtained in the force of the water on the riverbed and banks of a river. It is particularly powerful when the river is in flood. The force of the water results in the removal of sediment and other materials from the riverbed and banks of the river, causing erosion and other alterations. The most common type of hydropower plant is using a dam on a river to store water in a reservoir. When the water released from the reservoir flows through a turbine, a generator activates to produce electricity by spinning the turbine. Another type of hydropower plant called a pumped storage stores power. The power is sent from a power grid into the electric generators.

Geothermal energy

Geothermal power is energy generated by heat stored beneath the Earth's surface or the collection of absorbed heat derived from underground in the atmosphere and oceans.

Solar energy

Solar energy can be used to generate electricity, provide hot water, and to heat, cool, and light buildings. Photovoltaic (solar cell) systems convert sunlight directly into electricity. A solar or PV cell consists of semi-conducting material that absorbs the sunlight. The solar energy knocks electrons loose from their atoms, allowing the electrons to flow through the material to produce electricity.

Wind energy

Wind turbines capture the wind's energy with two or three propeller-like blades, which are mounted on a rotor, to generate electricity. The turbines sit high atop towers, taking advantage of the stronger and less turbulent wind at 100 ft (30 m) or more aboveground. Wind turbines can be used as stand-alone applications, or they can be connected to a utility power grid or even combined with a photovoltaic (solar cell) system. Stand-alone turbines are typically used for water pumping or communications.

In recent years, some studies have concentrated on energy planning and energy policy making. Kahraman and Kaya (2010) suggested a fuzzy multicriteria decision-making methodology based on the analytic hierarchy process (AHP) under fuzziness and allowed the evaluation scores from experts to be linguistic expressions, crisp or fuzzy numbers for the selection among energy policies for Turkey. Aydın et al. (2010) developed a decision support tool for site selection of wind energy turbines in the Geographic Information System environment using fuzzy decision making approach. This decision support tool enabled aggregation of individual satisfaction degrees of each alternative location for various fuzzy environmental objectives. Kahraman et al. (2010) suggested axiomatic design methodology for the selection among renewable energy alternatives under fuzzy environment. Kaya and Kahraman (2010) proposed a methodology based on fuzzy VIKOR and fuzzy AHP to determine the best renewable energy alternative for Istanbul. They also used the proposed methodology to selection among alternative energy production sites in Istanbul. Kucukali and Baris (2010) employed a fuzzy logic method to forecast the gross electricity demand of Turkey. Kahraman et al.

(2009) suggested two fuzzy multicriteria decision making methodologies for the selection among renewable energy alternatives. The first methodology was based on the fuzzy AHP which allowed the evaluation scores from experts to be linguistic expressions, crisp, or fuzzy numbers, while the second was based on AD principles under fuzziness which evaluated the alternatives under objective or subjective criteria with respect to the functional requirements obtained from experts. In the application of the proposed methodology the most appropriate renewable energy alternative was determined for Turkey. Afgan and Carvalho (2008) used sustainability assessment method for the evaluation of quality of the selected hybrid energy systems. They used the following indicators: economic indicator, environment indicator, and social indicator. Patlitzianas et al. (2008) presented an information decision support system, which consists of an expert subsystem, as well as a multi criteria decision making (MCDM) subsystem. The system supported the state toward the formulation of a modern environment, since it incorporates the ''new parameters'' of the energy market, namely the liberalization and the climate change. The system was successfully applied in the 13 accession member states of the European Union. Burton and Hubacek (2007) investigated a local case study of different scales of renewable energy provision for local government in the UK. They compared the perceived social, economic and environmental cost (SEE) of these small-scale energy technologies to larger-scale alternatives. In order to investigate whether the energy could have been generated at a lower SEE cost if large-scale projects had been available, a multi-criteria decision analysis (MCDA) methodology was used to compare the advantages and disadvantages of a number of different renewable energy technologies. They considered eight renewable energy technologies of differing scales: solar photovoltaic, micro-wind, microhydro, large-scale wind, large-scale hydro, energy from waste, landfill gas and biomass (wood chippings) based on the definition of renewable energy used by the UK government. Patlitzianas et al. (2007) presented an integrated multicriteria decision making approach, ordered weighted average, of qualitative judgments for assessing the environment of renewable energy producers in the fourteen different member states of the European Union accession. Afgan et al. (2007) presented an evaluation of the potential natural gas utilization in energy sector. They classified the criteria as economic, environmental, social and technological. Among the potential options of gas utilization following systems were considered: Gas turbine power plant, combine cycle plant, Combined Heat and Power (CHP) plant, steam turbine gas-fired power plant, fuel cells power plant. They also used multi-criteria method, general index of sustainability, for the assessment of potential options with priority given to the economic, environmental, social and technological criteria. Çam (2007) compared a conventional proportional integral controller and a fuzzy gain scheduled proportional integral controller for applying to a single area and a two area hydroelectric power plant, considering Turkey's several hydro power sources. Jebaraj and Iniyan (2007) developed a fuzzy-based linear programming optimal energy model that minimized the cost and determined the optimum allocation of different energy sources for the centralized and decentralized power generation in India. Polatidis et al. (2006) developed a methodological framework to provide insights regarding the suitability of multi-criteria techniques in the context of renewable energy planning. They created a comparative matrix with the various appropriate multi-criteria techniques and their performance for renewable energy planning. Ulutaş (2005) analyzed the appropriate energy policy problem which considers as a MCDM problem with interactive criteria and alternatives. She used the ANP to evaluate the alternative energy sources for Turkey's energy resources. Cavallaro and Ciraolo (2005) proposed a multicriteria method in order to support the selection and evaluation of one or more of the solutions to make a preliminary assessment regarding the feasibility of installing some wind energy turbines in a site on the island of Salina in Italy. They compared the four wind turbine configurations. They used a multicriteria methodology to rank the solutions from the best to the worst. Haralambopoulos and Polatidis (2003) described an applicable group decision-making framework for assisting with multi-criteria analysis in renewable energy projects, utilizing the PROMETHEE II outranking method to achieve group consensus in renewable energy projects. The proposed framework was tested in a case study concerning the exploitation of a geothermal resource, located in the island of Chios, Greece. Beccali et al. (2003) made an application of the multicriteria decision-making methodology to assess an action plan for the diffusion of renewable energy technologies at regional scale. They also carried out a case study for the island of Sardinia. They used ELECTRE-III method under fuzzy environment. Borges and Antunes (2003) presented an interactive approach to deal with fuzzy multiple objective linear programming problems based on the analysis of the decomposition of the parametric (weight) diagram into indifference regions corresponding to basic efficient solutions. The approach was illustrated to tackle uncertainty and imprecision associated with the coefficients of an input–output energy-economy planning model, aimed at providing decision support to decision makers in the analysis of the interactions between the energy system and the economy on a national level. Afgan and Carvalho (2002) presented the selection of criteria and options for the new and renewable energy technologies assessment based on the analysis and synthesis of parameters under the information deficiency method to define energy indicators used in the assessment of energy systems which met the sustainability criterion. They took into account energy resources, environment capacity, social indicators and economic indicators. Goumas and Lygerou (2000) extended a multicriteria method of ranking alternative projects, PROMETHEE, to deal with fuzzy input data. The proposed method was applied for the evaluation and ranking of alternative energy exploitation schemes of a low temperature geothermal energy.

The selection among renewable energy alternatives is a multicriteria decision making problem with many conflicting criteria. Hence, this problem should be solved by a multicriteria method. In this chapter, the most appropriate renewable energy alternative for Turkey is determined by using a multiple attribute utility model.

4 An Application: The Case of Turkey

The Republic of Turkey, located in Southeastern Europe and Southwestern Asia, has an area of about 780,580 sq km and a population of over 70 million. With its young population, growing energy demand per person, fast growing urbanization and economic development, Turkey has been one of the fast growing power markets of the world for the last two decades. Turkey is an energy importing country; more than half of the energy requirement has been supplied by imports. Turkey's primary energy sources include hydropower, geothermal, lignite, hard coal, oil, natural gas, wood, animal and plant wastes, solar and wind energy. In 2004, primary energy production and consumption has reached 24.1 million tonnes (Mt) of oil equivalent (Mtoe) and 81.9 Mtoe, respectively. Fossil fuels provided about 86.9% of the total energy consumption of the year 2004, with oil (31.5%) in first place, followed by coal (27.3%) and natural gas (22.8%). Turkey has not utilized nuclear energy yet. The Turkish coal sector, which includes hard coal as well as lignite, accounts for nearly one half of the country's total primary energy production (43.7%). The renewable collectively provided 13.2% of the primary energy, mostly in the form of combustible renewables and wastes (6.8%), hydropower (about 4.8%) and other renewable energy resources (approximately 1.6%) (Erdogdu, 2010; IEA, 2007).

Because of the increasing population and life standards in Turkey, fossil fuel consumption is increasing. As a result, fossil fuels are being depleted rapidly. Another important problem associated with fossil fuels is that their consumption has major negative impacts on the environment. Therefore, Turkey has to include renewable energy alternatives in their future energy plans so that they can produce reliable and environmentally friendly energy. For this aim, a multicriteria decision making methodology is used to determine the most appropriate renewable energy alternative for Turkey in this paper.

According to 1970-2006 data, Turkey produced 342,458 ktoe from its own renewable energies alternatives and consumed it's all. It is expected that by the year 2020, the renewable energy production will be 19,841.49 ktoe, while renewable energy consumption will be 19,841.49 ktoe (Republic of Turkey Ministry of Energy and Natural Resources, 2008). As fossil fuel energy becomes scarcer, Turkey will face energy shortages, significantly increasing energy prices, and energy insecurity within the next few decades. In addition, Turkey's continued reliance on fossil fuel consumption will contribute to accelerating the rates of domestic environmental quality and global warming. For these reasons, the development and use of renewable energy sources and technologies are increasingly becoming vital for sustainable economic development of Turkey.

In this chapter, the attributes in Table 1 will be used to evaluate renewable energy alternatives. They are briefly explained in the following (Kahraman et al., 2009):

Table 1 Main and sub criteria and their utility ranges to select the best renewable energy alternative

Main Criteria	Sub-Criteria	Utility Range	Utility Characteristics
C1: Technological	C11: Feasibility	0-10	The number of times tested successfully
	C12: Risk	20-0	The number of problems for failures in a tested case
	C13: Reliability	0-10	The number of times tested successfully
	C14: The duration of preparation phase	36- 12	Month
	C15: The duration of implementation phase	12- 6	Month
	C16: Continuity and predictability of performance	0-10	The number of times tested successfully
	C17: Local technical know how	0-20	Score
C2: Environmental	C21: Pollutant emission	200-50	g/km or g/km^2
	C22: Land requirements	100-1	km^2
	C23: Need of waste disposal	0-20	Score
C3: Socio-Political	C31: Compatibility with the national energy policy	0-20	Score
	C32: Political acceptance	0-20	Score
	C33: Social acceptance	0-20	Score
	C34: Labour impact	5-250	Manpower
C4: Economic	C41: Implementation cost	5,000,000 - 750,000	$/MW (Megawatt) per site
	C42: Availability of funds	0-20	Score
	C43: Economic value (PW, IRR, B/C)	750 - 40	MWh (Megawatt Hours)

Feasibility (C11): This criterion measures the secure of the possibility for implementation of the renewable energy. The number of times tested successfully can be taken into account as a decision parameter.

Risk (C12): The risk criterion evaluates the secure of the possibility for implementation of a renewable energy by measuring the number of problems for failures in a tested case.

Reliability (C13): This criterion evaluates the technology of the renewable energy. Technology may have been only tested in laboratory or only performed in pilot plants, or it could be still improved, or it is a consolidated technology.

The duration of preparation phase (C14): The criterion measures the availability of the renewable energy alternative to decrease financial assets and reach the minimum cost. The preparation phase is judgment by taking into accounts years or months.

The duration of implementation phase (C15): The criterion measures the applicability of the renewable energy alternative to reach the minimum cost. The cost of implementation phase is judgment by taking into accounts years or months of implementation.

Continuity and predictability of performance (C16): This criterion evaluates the operation and performance of the technology for renewable energy alternative. It is important to know if the technology operates continuously and confidently.

Local technical know how (C17): This criterion includes an evaluation which is based on a qualitative comparison between the complexity of the considered technology, and the capacity of local actors to ensure an appropriate operating support for maintenance and installation of technology for renewable energy alternative.

Pollutant emission (C21): The criterion measures the equivalent emission of CO_2, air emissions which are the results of combustion process, liquid wastes which are related to secondary products by fumes treatment or with process water, and solid wastes. The evaluation of the criterion includes type and quantity of emissions, and costs associated with wastes treatments. Also the electro-magnetic interferences, bad smells, and microclimatic changes for energy investment are taken into account in the evaluation of this criterion.

Land requirements (C22): Land requirement is one of the most critical factors for the energy investment. A strong demand for land can also determine the economic losses.

Need of waste disposal (C23): The criterion evaluates the renewable energy's damage on the quality of the environment. The renewable energy alternative can be evaluated to reduce damage on the quality of life and to increase sustainability by taking into account this criterion.

Compatibility with the national energy policy objectives (C31): The criterion analyzes the integration of the national energy policy and the suggested renewable energy alternative. It measures the degree of objectives' convergence between the government policy and the suggested policy. The criterion also takes into account the government's support, the tendency of institutional actors, and the policy of public information.

Political acceptance (C32): The criterion searches whether or not a consensus among leaders' opinions for proposed renewable energy alternative exists. Also it takes into account avoiding the reactions of the politicians and to satisfying of political leaders.

Social acceptance (C33): The criterion enhances consensus among social partners. Also it takes into account avoiding the reactions from special interest social groups for renewable energy alternatives.

Labour impact (C34): Renewable energy alternatives are evaluated by taking into account labour impact which is analyzed taking care of direct and indirect employment and the possible indirect creation of new professional figures are also assessed.

Implementation cost (C41): This criterion analyzes the total cost of the energy investment in order to be fully operational.

Availability of funds (C42): This criterion evaluates the national and international sources of funds, and economic support of government.

Economic value (PW, IRR, B/C) (C43): This criterion judges the proposed renewable energy alternative as economically by using one of the engineering economics techniques which are present worth (PW), internal rate of return (IRR), benefit/cost analysis (B/C), and payback period.

The hierarchical structure for the selection of the best renewable energy alternative is shown in Figure 2. The decision makers who are composed of one professor from Industrial Engineering Department, two professors from Institute of Energy, and two top managers of energy sector in Turkey, evaluate the selection process. In the first stage, the utility curves are obtained for the sub-criteria.

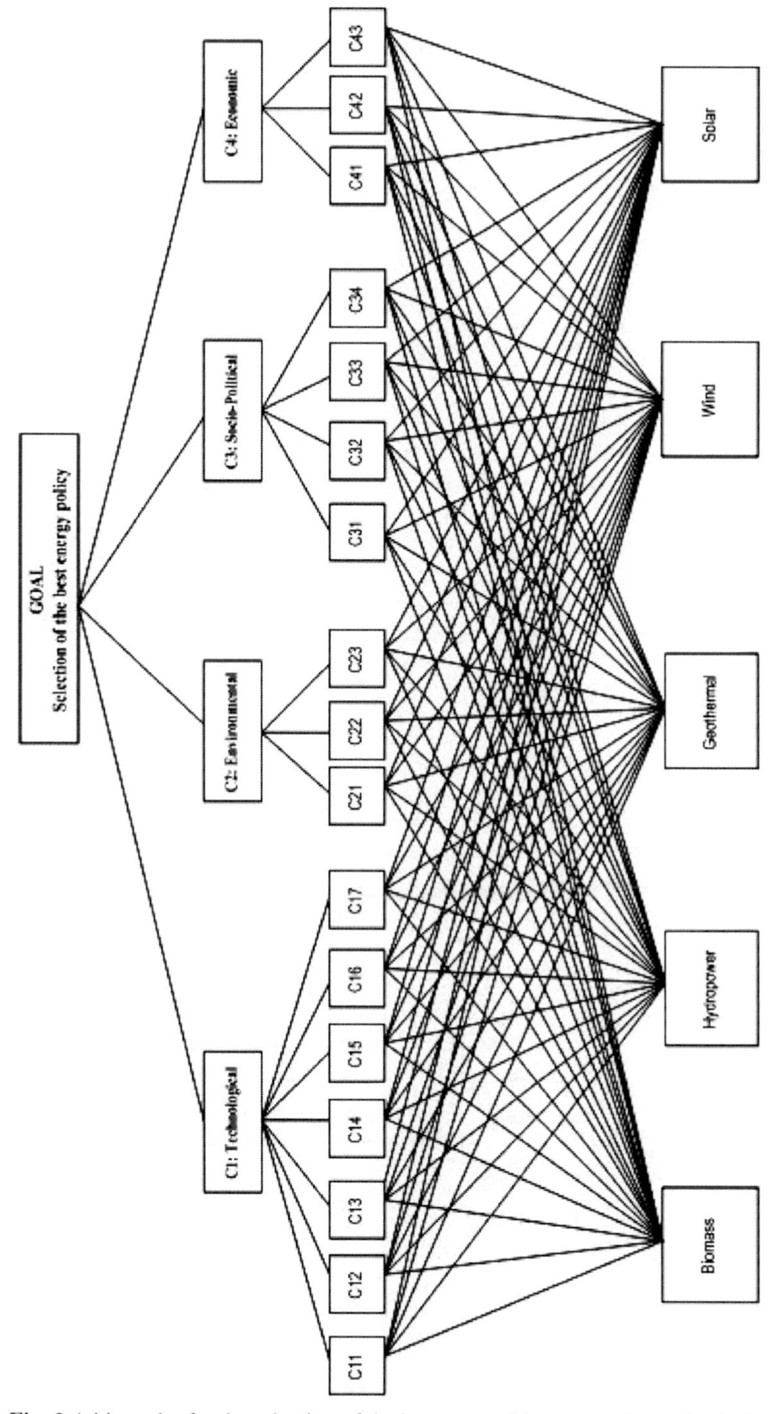

Fig. 2 A hierarchy for the selection of the best renewable energy alternative in Turkey

The utility curves for Technological, Environmental, Socio Political and Economic criteria are illustrated in Figures 3-6, respectively.

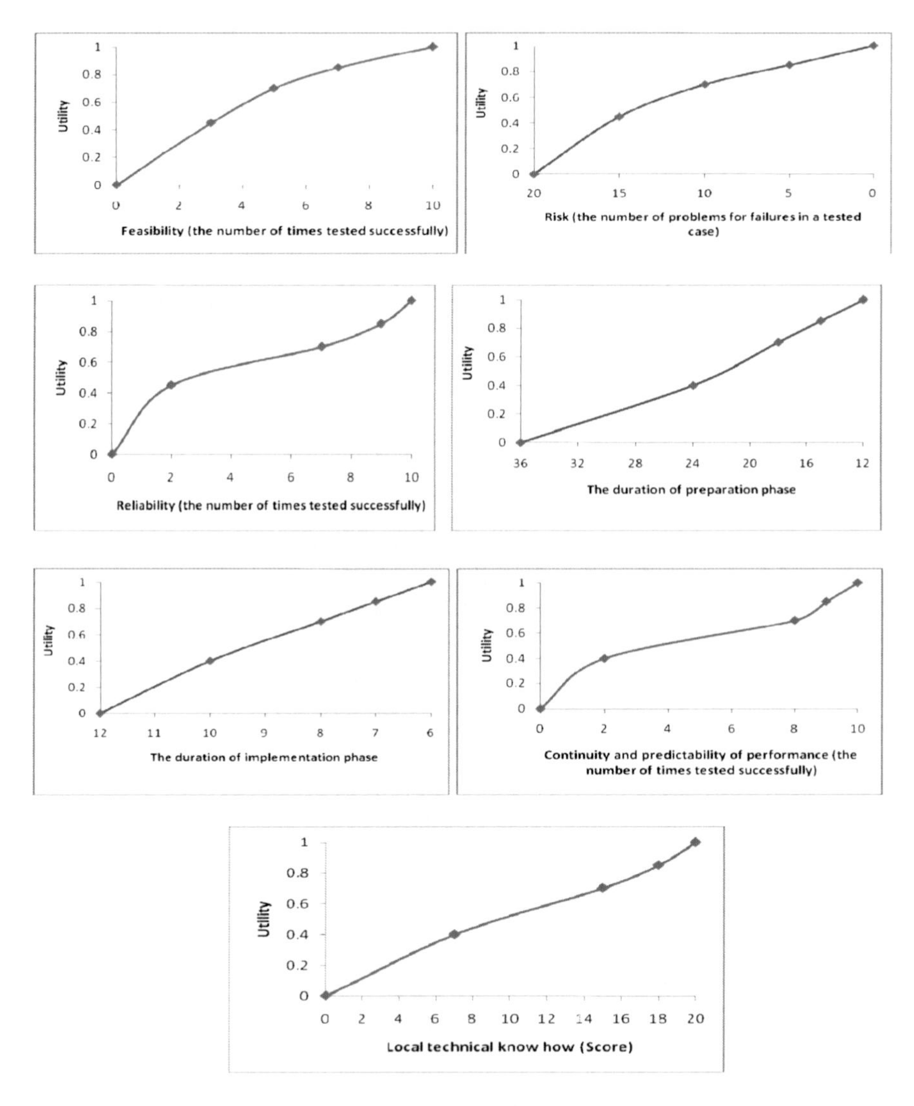

Fig. 3 The utility curves for Technological Criteria

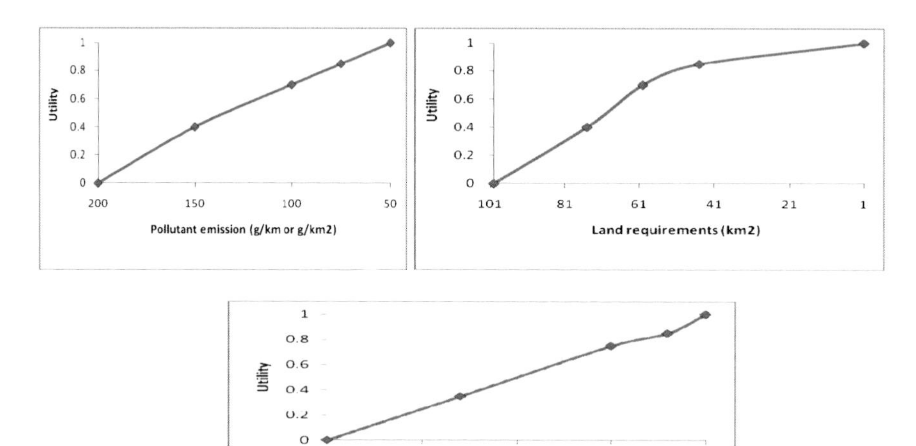

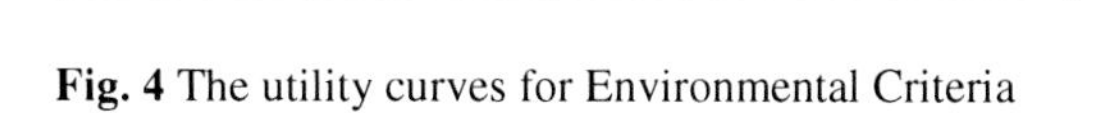

Fig. 4 The utility curves for Environmental Criteria

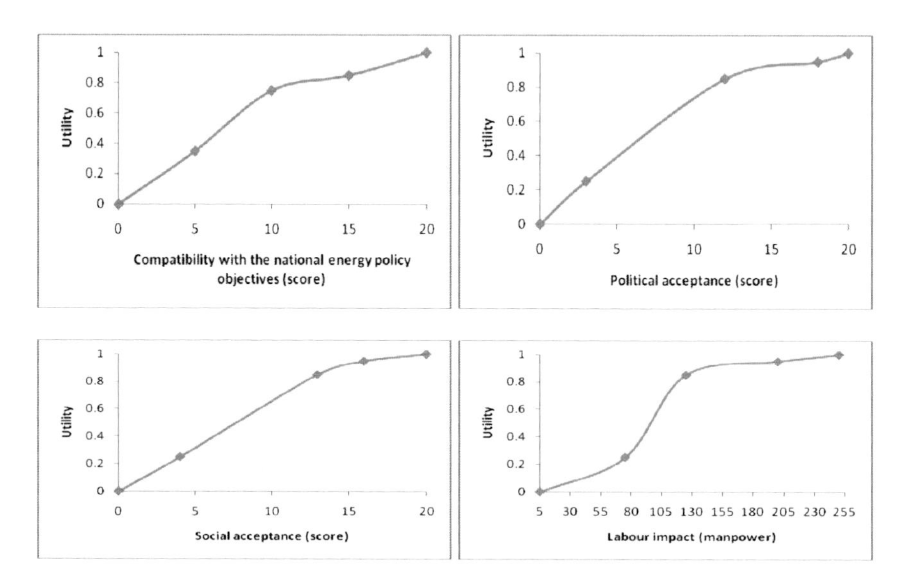

Fig. 5 The utility curves for Socio-Political Criteria

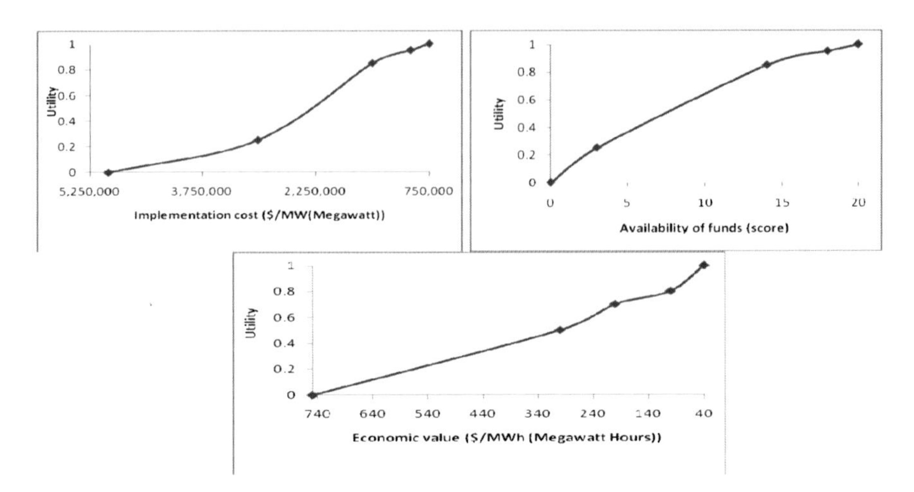

Fig. 6 The utility curves for Economic Criteria

The local and global weights and ranks of the selection attributes are obtained by using AHP. The obtained results are shown in Table 2.

Table 2 The local and global weights & ranks of attributes for renewable energy criteria

		Local		Global	
Main Attributes	Sub-Attributes	Weight	Rank	Weight	Rank
Technological (0.440)	Feasibility	0.106	5	0.046	7
	Risk	0.335	1	0.148	3
	Reliability	0.241	2	0.106	4
	The duration of preparation phase	0.043	6	0.019	12
	The duration of implementation phase	0.037	7	0.016	13
	Continuity and predictability of per-	0.108	4	0.048	6
	Local technical know how	0.130	3	0.057	5
Environmental (0.411)	Pollutant emission	0.507	1	0.208	1
	Land requirements	0.074	3	0.030	10
	Need of waste disposal	0.420	2	0.172	2
Economic (0.084)	Implementation cost	0.161	3	0.014	14
	Availability of funds	0.291	2	0.025	11
	Economic value (PW, IRR, B/C)	0.548	1	0.046	8
Socio-Political (0.064)	Compatibility with the national energy policy objectives	0.629	1	0.040	9
	Political acceptance	0.116	3	0.007	16
	Social acceptance	0.143	2	0.009	15
	Labour impact	0.111	4	0.007	17

It is now time to determine the scaling factors. For this aim, the two types of questions to obtain the scaling factors are asked: What probability P of all attribute outcomes at their best levels (C11*, C12*,...,C15*, C16*,...,C42*, C43*) versus probability $(1-P)$ of all attribute outcomes at their worst levels (C11^0, C12^0,...,C15^0,C16^0,...,C42^0, C43^0) would be as desirable as "pollutant emission" at its best level and all other attributes at their worst levels (C11^0, C12^0,...,C21*, C22^0,...,C42^0, C43^0)? The compromised answer from the decision makers was 0.05 $(k_1 = 0.05)$. The next step is to ask the second question: What "pollutant emission" level, given "need of waste disposal" at its worst, C23^0, (i.e. 5) would be as desirable as what "need of waste disposal", given the "pollutant emission" at its worst, C21^0, (i.e. 150 g/km or g/km^2)? The compromised answers from the decision makers were as follows:

$x_1 = 150$ for pollutant emission and $x_2 = 18$ for need of waste disposal. The scaling factor for need of waste disposal is calculated as follows:

$$k_1 U_1(x_1) = k_2 U_2(x_2)$$
$$0.05 \times U_1(150) = k_2 \times U_2(18)$$
$$0.05 \times 0.4 = k_2 \times 0.85$$
$$k_2 = 0.02$$

The scaling factors for the other attributes are obtained by the same way and they are shown in Table 3. Since $\sum_{i=1}^{n} k_i \neq 1.00$, the multiplicative utility model is suitable for renewable energy alternatives.Then the scaling constant K, is determined as $K = \text{-}0.000001$. The compromised outcomes from the decision makers for renewable energy alternatives and their utilities are obtained as shown in Table 3.

Table 3 The scaling factor values and utilities for renewable energy alternatives

			Biomass	Hydropower	Geothermal	Wind	Solar
Rank	Sub-Criteria	k_i	$U_i(x_i)$	$U_i(x_i)$	$U_i(x_i)$	$U_i(x_i)$	$U_i(x_i)$
1	C21	0.0500	0.75	0.75	0.7	0.98	0.98
2	C23	0.0235	0.85	0.7	0.9	0.9	0.9
3	C12	0.0097	0.9	0.85	0.8	0.95	0.95
4	C13	0.0051	0.85	0.85	0.85	0.95	0.95
5	C17	0.0026	0.85	0.85	0.95	0.85	0.3
6	C16	0.0064	0.8	0.8	0.83	0.9	0.9
7	C11	0.0053	0.8	0.8	0.8	0.8	0.45
8	C43	0.0023	0.75	0.75	0.75	0.85	0.85

Table 3 (*continued*)

9	C31	0.0015	0.8	0.7	0.8	0.95	0.95
10	C22	0.0019	0.7	0.7	0.7	0.9	0.9
11	C42	0.0024	0.75	0.9	0.74	0.75	0.75
12	C14	0.0025	0.8	0.8	0.8	0.8	0.8
13	C15	0.0031	0.83	0.8	0.83	0.65	0.5
14	C41	0.0013	0.75	0.72	0.75	0.7	0.7
15	C33	0.0003	0.85	0.8	0.7	0.9	0.9
16	C32	0.0004	0.85	0.8	0.7	0.95	0.95
17	C34	0.0001	0.73	0.7	0.7	0.7	0.7

The utility values for renewable alternatives are calculated and shown in Table 4.

Table 4 The utility values and their ranking for renewable alternatives

Alternatives	U(x)	Rank
Biomass	0.09440	3
Hydropower	0.09042	5
Geothermal	0.09243	4
Wind	0.10899	1
Solar	0.10527	2

According to Table 4, "wind energy" alternative is determined as the most appropriate alternative for renewable energy in Turkey when the alternative "hydropower energy" is clarified as the worst renewable energy alternative for Turkey. The ranking of renewable energy alternative is determined as follows: {Wind – Solar - Biomass - Geothermal - Hydropower}

5 Conclusion

Energy is considered one of the most important factors in the generation of wealth and also a key factor to show the economic development. The importance of energy in economic development has been recognized almost universally; the historical data attest to a strong relationship between the availability of energy and economic activity. It is well accepted that renewable energy alternatives have advantages over conventional energy systems in terms of environmental acceptability. Turkey is a rich country for the purposes of renewable energy and renewable energy investments have been increasing in Turkey. For this purpose, this study is based on the selection of the most appropriate renewable energy investment for Turkey.

The selection of the suitable renewable energy alternative is very important to plan future's energy consumption. Since the most forms of renewable energy alternatives are dependent on multicriteria decision making, this paper is concerned with using multiple attribute utility analysis to evaluate renewable energy resources for Turkey. Multiple attribute utility analysis is targeted in solving problems of trading off the achievement of some objectives against other objectives to obtain the maximum overall utility. Multiple attribute utility analysis is used to assess the decision-maker's preference structure and model it mathematically with a multiple attribute utility function. This multiple attribute utility function is then applied to help the decision maker reach an optimal decision.

In this chapter, MAUT is used to make a decision for selecting the best renewable energy for Turkey. The proposed MAUT methodology determines the most appropriate alternative based on utilities of criteria. The results of the proposed methodology suggest that "wind energy" as the best alternative, after considering four main criteria and 17 sub-criteria. Wind and Solar Energy alternatives are determined as the most suitable renewable alternatives, respectively for Turkey. This result confirms that wind energy causes no emissions and will be the most suitable alternative to resolve Turkey's energy problem in the future. One of the major contributions of wind energy to environmental protection is the decrease in CO2 emission. The ranking of energy alternatives is determined as {Wind – Solar - Biomass - Geothermal – Hydropower}.

In the future research, the fuzzy set theory can be used in MAUT to increase its flexibility and sensitivity . Also the using of fuzzy set theory will bring an advantage to obtain utility curves and to give flexibility for experts' evaluation.

References

Abouelnaga, A.E., Metwally, A., Nagy, M.E., Agamy, S.: Optimum selection of an energy resource using fuzzy logic. Nuclear Engineering and Design 239, 3062–3068 (2009)

Afgan, N.H., Carvalho, M.G.: Multi-criteria assessment of new and renewable energy power plants. Energy 27, 739–755 (2002)

Afgan, N.H., Carvalho, M.G.: Sustainability assessment of a hybrid energy system. Energy Policy 36, 2903–2910 (2008)

Afgan, N.H., Pilavachi, P.A., Carvalho, M.G.: Multi-criteria evaluation of natural gas resources. Energy Policy 35, 704–713 (2007)

Aydin, N.Y., Kentel, E., Duzgun, S.: GIS-based environmental assessment of wind energy systems for spatial planning: A case study from Western Turkey. Renewable and Sustainable Energy Reviews 14(1), 364–373 (2010)

Beccali, M., Cellura, M., Mistretta, M.: Decision-making in energy planning: Application of the ELECTRE method at regional level for the diffusion of renewable energy technology. Renewable Energy 28, 2063–2087 (2003)

Bell, D.V., Keeney, R.L., Raiffa, H.: Conflicting objectives in decisions. Wiley-Interscience, International Institute for Applied Systems Analysis. John Wiley, New York (1978)

Borges, A.R., Antunes, C.H.: A fuzzy multiple objective decision support model for energy-economy planning. European Journal of Operational Research 145, 304–316 (2003)

Burton, J., Hubacek, K.: Is small beautiful? A multicriteria assessment of small-scale energy technology applications in local governments. Energy Policy 35, 6402–6412 (2007)

Butler, J., Morrice, D.J., Mullarkey, P.W.: A multiple attribute utility theory approach to ranking and selection. Management Science 47(6), 800–816 (2001)

Çam, E.: Application of fuzzy logic for load frequency control of hydroelectrical power plants. Energy Conversion and Management 48(4), 1281–1288 (2007)

Canada, J.R., Sullivian, W.G.: Economic and multiattribute evaluation of advanced manufacturing systems. Prentice Hall, New Jersey (1989)

Cavallaro, F., Ciraolo, L.: A multicriteria approach to evaluate wind energy plants on an Italian island. Energy Policy 33, 235–244 (2005)

Cirtita, H., Ilieş, L.: Does structure influence performance in downstream supply chain? Review of International Comparative Management 10(1), 106–112 (2009)

De Melo Brito, A.J., De Almeida Filho, A.T., De Almeida, A.T.: Multi-criteria decision model for selecting repair contracts by applying utility theory and variable interdependent parameters. Journal of Management Mathematics 21, 349–361 (2010)

Demirbaş, A.: Importance of biomass energy sources for Turkey. Energy Policy 36, 834–842 (2008)

Erdogdu, E.: Natural gas demand in Turkey. Applied Energy 87(1), 211–219 (2010)

Goumas, M., Lygerou, V.: An extension of the PROMETHEE method for decision making in fuzzy environment: Ranking of alternative energy exploitation projects. European Journal of Operational Research 123, 606–613 (2000)

Haralambopoulos, D.A., Polatidis, H.: Renewable energy projects: structuring a multicriteria group decision-making framework. Renewable Energy 28, 961–973 (2003)

Hepbaslı, A., Ozalp, N.: Development of energy efficiency and management implementation in the Turkish industrial sector. Energy Conversion and Management 44, 231–249 (2003)

Hendrickson, C., Horvath, A., Joshi, S., Lave, L.: Economic input-output models for environmental life-cycle assessment. Policy Analysis 32(7), 184–191 (1998)

IEA (International Energy Agency) (2007) 2004 Energy Balances for Turkey, `http://www.iea.org/stats`

Jebaraj, S., Iniyan, S.: An optimal energy allocation model using fuzzy linear programming for energy planning in India for the year 2020. International Journal of Energy Technology and Policy 5(4), 509–531 (2007)

Jimenez, A., Mateos, A., Rios-Insua, S.: Missing consequences in multiattribute utility theory. Omega 37, 395–410 (2009)

Jimenez, A., Rios-Insua, S., Mateos, A.: A decision support system for multiattribute utility evaluation based on imprecise assignments. Decision Support Systems 36, 65–79 (2003)

Kahraman, C., Çebi, S., Kaya, İ.: Selection among renewable energy alternatives using fuzzy axiomatic design: The case of Turkey. Journal of Universal Computer Science 16(1), 82–102 (2010)

Kahraman, C., Kaya, İ.: A fuzzy multicriteria methodology for selection among energy alternatives. Expert Systems with Applications 37(9), 6270–6281 (2010)

Kahraman, C., Kaya, İ., Çebi, S.: A comparative analysis for multiattribute selection among renewable energy alternatives using fuzzy axiomatic design and fuzzy analytic hierarchy process. Energy The International Journal 34(10), 1603–1616 (2009)

Kainuma, Y., Tawara, N.: A multiple attribute utility theory approach to lean and green supply chain management. International Journal of Production Economics 101, 99–108 (2006)

Kaya, T., Kahraman, C.: Multicriteria renewable energy planning using an integrated fuzzy VIKOR & AHP methodology: The case of Istanbul. Energy 35(6), 2517–2527 (2010)

Kaygusuz, K.: Environmental impacts of energy utilisation and renewable energy policies in Turkey. Energy Policy 30, 689–698 (2002)

Kaygusuz, K.: Energy policy and climate change in Turkey. Energy Conversion and Management 44, 1671–1688 (2003)

Keeney, R.L., Raiffa, H.: Decisions with multiple objectives: preferences and value tradeoffs. Cambridge University Press, Cambridge (1993)

Kucukali, S., Baris, K.: Turkey's short-term gross annual electricity demand forecast by fuzzy logic approach. Energy Policy 38(5), 2438–2445 (2010)

Malakooti, B.: Ranking multiple criteria alternatives with half-space, convex, and nonconvex dominating cones: quasi-concave and quasi-convex multiple attribute utility functions. Computers & Operations Research 16(2), 117–127 (1989)

Malakooti, B., Subramanian, S.: Generalized polynomial decomposable multiple attribute utility functions for ranking and rating multiple criteria discrete alternatives. Applied Mathematics and Computation 106, 69–102 (1999)

Nishizaki, I., Katagiri, H., Hayashida, T.: Sensitivity analysis incorporating fuzzy evaluation for scaling constants of multiattribute utility functions. Central European Journal of Operations Research 18, 383–396 (2010)

Patlitzianas, K.D., Ntotas, K., Doukas, H., Psarras, J.: Assessing the renewable energy producers' environment in EU accession member states. Energy Conversion and Management 48, 890–897 (2007)

Patlitzianas, K.D., Pappa, A., Psarras, J.: An information decision support system towards the formulation of a modern energy companies' environment. Renewable and Sustainable Energy Reviews 12, 790–806 (2008)

Polatidis, H., Haralambopoulos, D.A., Munda, G., Vreeker, R.: Selecting an appropriate multi-criteria decision analysis technique for renewable energy planning. Energy Sources Part B 1, 181–193 (2006)

Republic of Turkey Ministry Of Energy and Natural Resources (2008) Statistics, `http://www.enerji.gov.tr/istatistik.asp`

Sohn, K.Y., Yang, J.W., Kang, C.S.: Assimilation of public opinions in nuclear decision-making using risk perception. Annals of Nuclear Energy 28, 553–563 (2001)

Streicher-Porte, M., Marthaler, C., Boni, H., Schluep, M., Camacho, A., Hilty, L.M.: One laptop per child, local refurbishment or overseas donations? Sustainability assessment of computer supply scenarios for schools in Colombia. Journal of Environmental Management 90, 3498–3511 (2009)

Ulutaş, B.H.: Determination of the appropriate energy policy for Turkey. Energy 30, 1146–1161 (2005)

Wang, M.L., Kuo, T.C., Liu, J.W.: Identifying target green 3C customers in Taiwan using multiattribute utility theory. Expert Systems with Applications 36, 12562–12569 (2009)

Xu, N., Huang, S.H.: Multiple attributes utility analysis in setup plan evaluation. Journal of Manufacturing Science and Engineering 128, 220–227 (2006)

Yang, Z.L., Bonsall, S., Wang, J.: Use of hybrid multiple uncertain attribute decision making techniques in safety management. Expert Systems with Applications 36, 1569–1586 (2009)

Zhang, H.: Multi-objective simulation-optimization for earthmoving operations. Automation in Construction 18, 79–86 (2008)

Zhang, H., Xing, F.: Fuzzy-multi-objective particle swarm optimization for time–cost–quality tradeoff in construction. Automation in Construction (2010), doi:10.1016/j.autcon.2010.07.014

A Novel Fuzzy-Based Methodology for Biogas Fuelled Hybrid Energy Systems Decision Making

Alexandre Barin[1], Luciane N. Canha[1], Karine M. Magnago[1], Manuel A. Matos[2], and Breno Wottrich[1]

[1] Federal University of Santa Maria / CEEMA- Brazil
alexandrebarin@hotmail.com, lncanha@ct.ufsm.br, kamagnago@gmail.com, b.wottrich@gmail.com

[2] Institute for Systems and Computer Engineering of Porto- Portugal
mmatos@inescporto.pt

Abstract. In response to the soaring energy crisis and the related pollution problems worldwide, it is essential to apply new technologies that use renewable energy sources in both an efficient and environmentally friendly manner. In this way, biomass offers one of the largest potential among renewable energy sources. The aim of this work is to demonstrate a novel fuzzy-based methodology for selecting hybrid energy systems fuelled by biogas. Fuzzy multi-rules and fuzzy multi-sets are used to evaluate the main operational characteristics of five types of renewable sources fuelled by biogas. The possibility of using the methodology for energy storage system evaluation is also assessed. The construction of the fuzzy multi-rules and fuzzy multi-sets is based on the following methods: *Mamdani* (fuzzification process), *Max-Min* (inference process), and *Center of Gravity* (defuzzification process). Several criteria are used: costs, efficiency, cogeneration, life-cycle, technical maturity, power application range, and environmental impacts. The methodology considers three different settings with two different constraints: costs and environment. One of the most relevant aspects presented by this work is about the previous classification of the criteria. It was created according to the different relevance observed among the attributes. The purpose of the proposed arrangement is to facilitate the understanding of the methodology and to increase the possibility of incorporating the decision makers' preferences on the decision-aid process. These aspects are essential to strengthen the final decision.

1 Introduction

Due to the current and predictable energy deficit and related environmental problems, the use of renewable energy sources has been attracting much attention

K. Gopalakrishnan et al. (Eds.): Soft Comput. in Green & Renew. Ener. Sys., STUDFUZZ 269, pp. 183–198.
springerlink.com

[Farret and Simões 2006]. To solve these constraints, several environmental-friendly fuels have been proposed as substitutes to conventional fossil fuels. In particular, the use of biogas is attractive, due to its high electrical efficiency and low environmental impact [Xuan et al 2008]. The production of biogas by anaerobic digesters has been shown to be an effective method for waste treatment. When exploiting the animal waste, like swine manure for example, highly concentrated hog farming operations often generate manure quantities too large to be applied to the surrounding land at agricultural rates [Mueller 2007]. In this way, one environmental and economically attractive solution is to integrate anaerobic digesters with hybrid technologies that may combine electrical power supply, energy storage and also cogeneration.

In order to increase the application of renewable technologies, there are governmental incentives in several countries. One remarkable example of this trend occurs in Brazil. The Federal Government approved the Decree 10438/02, which creates the Program for Alternative Sources of Energy - PROINFA, introducing incentives for hydro power, wind generation and biomass sources.

However, an effective methodology for the energy management is essential to guarantee the expansion of hybrid energy sources among users [Zopounidis and Doumpos 2002]. This methodology must be able to deal with economic, operational and environmental constrains. Accordingly, it is important to select a multicriteria method that best satisfy the management needs [Cormio et al 2003]. Several authors presented excellent results by using muticriteria decision aiding methodologies for energy management problems. *ELECTRE* [Beccali et al 2003], *PROMETHEE* [Belton and Stewart 2002], *MACBETH* [Bana e Costa and Vansnick 1999], *AHP* [Wedley et al 2001] and also *Fuzzy sets* [Ramírez-Rosado and Dominguez-Navarro 2004] are some examples.

The problem addressed in this work is the ordering problem. A set of alternatives – renewable energy technologies (RET) and energy storage systems (ESS) – must be ordered taking to account their attributes values and the preferences of the decision maker represented by the classification of the criteria. This paper therefore evaluates the characteristics of RET fuelled by biogas – three types of fuel cells (FC), one microturbine (MT) and one Otto Engine (OTTO) – and ESS – flywheel and conventional and flow batteries. To address this problem, a new type of fuzzy rule-based construction is proposed and it is presented the application of the fuzzy-based methodology for biogas fuelled hybrid energy systems decision making.

The paper is organized as follows: Section 2 includes the main characteristics about RET and the ESS in analysis. Section 3 presents the selected criteria, the decision makers' analysis and the classification of criteria under different perspectives. Section 4 introduces the fuzzy concepts and further methodological aspects. Subsequently, Section 5 outlines the application of Inference Systems concerning RET and ESS. Concluding remarks are discussed in Section 6.

2 Renewable Energy Technologies and Storage Energy Systems

Before presenting the main characteristics of the renewable energy technologies and energy storage system, it is essential to evaluate the availability of swine manure for electricity generation. In this paper, RET and ESS selection are achieved for the specific region under analysis, where it is possible to generate about 1 MW of power using biodigesters. With those digesters, there are stabilization ponds (anaerobic, facultative and maturation) connected in sequence and fed with sludge from a swine manure treatment unit. It is necessary to observe that the region under analysis is located in a tropical country, where temperatures typically range between 20°C and 35°C year-round - ideal for biogas.

A brief description of the energy technologies evaluated in this paper is presented below.

RET – Fuel cells and microturbines (MT) fuelled by biogas have been increasingly used in a wide range of applications. These technologies have attracted interest mainly due to their environmental advantages, with the possibility of combining high efficiencies with low greenhouse gas (GHG) emissions. In particular, three high-temperature fuel cells described in the literature were here evaluated. The main reason for this choice is the poor compatibility of low-temperature fuel cells with biogas (due to the poisoning possibility in the catalyst of the cell).

Otto engines are internal combustion engines supported by the Otto cycle. Otto engines fuelled by biogas are mature and low-cost technologies, but with high GHG emissions. The RET selected in this paper can also support the cogeneration to provide heat for a variety of applications.

ESS – Flywheels can accumulate and store mechanical energy in kinetic form. The stored energy depends on the inertia and speed of the rotating mass (rotor). The flywheel is a rotor placed in a vacuum enclosure to eliminate friction-loss from the air and mounted on bearings for a stable operation. A flywheel offers high density energy and high efficiency.

Batteries are the most common devices used to store electrical energy. Traditionally, they have been used for small scale applications. However, due to the liberalization of electricity markets, battery manufacturers have been used for large scale energy storage applications. Flow Batteries, also known as Regenerative Fuel Cells or Redox Flow Systems, are a new class of batteries that have been achieving substantial progress – technically and commercially. Flow Batteries present some features that make them especially attractive for utility-scale applications. The operational principle differs from classical batteries, since the latter store energy both in the electrolyte and the electrodes, while flow batteries store and release energy using a reversible reaction between two electrolyte solutions, separated by an ion-permeable membrane. Both electrolytes are stored separately in bulk storage tanks, whoze size defines the energy capacity of the storage system. The power rating is determined by the cell stack. Therefore, the power and energy rating are decoupled, which provides to the system designer an extra degree of freedom when structuring the system.

3 Selected Criteria and Proposed Methodology

A description of the criteria evaluated by the proposed methodology is described below. It is important to observe that the definition and evaluation of the selected criteria must take into account an actual database and the management needs for each specific case. After analyzing these aspects, it is possible to develop the methodology for RET and ESS selection. The criteria evaluated in this work are classified as qualitative and quantitative.

The qualitative criteria are expressed through scores stipulated by the decision maker (DM) – a group of researchers from The Federal University of Santa Maria – in the intervals from 0 to 1.0, with 1.0 being the highest score. These scores are defined according to the analysis of the actual database, taking into account social, political and economic aspects related to the particular region under analysis, e.g. RET and ESS installation in a specific region of Brazil. In addition, the experience of the selected decision makers is another key aspect in determining the scores.

The qualitative criteria considered in this work are:

- technical maturity (TM);
- environmental impacts (IMP) concerning: end-of-life disposal of ESS and GHG emissions from RET.

The quantitative criteria are expressed through rated data. The quantitative criteria evaluated in this study are:

- efficiency (EF) in %;
- efficiency of cogeneration (CO) in %;
- costs in US$/ kW;
- life-cycle (LC) in years;

The perspectives simulated in this study are evaluated by the prior classification of the criteria created through DM preferences. This classification was developed according to the different relevance observed among the attributes. The purpose of the proposed arrangement is to facilitate the understanding of the methodology and to improve the DM interaction over the decision making process.

The classification defined for RET analysis according to each proposed perspective is:

- Environmental Perspective: 1st environmental impacts, 2nd efficiency, 3rd life cycle, 4th cogeneration, 5th costs.
- Costs Perspective: 1st costs, 2nd cogeneration, 3rd efficiency, 4th life cycle, 5th environmental impacts.
- Environmental-costs Perspective: 1st environmental impacts, 2nd costs, 3rd efficiency, 4th life cycle, 5th cogeneration.

The classification defined for ESS analysis according to each proposed perspective is:

- Environmental Perspective: 1st environmental impacts, 2nd efficiency, 3rd life cycle, 4th technical maturity, 5th costs.

- Costs Perspective: 1st costs, 2nd technical maturity, 3rd efficiency, 4th life cycle, 5th environmental impacts.
- Environmental-costs Perspective: 1st environmental impacts, 2nd costs, 3rd efficiency, 4th life cycle, 5th technical maturity.

By using the prior classification of the criteria is possible to produce fuzzy models with both a small number of interpretable rules and high precision. The proposed fuzzy-based methodology allows the decision maker to interact with the methodology, modifying the criteria rank in the classification of criteria and making changes in the set of rules.

4 Fuzzy-Based Methodology

4.1 Initial Concepts

Fuzzy Logic was proposed by Zadeh. Fuzzy Logic is considered one of the most powerful methods encompassing many fields of application [Siler and Buckley 2005]. Indeed, fuzzy rule-based expert systems can improve the interpretability of results and increase the interaction of DM on the decision making process.

For the inference fuzzy process there are two well-established classes of fuzzy controllers: *Mamdani* and *Takagi-Sugeno*. The most fundamental difference between *Mamdani* and *Sugeno* is the way the crisp output is generated from the fuzzy inputs [Hamam e Georganas 2008]. While *Mamdani* uses the technique of defuzzification of a fuzzy output, *Sugeno* uses weighted average to compute the crisp output. Therefore in *Sugeno* the defuzzification process is bypassed. The difference between the controllers is illustrated in Figure 1. The expressive power and interpretability of the *Mamdani* output are lost when using *Sugeno*, since the consequents of the rules are not fuzzy [Yusoff et al 2007].

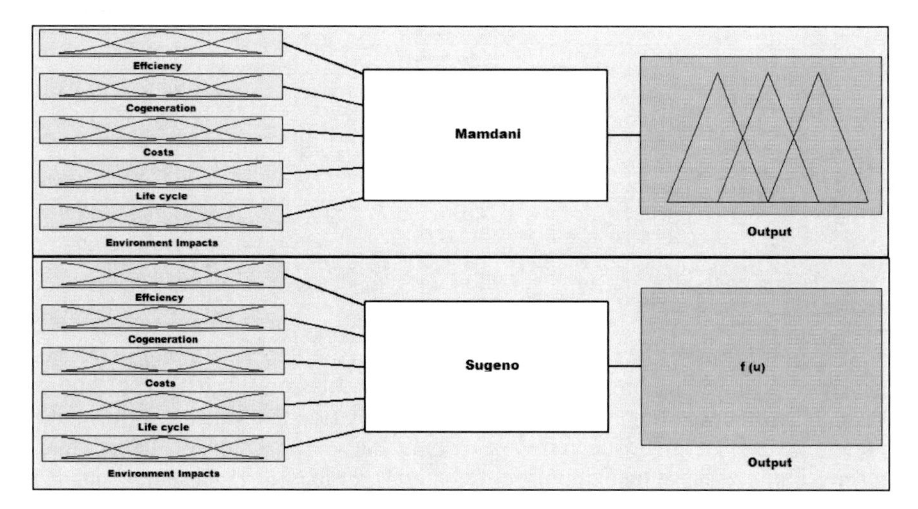

Fig. 1 Mamdani controller and Takagi-Sugeno controller (MATLAB® Software)

In those controllers, the inference processes are based on a set of rules like:

- IF A is High and B is Medium THEN Z is Medium (Mamdani FIS).
- IF A is High and B is Medium THEN $Z = \alpha + A\beta + B\gamma$ (Sugeno FIS).

In a control system, input variables A and B could be the state variables, while output variable Z would be the control variable (α, β, γ are constants) [Matos 2002]. In this paper, the input variables will be the criteria difference represented by membership functions and the output variable Z will be the value difference.

The most common fuzzy system is the *Mamdani* system, which is used in this paper. The choice of *Mamdani* controller it is related to the following aspects [Li-Xin Wang 1993]:

- it is suitable for engineering systems because its inputs and outputs are real-valued variables;
- it provides a natural framework for incorporating fuzzy rules from human experts;
- there is much freedom for the choices of fuzzifier, fuzzy inference engine, and defuzzifier;
- it provides an effective framework to integrate numerical and linguistic information.

Mamdani controller performs three major steps: fuzzification of the input variables; inference (rule evaluation and implication plus aggregation); and defuzzification – as illustrated in Figure 2.

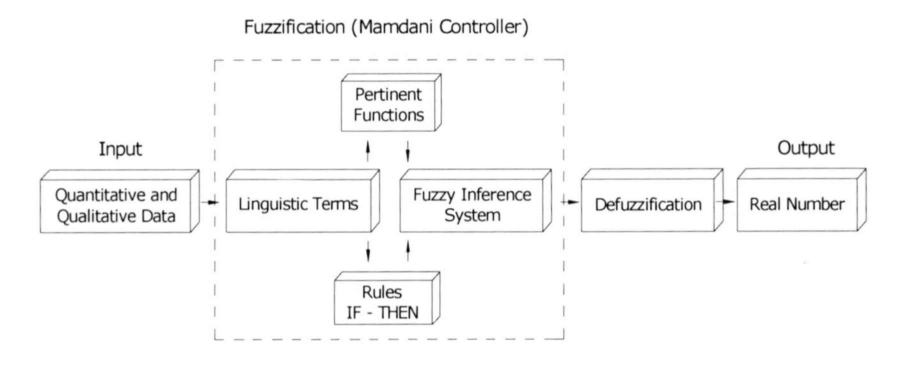

Fig. 2 Basic steps on the Mamdani controller
** some authors connect the aggregation step together with the defuzzification.*

A basic *Mamdani* fuzzy system accepts numbers as input, and then translates the input numbers into linguistic terms, such as low, medium, high (fuzzification). Rules map the input linguistic terms, which are represented in membership functions, into similar linguistic terms describing the output linguistic terms (inference). Finally, the output linguistic terms are translated into an output number (defuzzification).

4.2 Fuzzification

4.2.1 Linguistic Terms and Membership Functions

The membership functions are represented in fuzzy sets with a certain shape. It is popular to use trapezoidal or triangular fuzzy sets due to their computational efficiency [Zimmermann 2001]. In this paper, inputs and the output are arranged in five linguistic terms - very low (VL), low (L), medium (M), high (H) and very high (VH) - represented by five membership functions applied in each fuzzy variable, as shown in Figure 3. The number of membership functions used in the fuzzy set is determined to maintain a good accuracy for the analysis concerning the three different perspectives. The same fuzzy set is applied for all selected criteria, which enables assessing any perspective that could be suggested by the decision makers.

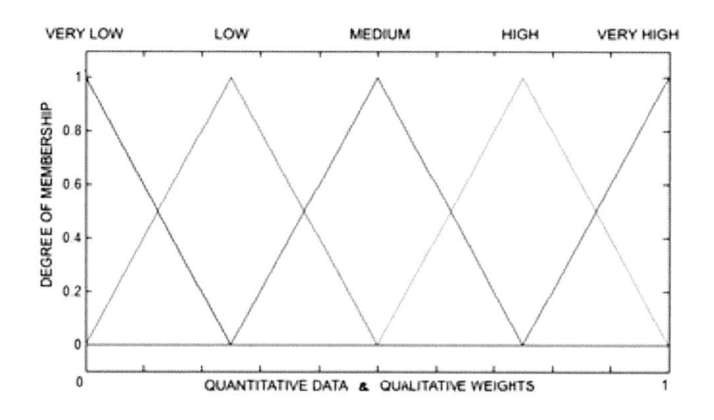

Fig. 3 Fuzzy varaiable defined for each criterion (MATLAB® Software)

4.2.2 Multi-rules-Based Construction

The multi-rules-based used in this work consists of a collection of if-then propositions. The number of fuzzy rules grows exponentially with the number of input criteria and with the number of membership functions used to represent each criterion. The improvement of rules has attracted much attention for a long time in the fuzzy community. In the literature, different aspects and various techniques have been studied, such as *hierarchical fuzzy* and *genetic algorithms* [Alcalá 2007], *artificial neural networks* [Mantas et al 2006], among others. These techniques try to reduce the number of rules, while keeping a good accuracy.

In fact, certain states can be neglected in most applications either because they are impossible or because a control action would not be helpful. It is therefore sufficient to write rules that cover only parts of the state space. In the presented paper, this aspect is considered for the construction of the classification of criteria.

The development of these rules for the proposed methodology is established according to three important definitions, which follow the classification of the criteria. These definitions are presented below:

- Definition of the Logical Operations – only the Logical Operation "and" is used. The application of only one logical operation is optional; it is possible to use the "or" logical operation instead or together with "and" in the same rule. In this case, the logical operation "and" was applied according to DM preferences.
- Basis Definition following a standard rule, for example:

  IF A = L and B = L and C = L and D = L THEN OUTPUT = L.
  IF A = M and B = M and C = M and D = M THEN OUTPUT = M.
  IF A = H and B = H and C = H and D = H THEN OUTPUT = H.

- Definition of Construction level increasing the value of the pertinent function after three equal outputs (S), e.g.

  IF A= VL and B= VL and C = VL and D = VL, THEN OUTPUT = VL.
  IF A= VL and B= VL and C = VL and D = L, THEN OUTPUT = VL.
  IF A= VL and B= VL and C = VL and D = M, THEN OUTPUT = VL (last VL).
  IF A= VL and B= VL and C = VL and D = H, THEN OUTPUT = L (first L).

- Definition of Start/End point establishing the initial point (start) of the Construction level. This definition is applied to attribute some relevance to the most important criterion (criterion A in this case), following the classification of the criteria, e. g.

  IF A = VL, THEN OUTPUT = VL. Thus, the start point is:
  IF A = L and B= VL and C = VL and D = VL, THEN OUTPUT = VL.
  IF A = VH, THEN OUTPUT = VH. Thus, the end point is:
  IF A = H and B= VH and C = VH and D = VH, THEN OUTPUT = VH.

It is necessary to observe that application of this definition is optional. In this case, it was applied according to DM preferences.

- Definition of the criterion for increasing value. This step is complementing the Definition of Construction level. For this definition, the criterion E (less important in the classification of the criteria) is used just to increase the value of the Output from Medium to High in the membership function, for instance:

  1.IF A= L and B= L and C = VH and D = VL, THEN OUTPUT = L.
  2.IF A= L and B= L and C = VH and D = L, THEN OUTPUT = M.
  3.IF A= L and B= L and C = VH and D = M, THEN OUTPUT = M.
  4.IF A= L and B= L and C = VH and D = H, THEN OUTPUT = M.
  IF A= L and B= L and C = VH and D = VH, THEN OUTPUT = M^{*H},
  and for M^{*H} :

5.IF A= L and B= L and C = VH and D = VH and E = VL, THEN OUTPUT = M;(never decreasing the value of Output).
6.IF A= L and B= L and C = VH and D = VH and E = L, OUTPUT = M;
7.IF A= L and B= L and C = VH and D = VH and E = M, OUTPUT = M;
8.IF A= L and B= L and C = VH and D = VH and E = H, OUTPUT = M;
9. IF A= L and B= L and C = VH and D = VH and E = VH, OUTPUT = H (increasing the value of Output).

Or, for example:

1.IF A= L and B= M and C = VH and D = VL, THEN OUTPUT = M.
2.IF A= L and B= M and C = VH and D = L, THEN OUTPUT = M.
3.IF A= L and B= M and C = VH and D = M, THEN OUTPUT = M.
IF A= L and B= M and C = VH and D = H, THEN OUTPUT = M^{*H},
and for M^{*H} :
4.IF A= L and B= M and C = VH and D = H and E = VL, OUTPUT = M;
5.IF A= L and B= M and C = VH and D = H and E = L, OUTPUT = M;
6.IF A= L and B= M and C = VH and D = H and E = M, OUTPUT = M;
7..IF A= L and B= M and C = VH and D = H and E = H, OUTPUT = M;
8.IF A= L and B= M and C = VH and D = H and E = VH, OUTPUT = H (increasing the value of Output).

IF A= L and B= M and C = VH and D = VH, THEN OUTPUT = $M^{\#H}$,
and for $M^{\#H}$:

9.IF A= L and B= M and C = VH and D = VH and E = VL, OUTPUT = M;
10.IF A= L and B= M and C = VH and D = VH and E = L, OUTPUT = M;
11.IF A= L and B= M and C = VH and D = VH and E = M, OUTPUT = M;
12.IF A= L and B= M and C = VH and D = VH and E = H, OUTPUT = H (increasing the value of Output);
13.IF A= L and B= M and C = VH and D = VH and E = VH, OUTPUT = H (increasing the value of Output).

Table 1 presents the complete set of rules developed by using the classification of criteria and the proposed definitions. To understand the table construction, it is necessary to analyze each one of the outputs separately. For example, considering the first cell in bold, inside the first line of the Table 1:

IF A=L and B=L and C= VL and D=VL THEN ***OUTPUT*** = **VL;** or considering A=M in this same cell - just skipping the first column for criterion B:
IF A=M and B=VL and C= VL and D=VL THEN ***OUTPUT*** = **VL.**
Now observing the second cell in bold located in the fourth line:
IF A=H and B=VH and C= VL and D=H THEN OUTPUT = **M** ($M*^{H}$);
But considering, in this case, the definition of the criterion for increasing value, the output is given by:

IF A=H and B=VH and C= VL and D=H and E=VH THEN OUTPUT = **H.**

Table 1 Complete set of rules follwing the linguistic terms - very low (VL), low (L), medium (M), high (H) and very high (VH).

		$A = L$					$A = M$	$A = H$
		$B=VL$	$B=L$	$B=M$	$B=H$	$B=VH$	$B=VH$	$B=VH$
$C=VL$	$D=VL$	VL	**VL**	VL	L	L	L	M
	$D=L$	VL	VL	L	L	L	M	M
	$D=M$	VL	L	L	L	M	M	M
	$D=H$	L	L	L	M	M	M	**M*^{H}**
	$D=VH$	L	L	M	M	M	M*^{H}	M$^{\#H}$
$C=L$	$D=VL$	VL	VL	L	L	L	M	M
	$D=L$	VL	L (BASIS)	L	L	M	M	M
	$D=M$	L	L	L	M	M	M	M*^{H}
	$D=H$	L	L	M	M	M	M*^{H}	M$^{\#H}$
	$D=VH$	L	M	M	M	M*^{H}	M$^{\#H}$	H
$C=M$	$D=VL$	VL	L	L	L	M	M	M
	$D=L$	L	L	L	M	M	M	M*^{H}
	$D=M$	L	L	M	M (BASIS)	M	M*^{H}	M$^{\#H}$
	$D=H$	L	M	M	M	M*^{H}	M$^{\#H}$	H
	$D=VH$	M	M	M	M*^{H}	M$^{\#H}$	H	H
$C=H$	$D=VL$	L	L	L	M	M	M	M*^{H}
	$D=L$	L	L	M	M	M	M*^{H}	M$^{\#H}$
	$D=M$	L	M	M	M	M*^{H}	M$^{\#H}$	H
	$D=H$	M	M	M	M*^{H}	M$^{\#H}$	H (BASIS)	H
	$D=VH$	M	M	M*^{H}	M$^{\#H}$	H	H	H
$C=VH$	$D=VL$	L	L	M	M	M	M*^{H}	M$^{\#H}$
	$D=L$	L	M	M	M	M*^{H}	M$^{\#H}$	H
	$D=M$	M	M	M	M*^{H}	M$^{\#H}$	H	H
	$D=H$	M	M	M*^{H}	M$^{\#H}$	H	H	H
	$D=VH$	M	M*^{H}	M$^{\#H}$	H	H	H	H *^{VH}

The cells defined as "BASIS" inside Table 1 represent the output of the Basis Definition. As stated before, the complete set of rules is very flexible, allowing the DM to increase or decrease the importance of each criterion. The total number of rules created in this study case was 625. Ming-Ling Lee et al [2003] established that a single-output fuzzy logic system with n input criteria and m membership functions defined for each input variable is composed by $x = m^n$fuzzy rules. This study evaluates five criteria n, each one represented by one variable with five membership functions m. According to [Ming-Ling Lee et al 2003] the total number of rules would follow equation 1.

$$x = m^n = 5^5 = 3125 \quad (1)$$

In this case, the total number of rules is greatly reduced under the application of the classification of criteria.

4.3 Inference (Implication and Aggregation)

The *Maximum of minimum* method – maximum aggregation of the minimum implication - is here used as the inference process. It is the most commonly used inference process found in the literature [Gegov e Gobalakrishnan 2007], [Yusoff et al 2007], [Hamam e Georganas 2008], Moreover, Kiszka et al [1985] calculated the Medium Square Error for three different inference methods, in which the Max-Min method achieved the highest performance. The *Max-Min* method is represented in Figure 4.

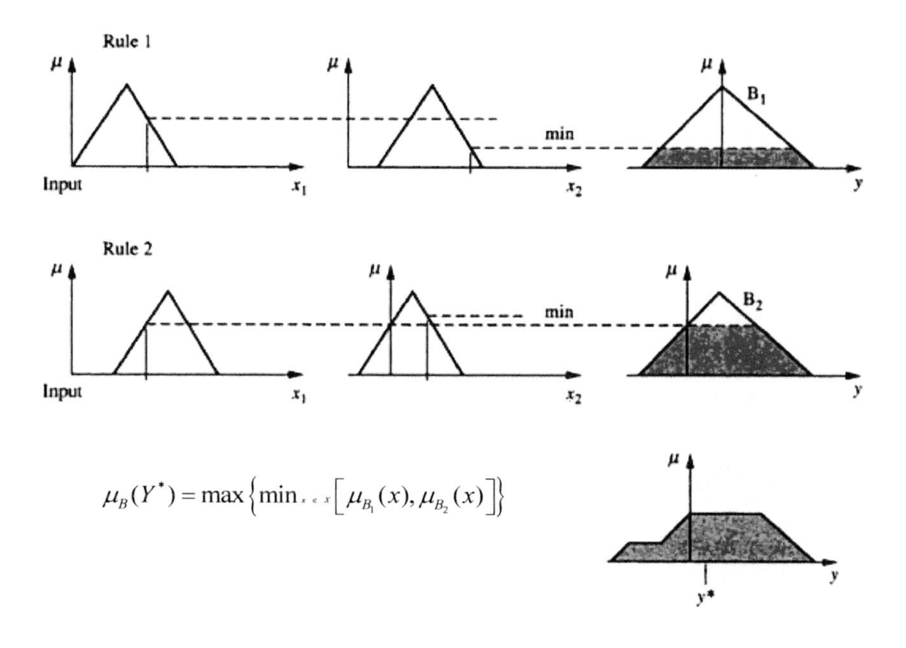

Fig. 4 Maximum of minimum method used as inference process

4.4 Defuzzification Process

Regarding the defuzzification process, there are several possible choices to be made and many different methods have been proposed [Barros and Bassanezi 2006]. This study applied the so-called *Center of Area* (COA) or *Center of Gravity* (COG) method, as illustrated in Figure 5. This method chooses the control action that corresponds to the center of the area with membership greater than zero. The area is weighted with the value of the membership function. The solution is a compromise, due to the fuzziness of the consequences. The choice for COG is justified in [Driankov et al 1996], who suggested some requirements that should be satisfied by an ideal defuzzification method. In conclusion, the COG application satisfies the three major requirements analyzed by [Driankov et al 1996] – continuity, disambiguity and plausibility. It is necessary to observe that the dezzuzification methods *Center of sums* and *Height* also satisfy these requirements. It could be therefore possible to apply any of these three methods in the proposed study.

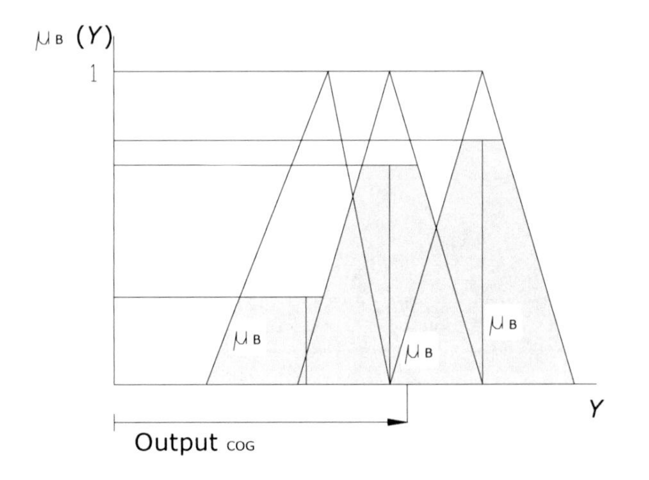

Fig. 5 Center of Gravity method (COG) used as defuzzification process [Virant, 2000]

5 Fuzzy Practical Analysis

The methodology is tested using the MATLAB® Software under multi-rules-based decision and multi-sets considerations. A brief description of fuzzy modeling used in this practical analysis is presented below:

- Fuzzification process – *Mamdani Controller*;
- Rule evaluation – logical operator used in the development of rules to obtain a single number: *AND Operator*;
- Implication – evaluation of each rule generating a single output: *Method of Minimum* (MIN);
- Aggregation – unification of the output of all rules: *Method of Maximum* (MAX).
- Defuzzification – the output linguistic value is translated into an output number: *Center of Gravity*.

Moreover, the arrangement modeled for RET and ESS analyses is described with five inputs (criteria in analysis) and only one output (final score). The main fuzzy variable used to characterize each criterion was described in Figure 1. The multi-rules-based used in this work consists of a collection of if-then propositions taking as basis the prior classification of the criteria, defined for the three different perspectives. The values that represent the rated data and the weights stipulated by the DM are presented in Table 2 - RET and Table 3 - ESS. The values are defined in the interval from 0 to 1.0, considering the scores for the qualitative criteria and the normalized values *(NV)* for the quantitative criteria *(QC)* according to equation 2.

$$NV = \frac{QC}{\max QC} \tag{2}$$

Table 2 Database used in the proposed methodology for RET analysis *(The higher the better)*

RET	EF	EF cog	$	LC	EI
MT	0.56	0.85	0.54	1.00	0.80
OTTO	0.70	1.00	1.00	0.50	0.40
PA	0.90	1.00	0.30	0.40	0.70
MC	1.00	1.00	0.24	0.25	0.90
SO	1.00	0.93	0.20	0.50	0.90

Table 3 Database used in the proposed methodology for ESS analysis *(The higher the better)*

ESS	EF	$	LC	EI	TM
FLY	1.00	0.60	1.00	0.90	0.80
C. BAT	0.83	1.00	0.40	0.60	0.80
F. BAT	0.94	0.16	0.80	0.50	0.40

The comparisons of the final classifications (CL) obtained by using the proposed fuzzy methodology for the different perspectives under analysis are presented in Tables 4 and 5.

Table 4 Final Classification for RET analysis acoording the perspectives: costs, environment impacts (EI) and environment impacts-costs

RET	Costs	CL	EI	CL	EI-Costs	CL
MT	0.750	2nd	0.771	2nd	0.757	1st
OTTO	0.920	1st	0.683	3rd	0.683	3rd
PAFC	0.645	3rd	0.567	4th	0.645	5th
MCFC	0.629	4th	0.797	1st	0.665	4th
SOFC	0.587	5th	0.797	1st	0.699	2nd

Table 5 Final Classification for ESS analysis acoording the perspectives: costs. environment impacts (EI) and environment impacts-costs

RET	Costs	CL	EI	CL	EI-Costs	CL
FLY	0.771	2nd	0.797	1st	0.910	1st
C. BAT	0.920	1st	0.645	2nd	0.645	2nd
F. BAT	0.395	3rd	0.567	3rd	0.500	3rd

By observing the data presented above, the most appropriate hybrid systems selected by the proposed methodology for each perspective are: the Otto engine with conventional battery for the costs setting; the SOFC or the MCFC with the flywheel for the environmental perspective; the microturbine with the flywheel for the environmental-costs scenario.

To conclude, it is important to emphasize that this novel fuzzy-based methodology may consider several criteria and perspectives by simply adjusting the fuzzy multi-rules and multi-sets, in accordance to each specific case.

6 Summary and Conclusions

This paper presented a study addressing the problem of finding appropriate renewable hybrid systems using biogas by anaerobic digesters, according to different perspectives. To achieve this goal, a methodology incorporating fuzzy multi-rules and fuzzy multi-sets was developed. A prior classification of the criteria relevance was defined in relation to each perspective. This arrangement facilitates the understanding of the methodology and increases the possibility of incorporating the DM preferences on the decision making process.

With relation to the RET scenarios, clean technologies such as fuel cells were not selected as most appropriate choice for the environment-costs perspective, mainly because their costs are still higher than other options under study. However, the microturbine using biogas appeared as a promising renewable energy source.

Regarding ESS analysis, the flywheel was selected as the most appropriate technology for the environment perspective and environment-costs perspective. This result is understandable, once the flywheel presents a high energy density, a high efficiency, a high life cycle, and it does not offer any kind of negative environmental impact.

To summarize, the final results illustrate the use of the novel fuzzy-based methodology for biogas fuelled hybrid energy systems decision making. It takes into consideration not only operational characteristics, but also social, economic and environmental aspects. Further improvements to the criteria evaluation are recommended - by using ISO standard 14040 - especially for analyzing and comparing different types of environmental impact categories. The ISO standard 14040 application would provide more credibility to the methodology outcome.

Acknowledgment

The authors would like to thank CAPES and PDEE Program for their financial support.

References

Alcalá, R., Alcalá-Fdez, J., Herrera, F., Otero, J.: Genetic learning of accurate and compact fuzzy rule based systems based on the 2-tuples linguistic representation. International Journal of Approximate Reasoning 44, 45–64 (2007)

Bana e Costa, A., Vansnick, J.: Applications of the MACBETH Approach in the Framework of an Additive Aggregation Model. In: Meskens, N., Roubens, M. (eds.), pp. 131–157 (1999)

Barros, L.C., Bassanezi, R.C.: Tópicos de lógica fuzzy e biomatemática. Unicamp – Imecc, São Paulo (2006)

Beccali, M., Cellura, M., Mistretta, M.: Decision-making in energy planning-application of the ELECTRE method at regional level for the diffusion of renewable energy technology. Renewable Energy 28, 2063–2087 (2003)

Belton, V., Stewart, J.: Multiple criteria decision analysis: an integrated approach, p. 372. Springer, Heidelberg (2002)

Cormio, C., Dicorato, M., Minoia, A., Trovato, M.: A regional energy planning methodology including renewable energy sources and environmental constraints. Renewable and Sustainable Energy 7, 99–130 (2003)

Driankov, D., Hellendoorn, H., Reinfrank, M.: An introduction to fuzzy control, p. 316. Springer, Heidelberg (1996)

Farret, F.A., Simões, M.G.: Integration of Alternative Sources of Energy. John Wiley & Sons, USA (2006)

Gegov, A., Gobalakrishnan, N.: Advanced inference in Fuzzy Systems by Rule Base Compression. Mathware & Soft Computing 14, 201–216 (2007)

Hamam, A., Georganas, N.D.: A Comparison of Mamdani and Sugeno Fuzzy Inference Systems for Evaluating the Quality of Experience of Hapto-Audio-Visual Applications. In: International Workshop on Haptic Audio Visual Environments and their Application, Canada (2008)

Klir, G.J., Yuan, B.: Fuzzy Sets and Fuzzy Logic: Theory Applications, p. 592. Prentice-Hall, London (2003)

Kiszka, J.B., Kochanska, M.E., Sliwinska, D.S.: The influence of some fuzzy implications operators on the accuracy of fuzzy model. Fuzzy Sets and Systems 15, 111–128 (1985)

Mantas, C.J., Puche, J.M., Mantas, J.M.: Extraction of similarity based fuzzy rules from artificial neural networks. International Journal of Approximate Reasoning 43, 202–221 (2006)

Matos, M.A.: Eliciting and aggregation preferences with fuzzy inference systems. In: 56th Meeting of the European Working Group, Coimbra, pp. 213–227 (2002)

Mueller, S.: Manure's allure: Variation of the financial, environmental, and economic benefits from combined heat and power systems integrated with anaerobic digesters at hog farms across geographic and economic regions. Renewable Energy 32, 248–256 (2007)

Ramírez-Rosado, I.J., Dominguez-Navarro, J.A.: Possibilistic model based on fuzzy sets for the multi-objective optimal planning of electric power distribution networks. Transactions on Power Systems 19, 1801–1810 (2004)

Siler, W., Buckley, J.: Fuzzy Expert Systems and Fuzzy Reasoning. John Wiley & Sons, USA (2005)

Xuan, J., Leung, M.K.H., Leung, Y.C., Ni, M.: A review of biomass-derived fuel processors for fuel cell systems. Renewable and Sustainable Energy Reviews 13, 1301–1313 (2008)

Virant, J.: Design Considerations of Time in Fuzzy Systems, p. 512. Springer, Heidelberg (2000)

Wang, L.-X.: Adaptive Fuzzy Systems and Control: Design and Stability Analysis, p. 352. Prentice Hall PTR, Englewood Cliffs (1993)

Wedley, W.C., Ung Choo, E., Schoner, B.: Magnitude adjustment for AHP ben-efit/cost ratios. European Journal of Operational Research 133, 335–342 (2001)

Yusoff, M., Mutalib, S., Rahman, S.A., Mohamed, A.: Intelligent Water Disper-sal Controller: Comparison between Mamdani and Sugeno Approaches. In: Proceedings of the International Conference Computational Science and its Applications. ICCSA, Malaysia (2007)

Zimmermann, H.: Fuzzy Set Theory and Its Applications. Springer, Heidelberg (2001)

Zopounidis, C., Doumpos, M.: Multicriteria classification and sorting meth-ods: a literature review. European Journal of Operational Research 138, 229–246 (2002)

Two New Applications of Artificial Neural Networks: Estimation of Instantaneous Performance Ratio and of the Energy Produced by PV Generators

Florencia Almonacid, Catalina Rus, Pedro Pérez-Higueras, and Leocadio Hontoria

Research Group "IDEA"
Department of Electronics Engineering, Polytechnics School of Jaén,
University of Jaén, 23071- Jaén, Spain
hontoria@ujaen.es

Abstract. The growth of photovoltaic (PV) for electricity generation is one of the highest in the field of the renewable energies and this tendency is expected to continue in the next years. As an obvious consequence, an increasing number of new PV components and devices, mainly arrays and inverters, are coming into the PV market. The need for PV arrays and inverters to be characterized has then become a more and more important aspect. Due to the variable nature of the operating conditions in PV systems, the complete characterization of these elements is quite a difficult issue.

One aspect that can help to achieve this goal is to improve methods for estimating the energy produced by photovoltaic generators. Overall, the annual energy provided by a PV generator is directly proportional to the annual radiation incident on the plane of PV generator and the installed nominal power or peak power. However, there are a number of reasons that cause a decrease in the expected energy and include; mismatch losses, dirt and dust, ohmic losses and many more. In this chapter we present two new studies in the PV field. The first one concerns the application of the Artificial Neural Networks (ANN) for estimating the instantaneous Performance Ratio, which is the fundamental parameter in the characterization of PV systems. The second study aims to compare the results of several methods for estimating the annual energy produced by a PV generator, three classical and one based on artificial neural networks, in different types of systems with different settings and types of modules.

Therefore, in this chapter the classical methods for estimating the energy provided by a PV generator, are compared with a method developed by University of Jaén based on artificial neural networks (ANN). While classical methods take into account only temperature losses, the method based on ANN take into account in addition to temperature losses the low irradiation losses, spectral and angular losses, and some other losses as nominal power losses. Additionally, as it was mentioned previously a study on the Performance Ratio of PV systems will be presented.

K. Gopalakrishnan et al. (Eds.): Soft Comput. in Green & Renew. Ener. Sys., STUDFUZZ 269, pp. 199–232.
springerlink.com

Nomenclature

C_A Generator Capacity (kWh)
C_S Accumulator Capacity (kWh)
E_{DC} Annual energy provided by a PV generator (Wh)
Eac Annual energy injected into the grid in kWh
FF Form Factor
G Irradiance (W/m^2)
G_{STC} Incident irradiance at standard test conditions (1000 W/m^2)
H_A Annual irradiation incident on the plane of PV generator (Wh/m^2)
H_d Daily solar irradiation (Wh/m^2)
H_{dm} Monthly mean daily solar irradiation (Wh/m^2)
I Current (A)
I_G PV generator current (A)
I_m Cell current at the maximum power point (A).
I_{MAXG} PV generator current at the maximum power point (A).
$I_{m,STC}$ Module current at maximum power in STC (A).
I_{sc} Cell short circuit current (A).
$I_{sc,\ STC}$ Cell short circuit current in STC (A).
K_T Daily clarity index
k_t Hourly clarity index
k_{t-p} Predicted value of k_t
k_{t-ave} Average value of k_t
K_{TDY} Yearly Clarity Index
LOLP Loss Of Load Probability
LLP Loss of Load Probability
N Number of samples
N_{cp} Parallel cells number of PV module
N_{cs} Series cells number of PV module
N_{mp} Parallel modules number of PV generator
N_{ms} Series modules number of PV generator
NOCT Nominal Operating Cell Temperature (ºC)
P_{DC} PV generator output power (W)
P_{AC} Output inverter power (W)
P_{FV} PV generator Peak Power generator (W)
P_{gen} Maximum power of the generator in kWp
P_m Cell maximum power (W)
P_{MAXG} PV generator maximum power (W)
$P_{m,STC}$ Cell maximum power in STC (W)
PR Performance Ratio
PR_{INST} Instantaneous Performance Ratio
PR_{ANNUAL} Annual Performance Ratio
P_{Oi} Measured power of PV module (W)
P_{Ti} Theoretical power of PV module (W)
r_s Standard resistance

R_s Series resistance of a photovoltaic cell module (Ω)
T_{am} Ambient temperature (ºC)
T_c Cell (Module) Temperature (ºC)
V Voltage (V)
V_G PV generator voltage (V)
V_m Cell voltage at the maximum power point (V)
V_{MAXG} PV generator voltage at the maximum power point (V)
$V_{m,STC}$ Module voltage at maximum power in STC (V)
V_{oc} Cell open circuit voltage (V)
$V_{oc,\,STC}$ Cell open circuit voltage in STC (V)
V_t Thermal voltage (V)
η_i The inverter efficiency
γ Cell maximum power temperature coefficient (ºC^{-1})

Abbreviations

AC Aguiar and Collares-Pereira
AIL Accredited Independent Laboratory
ANN Artificial Neural Network
ARMA Autoregressive Moving Average
GH Graham and Hollands
GCPVS Grid-Connected Photovoltaic System
IES Instituto Energía Solar (Solar Energy Research Centre)
LOLP Loss of Load Probability
MLP MultiLayer Perceptron
NN Neural Network
PV PhotoVoltaics
PVPSP PhotoVoltaics Power System Program
STC Standard Test Conditions
TSP Time Series Prediction

1 Introduction

Artificial Neural Networks (ANN) has been used in the Solar Energy field for solving several problems during the last decades with very good results. There are a lot of ANN applications in this field. The Research & Development Group for Solar Energy (IDEA Group) at the University of Jaén has applied different ANN for solving problems in various Solar Energy fields such as the generation of synthetical series of solar radiation, drawing solar radiation maps, characterization of PV modules, etc.

In this chapter of the book, we are going to present our last two research works with ANN; **the obtainment of instantaneous Performance Ratio and the estimation of the energy provided by PV generators**, developed by our R & D

Group. In order to present the ANN developed and used, initially we are going to present a summary of the evolution of our research. Our first investigations where done in 1999 and from that date we have improved an ANN, particularly the Multilayer Perceptron (MLP) methodology for our purposes.

This chapter is developed in this way: the first section is this introduction. In section 2 a summary of the evolution of our investigations in the application of ANN to solve different problems in Solar Energy field is presented. Our investigations in this field are quite wide since we have developed methods using ANN for such applications as prediction of solar radiation and drawing solar radiations maps, sizing photovoltaic systems or characterisation of photovoltaic solar modules. Nevertheless, the two main objectives of this chapter are: the obtainment of the instantaneous performance ratio, which is presented in section 3 and the study on the energy produced by three PV generators using ANN, which is described and explained in section 4. At the end we present some conclusions.

2 Previous Research Work

2.1 Artificial Neural Networks and Time Series Prediction

Our first work with ANN was on the prediction of solar radiation data (Time Series Prediction, TSP). We developed a MultiLayer Perceptron (MLP), which produced very good results when it was compared with other classical methods for predicting solar radiation data. This first work was done in 1999 and presented in [Zufiria et al. 1999]. Here we are going to summarize these first results.

The design and analysis of photovoltaic systems is usually performed via numerical simulations which require as input data large time sequences of hourly or daily irradiation values [Graham and Hollands 1988, 1990, Aguiar and Collares-Pereira 1991, 1992]. Nevertheless, these historic radiation measurements do not exist in most of the world countries, and, if any, their quality is questionable or they have plenty of missing values [Knight et al. 1991]. The synthetic generation of hourly or daily solar irradiation values is often the only practical way to obtain radiation data for any given location.

Several mathematical radiation models and methods have been developed [Aguiar and Collares-Pereira 1992, Balouktsis and Tsakides 1986, Graham and Hollands 1988, 1990, Goh and Tan 1977, Knight et al. 1991, Mustachi et al. 1979] to generate sequences of values, which try to preserve the same statistical properties such as average, variance, probability density function [Aguiar and Collares-Pereira 1991, Hollands and Hughet 1983, Liu and Jordan 1960] and sequential characteristics as those of the historical records (i.e., those observed in nature). The output of these models may be used for example for the construction of typical meteorological years [Knight et al. 1991] or to provide computer-generated data sequences for the analysis and design of photovoltaic (PV) converters, which is usually performed using numerical simulation tools [Graham and Hollands 1990]. For the study of PV systems with a high storage capacity daily radiation data will usually suffice as the storage attenuates the effects of hourly variations; but for PV systems with one or two-hour response time, such as peak plants or PV plants which return energy to the network at maximum charge instants, hourly series are required.

The models proposed by Aguiar and Collares-Pereira [1992] and by Graham and Hollands [1988, 1990] referred from now on as AC and GH respectively for short, may be considered as paradigms in the field of hourly [Aguiar and Collares-Pereira 1991, Graham and Hollands 1990] radiation modelling. They are auto-regressive first-order models [Priestley 1988], based on a stochastic disaggregation methodology, that generate hourly irradiation series making use of daily values. These values can be obtained from historic records (which are more common than hourly records) or using daily generation methods [Graham and Hollands 1988] in turn (which are more validated than hourly methods). Complex empirical expressions are proposed to relate the hourly and daily values, obtaining the parameters in the formulae via regression analysis [Werbos 1974] over historical data.

The methodology presented for Times Series Prediction and system identification via MLP defines the framework of the method developed for the generation of hourly clarity index series $\{k_t\}$. For the computational experiments, a set of hourly clarity index k_t values measured in several Spanish locations and its corresponding daily K_T values were used. As a first approach, in order to evaluate the quality of the generated series, the first years were considered as a *training set* and the last year was employed for testing the validity of the generated series, in each location.

The proposed method has been developed via a step by step inclusion of the available associated information. The greatest advantage of the MLP-based methodology is that explicit knowledge of the relationship among all the information sources is not needed. Such information sources can be progressively incorporated in different steps upon the proposed method. The details of this step-by-step procedure can be found in [Zufiria et al. 1999]. The final procedure carried out by employing a MLP in a mixed feedback-feedforward configuration, is shown in Fig. 1.

The seven inputs required are [Hontoria et al. 1999, Zufiria et al. 1999]:

- s: this is an input, taking values $\{0,1\}$ that indicates whether an hour is between sunrise and sunset (k_t should be different from 0) or not (k_t has to be 0),
- d_n: distance (days) between the value to be generated and the day with maximum value in the $\{k_t\}$ annual distribution. This is used in order to keep the hourly global irradiation monthly periodicity and seasonal nature,
- h_n: indicates the hour order number of the k_t value (ensuring the hourly global irradiation periodicity and seasonal nature). Both inputs (d_n and h_n) are normalised to the range [0,1],
- K_T : daily clarity index,
- $k_t(h\text{-}3)$, $k_t(h\text{-}2)$, $k_t(h\text{-}1)$: hourly clarity index of the three previous hours.

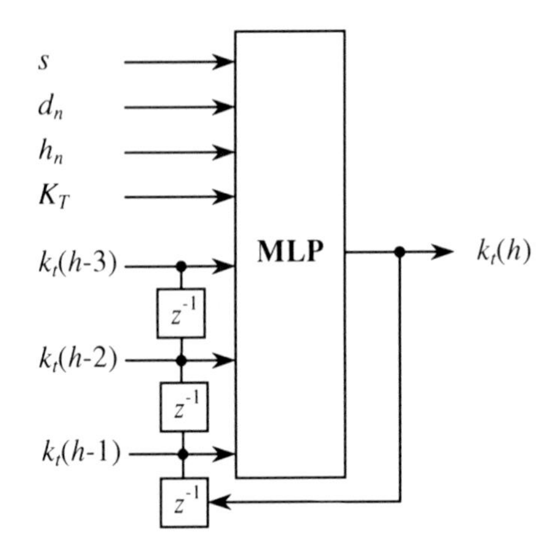

Fig. 1 MLP Architecture for predicting k_{tm}

The training was done using the backpropagation algorithm (with momentum and random presentations), which is eventually combined with second order optimisation methods during the last few epochs. The hidden layer has fifteen nodes. For the training of the MLP and for the subsequent validation it is necessary to use real solar radiation series. For this objective several Spanish locations with different latitudes and climates were used.

In Fig. 2 we present an example of the artificial series of solar radiation obtained by our MLP (Neural in the figure), the Aguiar-Collares Method (AC in the figure), the Graham-Hollands Method (GH) and the real data.

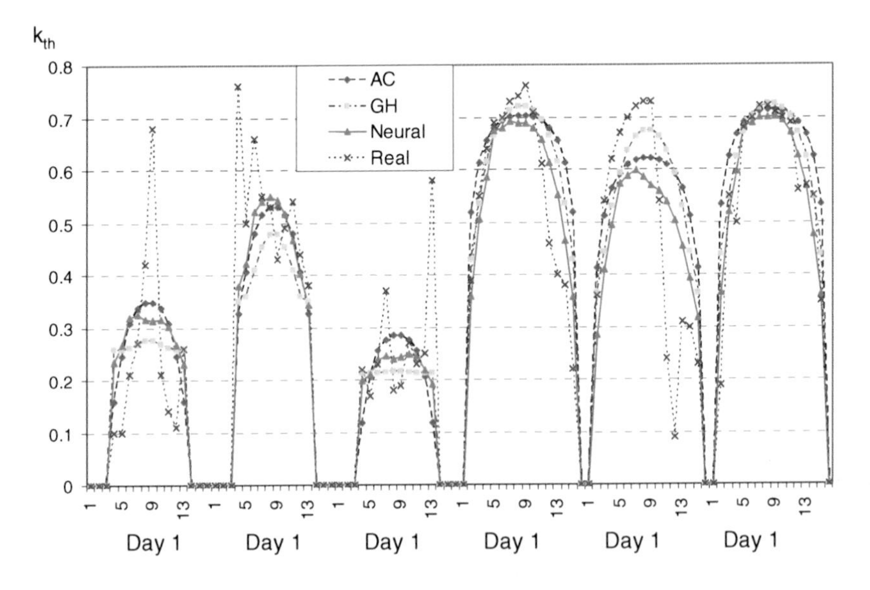

Fig. 2 Artificial Series obtained by different methods and Real data

We have calculated the Root Mean Square Error (RMSE) among the real data and the data obtained by the three other methods (AC, GH and Neural) and the results shown a better approximation of the data provided by the Neural Networks (MLP) method.

2.2 Artificial Neural Networks and Solar Maps

A very important application of the methodology proposed using the MLP in the solar radiation field is the drawing of solar radiation maps. The availability of information about the solar radiation in the location where a solar system is going to be installed is necessary for the designer of solar systems. This information, in case that exists, can be available in several ways. The most common one is by means of several tables with a lot of very useful information (usually large solar radiation sequences), but they are extremely difficult to handle. Nevertheless, another way to present this information is by means of different solar radiation maps (one for each month) of the area where the installation is going to be made. This second way is usually more efficient, easy to handle and preferable by the designers, and can be used at the initial stage of the solar system design and sizing.

As part of this work we used the MLP for drawing solar radiation maps for Spain [Hontoria et al. 2000]. In this case the MLP was slightly different than the MLP of our first work, and included two new input parameters: latitude and altitude of the site, to produce the variation needed for the Time Series Prediction of solar radiation for different points of the solar radiation map. In Fig. 3, the MLP used is shown, and two examples (Figs. 4 and 5) of solar radiation maps obtained with this method are presented.

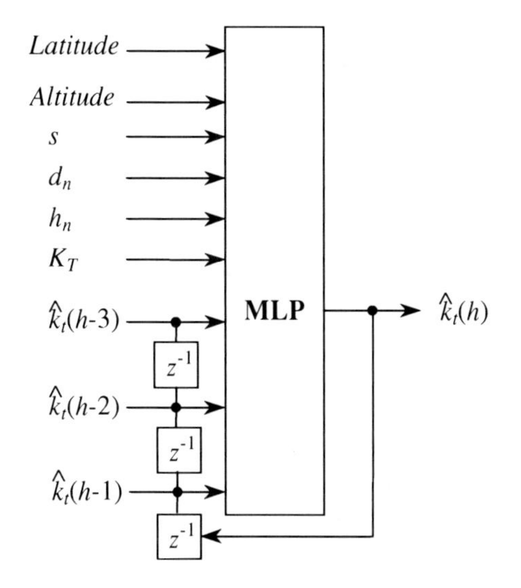

Fig. 3 MLP Architecture for clarity indexes prediction, including latitude and altitude

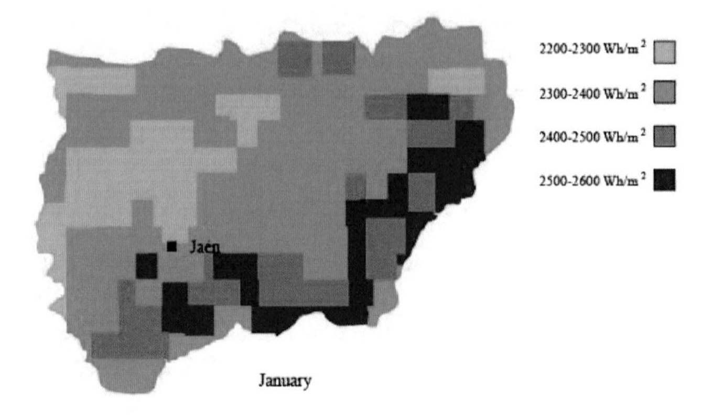

Fig. 4 Solar radiation map of Jaén in winter

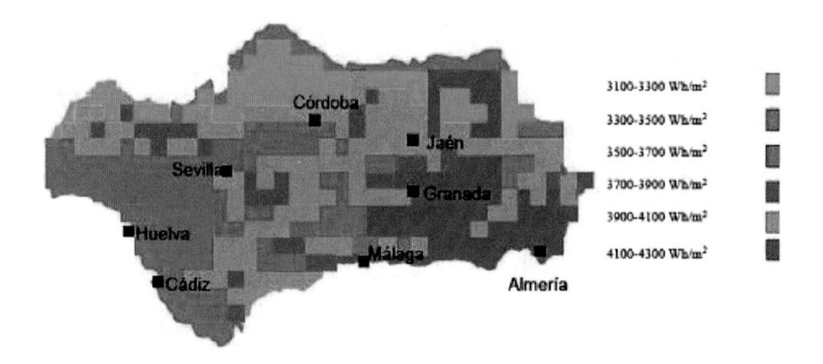

Fig. 5 Solar radiation map of Andalucía in autumn

2.3 *Artificial Neural Networks and Stand Alone Photovoltaic Systems*

Once we have developed the MLP for applications in the Solar Radiation field, the next step was to try to apply our knowledge in the Photovoltaic (PV) field. There are mainly two types of Photovoltaic Systems: the Stand Alone Photovoltaic Systems and the Grid Connected Photovoltaic Systems. We have used ANN methodology in both types of systems.

The first study we carried out in the PV field was done in order to apply the MLP for establishing a relation between some variables needed for sizing stand alone PV systems. There are many methods available for sizing stand alone photovoltaic systems. The most important are methods which use equations to describe the PV system size as a function of reliability. These are called analytical methods. They allow the designer to optimise the energy and economic cost of the PV system. Many of the analytical methods employ the concept of reliability of the system or the complementary term, i.e., Loss of Load Probability (LOLP). The

LOLP represents the percentage of time that the PV/storage system will not be able to satisfy the load. Several authors use this type of methodology for sizing stand alone PV systems [Chapman 1987, Egido and Lorenzo 1992, Lorenzo et al. 1994, Sidrach-de-Cardona and Lopez 1998, 1999, Barra et al. 1984, Bartoli et al. 1984]. With these methods the relations between the generator capacity, the accumulator capacity and the LOLP lead to curves called isoreliability curves or LOLP curves. Let CA be the generator capacity, which is defined as the ratio of the average energy output of the generator in the month with worst solar radiation input divided by the average consumption of the load (assuming a constant consumption of load for every month). The battery storage system is related to the term CS, the accumulator capacity, which is defined as the maximum energy that can be extracted from the accumulator divided by the average daily consumption of the load.

With the aim to establish relations between CA, CS and LOLP the numerical methods use system simulations while the analytical methods use equations. For instance, Egido and Lorenzo [1992] presented a method consisting in creating reliability maps for each LOLP value considered, (isoreliability lines). They proposed very simple equations to determine the isoreliability lines for many Spanish locations.

Once the LOLP curves are obtained, it is very simple to design both the capacity of the generator (CA) and the accumulator capacity (CS). Depending on the reliability needed for the PV system design, a specific value of the LOLP will be considered. The problem is to obtain the LOLP curves. A new approach for obtaining LOLP curves was developed by our Research Group [Hontoria et al. 2003]. It was based on a Multilayer Perceptron (MLP) neural network. Particularly, the method is based on the MLP's ability to extract, from a sufficiently general training data set, the existing relationships between variables whose interdependence is not known *a priori*. As was seen with a very simple structure of the MLP [Hontoria et al. 2001, 2002], the solar radiation series in locations where data are not available can be generated. For the case of LOLP curves prediction, a new architecture for the MLP is used. The MLP is trained with LOLP data (different LOLP values, i.e., 0.1, 0.01, 0.05,...) and radiation data from different Spanish locations (first stage). After training, the MLP is able to generate as many LOLP curves as needed, for a particular site (second stage). The main advantage of this tool is that the effort required in the first stage is similar to that required in other methods, however in the second stage, when the MLP is trained, it is extremely simple to obtain new LOLP curves.

Summing up, in this work, we trained several MLPs (one for each location considered) with many pairs of (CS, CA) data but for very few LOLP curves (just the values 0.01, 0.05 and 0.1) and after the training process each MLP was able to generate the LOLP for other values (for instance the value 0.02 or 0.25). The results obtained were very satisfactory.

The structure of the Multi Layer Perceptron (MLP) proposed for the generation of LOLP curves is shown in Fig. 6. It consists of three layers. The first one or input layer has three inputs as follows:

CS: Accumulator Capacity.
LOLP: Loss of Load Probability.
K_{TDY}: Yearly Clarity Index.

The second layer, also called the hidden layer, has nine neurons or nodes. Different tests have been done in order to choose the number of neurons and the actual number selected produced the best results. Finally the last layer has only one node, the value of the generator capacity (CA) to be estimated.

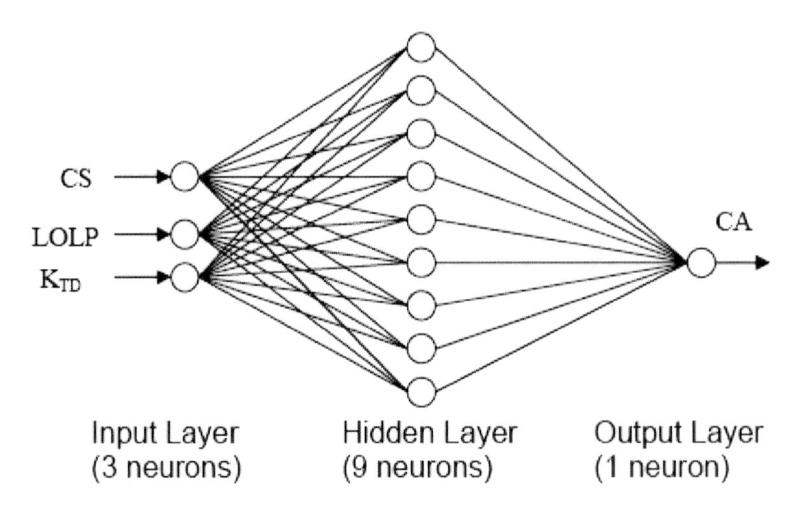

Fig. 6 MLP Architecture for the obtaining of the LOLP curves

The training was done with the backpropagation algorithm [Rumelhart and MacClelland 1986, Weigend et al. 1990].

By applying the proposed methodology many different LOLP curves have been calculated. Figure 7 shows one such example for Santander, a city in the North of Spain. As can be seen the LOLP curve obtained by the MLP is much closer to the real curve whereas the curve obtained with the simple IES method, developed by the instituto de Energia Solar Madrid, is not as good. As it was mentioned before the generation of the LOLP curves has been carried out not only for the locations used for the training, but also for those used for the validation. As a quality measure the quadratic error between the real curves and those obtained via the MLP has been calculated. In [Hontoria et al. 2003, 2005a], the complete methodology is described.

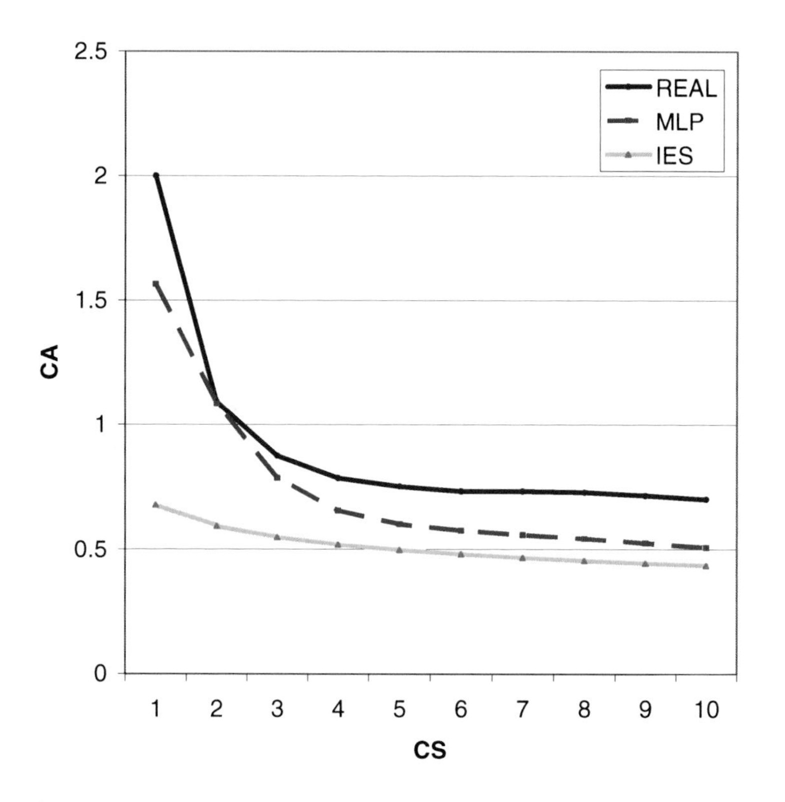

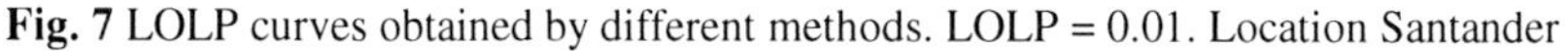
Fig. 7 LOLP curves obtained by different methods. LOLP = 0.01. Location Santander

2.4 Artificial Neural Networks and Characterization of PV Modules

A new evolution in our research with the MLP was to characterize PV modules. We applied the ANN methodology developed by our group for the characterization of PV modules of different technologies, for instances:

a) In [Almonacid et al. 2009, Hontoria et al. 2005b] we characterized Si-Crystalline PV modules with the MLP methodology developed.
b) In [Almonacid et al. 2010] the modules in study where CIS modules.
c) Additionally, thin film modules were characterized by our methodology as it can be seen in [Almonacid et al. 2007].

3 Artificial Neural Networks and Maintenance of Grid Connected Photovoltaic Systems

3.1 Introduction

In order to make a proper maintenance of a Grid-Connected Photovoltaic System (GCPVS) and to optimize its rate and payback, the parameter called Performance Ratio (PR) is usually employed. It is considered that a GCPVS is working properly if the value of its annual Performance Ratio has a certain value depending on the site where the system is located.

The Photovoltaic Power Systems Programme (PVPS) of the International Energy Agency (IEA) collects and analyses performance data of PV power plants in various system techniques and disseminates suitable information on the performance, long-term reliability and the technical and economic output of PV systems. In Table 1, the Annual Performance Ratio, for three years of the PV systems monitored, is shown.

Table 1 International Energy Agency (IEA) data of PR [EA 2007]

	Annual Performance Ratio (in %)
United States	*78*
Germany	*76*
Japan	*75*
Israel	*70*
United Kingdom	*63*
Spain	*60*
Norway	*59*
Austria	*23*

A tool to calculate the instantaneous PR, was designed using Artificial Neural Networks. Afterwards, a comparative study between the instantaneous PR obtained by the ANN and the one obtained with the measurements monitored by the system was carried out. As the comparative study performed indicated that our developed tool was adequate, it is possible to know how the PV system is working in a quick and efficient way [Rus et al. 2009].

Here we explain our investigations done in this work.

3.2 System Description

The Univer generator [Drift et al. 2007] consists of four photovoltaic subgenerators connected to the low voltage grid at Jaén University Campus (37°73'N, 3°78'W) in the south of Spain with a total power of 200 kWp. In Jaén, average yearly peak solar hours are 4.9h per day.

The installation presents two particular aspects: the **system location** and the **use of different PV integration**. The system is located at a crowded public building. In the whole PV integration plant, different traditional architectural solutions were used: parking canopies, pergolas and façade. The analysis was done with the data from the generator *"Pergola PV system"*.

The Pergola PV system, has an array of 180 semi-transparent Isofotón I-106 modules (characteristics shown in Table 2), with a total power of 19.08 kWp. The PV array is divided into 9 sub-arrays of 20 modules each. Let N_{mp} be the Parallel modules number of the PV generator sub-array, and N_{ms} the Series modules number of PV generator sub-array, the final distribution of the 20 modules in each array is: N_{ms} x N_{mp} = 10 x 2 modules. In the performance analysis, it was considered that all 9 subsystems operate with the same performance. In Fig. 8 a photograph of the Pergola PV system is presented and in Fig. 9 the schematic diagram of the modules connection is shown.

Fig. 8 Photography of "Pergola PV system"

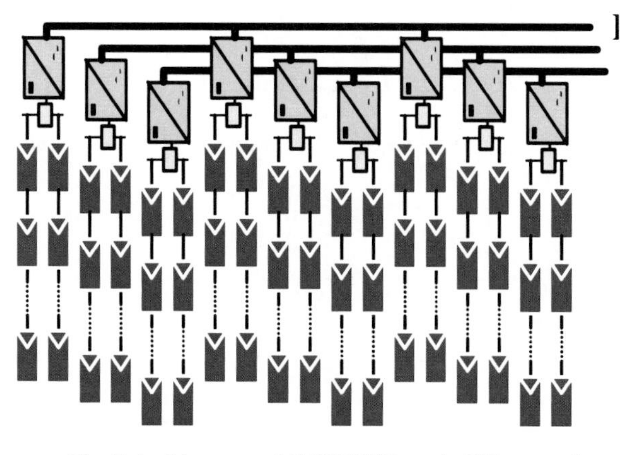

Fig. 9 Architecture of GCPVS "Pergola PV system"

In Table 2 the electric specifications of the modules used in the Pergola PV system are presented. All data are referred at Standard Test Conditions (STC), that is: 100 mW/m^2, AM 1.5 spectrum, and 25ºC cell temperature.

Table 2 Electrical values of the module

Electric Specifications	**I-106**
Maximum Power (Pmax)	106W
Short Circuit Current (Isc)	6.54A
Open Circuit Voltage (Voc)	21.6V
Maximum power current (Imax)	6.10A
Maximum power voltage (Vmax)	17.4V
Standard Test Conditions	

Additionally, in Table 3 the generator and modules parameters are presented, again at STC.

Table 3 Generator and module parameters

MODULES		**GENERATOR**	
Model	Isofotón I – 106	Orientation	52º SE
Material	Si-monocrystalline	Tilt	13º
Series Cells	36	Series modules	10
Parallel Cells	2	Parallel Modules	2
Total Cells	72	Total Modules	20
Standard Test Conditions			

The maximum power of each module is **106 Wp**, so the maximum power of the generator is **2120 Wp**.

3.3 MultiLayer Perceptron Developed

The MLP that we used and developed for obtaining the Performance Ratio of the system is presented in Fig. 10.

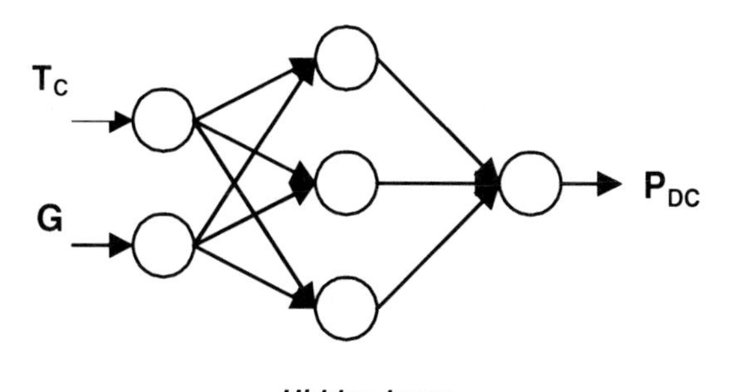

Fig. 10 Proposed architecture of the MLP

The MLP has three layers. The first one or *input layer* has two neurons or nodes, which are the following:

G: Irradiance (W/m^2)
T_c: Module Temperature (ºC)

The second layer (*hidden layer*), has three nodes. And finally the last layer (output layer) has only one node: the Power (in DC) P_{DC} of the PV generator in study.

The grid connected PV installations of Univer generator are fully monitored to assess the potential of PV technology and performance of this kind of systems. The monitoring system was designed according to the European Guidelines and IEC 61724 [IEC 1993].

This system includes also a data acquisition system, connected to a computer and the measured data is recorded every 10 min. The monitoring parameters for each PV system are: the ambient temperature T_{am} (ºC), the in-plane irradiance G (W/m^2), the array voltage V_{dc} (V), the array current I_{dc} (A) and the output inverter power P_{ac} (W).

3.4 Performance Ratio-Definitions and Calculations

In order to determine whether a GCPVS is working properly, one of the most common parameter used is the yearly value of the performance ratio (PR_{anual}). This parameter can be obtained using the following expression:

$$PR_{annual} = \frac{E_{ac}(kWh/year)}{P_{gen}(kWp)\cdot H\left(kWh/m^2 year\right)} \tag{1}$$

Where:
E_{ac}: annual energy fed into the grid in kWh,
P_{gen}: maximum power of the generator in kWp and
H: annual irradiation received by the generator in kW/m^2.

To determine, in an instantaneous way, the running of equipment PR_{anual} is not valid, so that a new parameter must be used. The instantaneous PR is defined as the relation between the irradiation received by the system, the power that is fed into the grid and the peak power of the generator. It is expressed as follows:

$$PR_{inst} = \frac{P_{ac}(kW)}{P_{gen}(kWp) \cdot G\left(kW/m^2\right)} \tag{2}$$

It can be checked out that there is not any value of the PR_{inst} which allows us to determine if the system works in the right way directly. This happens due to the fact that the value of the PR changes every moment of the day in which the system is being analyzed. Our aim is to show the way we can track a system through the monitoring of its PR_{inst}, and based on this, to classify whether the working state of our GCPVS is good or bad. This permits the early detection of any deviation from the system. In Fig. 11, the schematic diagram of the algorithm proposed is shown.

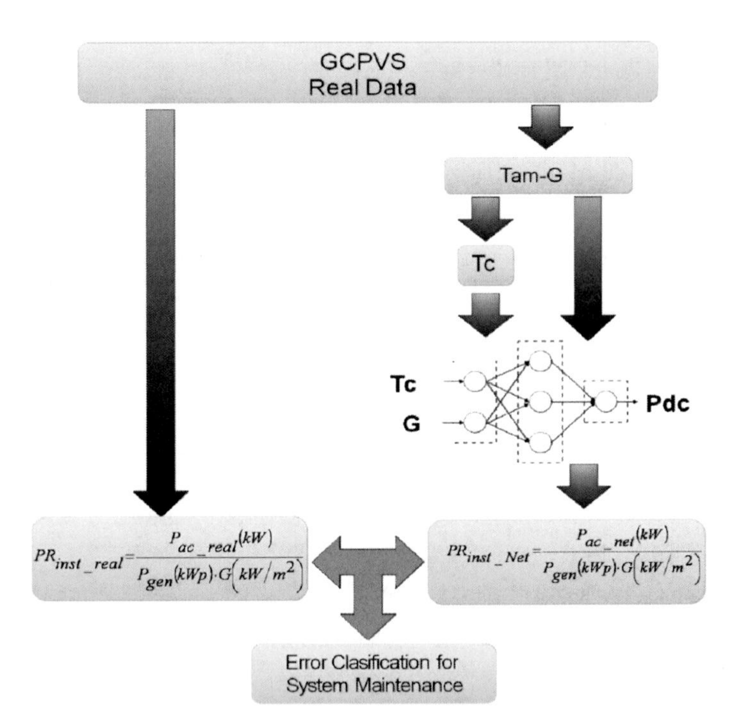

Fig. 11 Diagram of the algorithm proposed

To apply the algorithm proposed, the methodology we applied can be summarised as follows:

- First of all we need some entrance parameters. In this case the monitoring parameters used are:

The ambient temperature T_{am} (ºC),
The in-plane irradiance G (W/m^2),
The output inverter power P_{ac_real} (W).

Here we made a slight change, as we need the module (or cell temperature T_c) and we were monitoring the ambient temperature (T_{am}) as mentioned previously. The change is done using the following equation:

$$T_c = T_{am} + G \times \frac{NOCT - 20}{800} \tag{3}$$

Where:

T_{am}: ambient temperature, ºC
T_c: cell module temperature, ºC
$NOCT$: nominal operating cell temperature (for I-106 module this parameters is 47ºC)
G: irradiance for the ambient temperature, W/m^2

- Then the values of irradiance and temperature are presented to the net, obtaining Module Maximum Power (P_M) and Generator Power PM (P_{dc}) using series cells number (N_s), parallel cells number (N_p).
- The inverter efficiency, η_i, quantifies how well the DC input power is converter in useful AC power. It is defined as the ratio between AC output power (P_{ac}) and DC input power (P_{dc}). Among the different approaches that can be adopted, we have used a descriptive model that provides a good equilibrium between precision and simplicity [Pérez et al. 2004].

$$\eta_i = \frac{P_{ac}}{P_{dc}} = \frac{p_{in} - \left(b_0 + b_1 + b_2 \left(p_{in}\right)^2\right)}{p_{in}} \tag{4}$$

Where:

$$p_{in} = \left(\frac{P_{dc}}{P_{nom\_inv}}\right) \tag{5}$$

This model has the advantage that parameters b_0, b_1 and b_2 have physical meaning according the type of losses involved in the power conversion:

- b_0 represents the self-consumption losses
- b_1 represents the losses that varies linearly with the current (e.g., diodes)

- b_2 represents the Joule losses, which are proportional to the square of the current (voltage drops across the wiring, transformers, switching semiconductors, etc.).
- Then the P_{ac} is calculated.
- Calculation of Instantaneous Performance Ratio (both ANN and measured).
- Finally, a comparative study between different PR_{inst} is done with error classification for system maintenance.

3.5 Accuracy of the Methodology Developed

As we have explained in the previous subsection with the MLP developed we were able to calculate the instantaneous performance ratio, that we called net instantaneous performance ratio (PR_{inst_net}). In our investigations we did a comparative study between the two instantaneous performance ratios, the one obtained via the ANN methodology and the one obtained with the real data. In Fig. 12 we can observe that the two performance ratio are quite similar, which indicates that our method for obtaining the performance ratio is good.

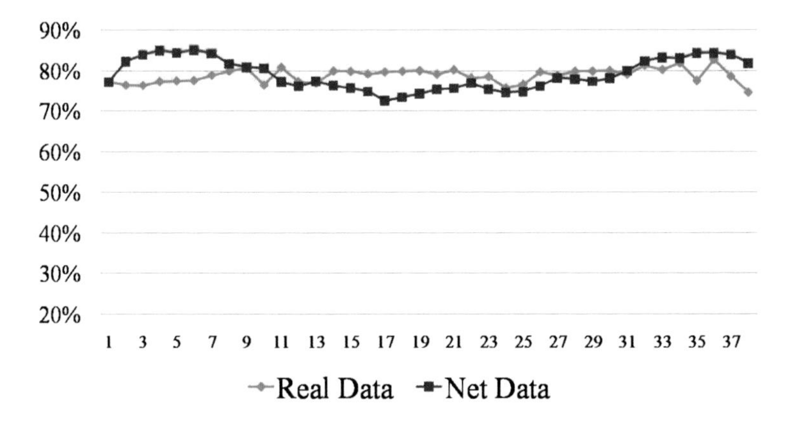

Fig. 12 PR_{inst} obtained by different methods

3.6 Behaviour of the System

Once we have calculated the instantaneous performance ratio it is possible to determine if the system is working properly or not thanks to the error system. The **error system** is defined as the difference between the real instantaneous performance ratio ($PR_{inst\text{-}real}$) and the instantaneous performance ratio obtained by the ANN (PR_{inst_net}). Initially we have to decide when the system is working in a good way and when it is working badly. The hypothesis used is that if the error system is not higher than 10% the systems is working properly. An example of the results we have obtained is presented in Fig. 13. In this figure the system error (remember that our time interval is ten minutes) is presented. As it can be observed the systems is always working properly, as the system error is under 10% always.

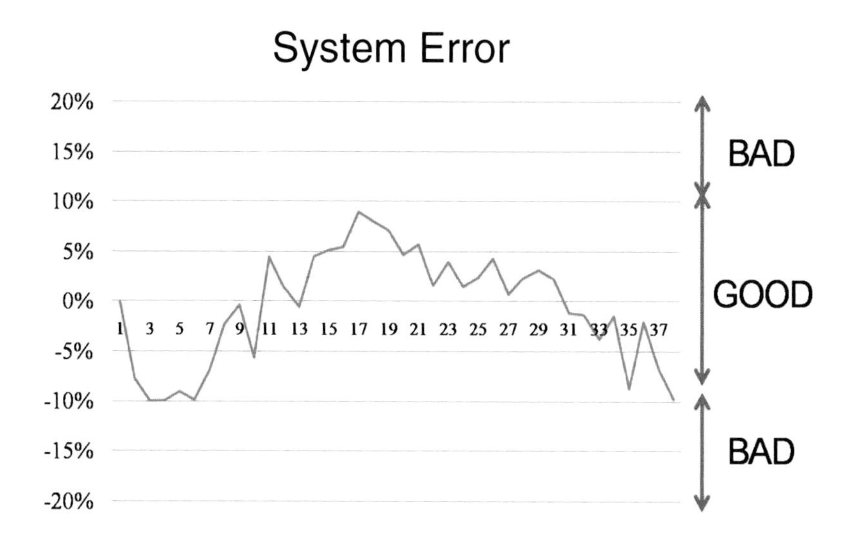

Fig. 13 System Error classification.

4 Estimation of the Energy Produced by pv Generators Using ann

4.1 Introduction

Photovoltaic (PV) systems have shown themselves to be one of the most promising applications for dealing with solar electricity generation. Because of this, in the last few years, the PV market has changed drastically. There has been a substantial market growth in the last years, with an ongoing trend in grid-connected applications. Countries enhance the international collaboration efforts which accelerate the development and deployment of photovoltaic solar energy as a significant and sustainable renewable energy option.

Nevertheless there is one difference between the theorical installed power and the power that actually, the PV system provides. This difference can be due to energetic losses introduced by the different factors present in every installation [Alonso-Abella and Chenlo 2004] such as:

- Angular and spectral effects
- Module temperature
- Low irradiance losses
- Mismatch losses
- Dirt and dust
- Ohmic loses
- Manufacturer warranted PV
- Nominal power

Determining the exact value of each type of loss for a particular system is a very complicated task, as it involves the use of complex mathematical models that do not always provide favourable results. Therefore, the values used for this purpose are the annual average values of energy losses. These annual values of losses are based on the experience acquired through the study of systems already installed and they are essentially statistical. These values can fluctuate from 11% to 45% (Table 4), which implies a wide range of variation, and this can lead to highly diverse estimations of energy depending on the values chosen.

Table 4 Typical maximal and minimum values used in the estimation of annual energy losses

	Minimum	Maximum
Temperature	5 %	15 %
Low irradiance	0,5 %	3 %
Angular and spectral losses	0,5 %	7 %
Tolerance	2 %	5 %
Mismatch	2 %	4 %
Dirty and dusk	0,5 %	4,5 %
Ohmic losses	0,5 %	1,5 %
Shading	0 %	5 %
Total	**11%**	**45 %**

On that account, it is interesting to develop models able to objectively consider most of these losses and to reduce the number of values that must be chosen in a subjective way, based on the experience of the person conducting the study, in order to bring down the uncertainty in estimating the energy produced by the PV generator.

4.2 System Description

The Univer Project [Drift et al. 2007] consists of the installation of a grid-connected photovoltaic system, with a total power of 200 kWp, in Jaen University Campus. We can find four subgenerators with a similar configuration, changing only the generator power, and three different architectural solutions: Two subgenerators which are part of the University parking covers and two subgenerators embedded in the building where the Transformation Center and the inverters are located. The system is made up by two subsystems based on the 60 kW-inverters and twenty-four subsystems with 2 kW-inverters.

Photovoltaic Systems 1&2 "Parking"

System "Parking 1" is integrated in one of the parking covers at the University Campus. It consists of a photovoltaic generator with 68 kWp nominal power and a 60 kW three-phase inverter. The photovoltaic generator consists of 640 modules of model ISOFOTON I-106, 80 modules connected in series, and 8 parallel arrays. For the integration of the photovoltaic generator, we use the existing parking covers at this University Campus, which are totally free from shadows, with a 30º southeast orientation and tilted at 7.5º. System "Parking 2" is the same as "Parking 1", and it is located in a parallel cover in the same parking area (see Fig. 14).

Fig. 14 Photograph of the 'Parking' system.

Photovoltaic System 3 "Pergola"

This PV generator is integrated in the Connection and Control Building of the project. In this building the inverters, data acquisition system and the safety and protection systems are located. The PV system consists of a photovoltaic generator with a nominal power of 20 kWp, made up by 9 subgenerators (2 kWp) and string oriented inverters. One of the aims of this integration is to get a shaded area that improves the environmental climatic conditions, which is very necessary in this part of Spain, and at the same time, to be useful as a PV application demonstration for students (see Fig. 15).

Fig. 15 Photograph of the 'Pergola' system

Photovoltaic System 4 "Façade"

This PV generator is integrated in the south façade of the building, which is located close to the Connection and Control Building. It consists of 15 subgnerators with a total of 40 kWp capacity PV polycrystalline modules and a 40 kW string oriented inverter (see Fig. 16).

Fig. 16 Photograph of the 'Façade' system

4.3 Methodology Developed for the Estimation of the Energy

In this new application, the methodology developed by our Research Group is based on an artificial neural network developed by at the University of Jaen, which allows the electrical characterisation of the PV modules [Almonacid et al. 2006, 2007, 2009, 2010]. This study uses the advantages of an artificial neural network of the Perceptron Multilayer type which does not required any knowledge of internal system, has less computational effort and offers a compact solution to obtain the V-I curves of si-crystalline modules.

The structure of the neural network (see Fig. 17) consists of three layers. The first one or ***input layer*** has two neurons or nodes. The second layer (***hidden layer***), has three nodes. And finally there is a last layer called the ***output layer***. This network has been properly trained and validated using V-I curves measured outdoor for different radiation and temperatures. The result is an artificial neural network which can obtain the V-I curve of a PV module for a pair of irradiance [G] and cell temperature [T_C] values.

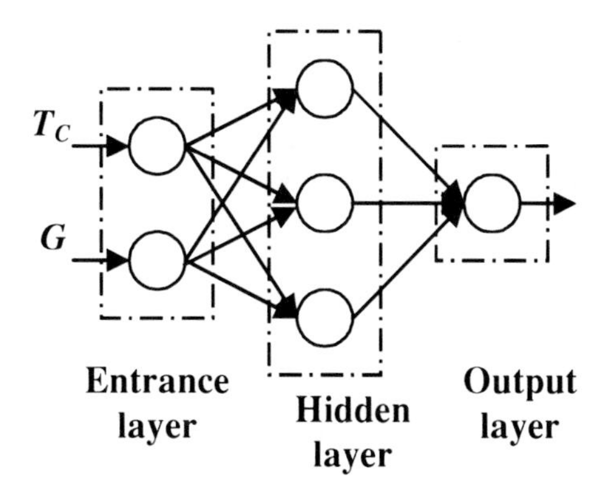

Fig. 17 Architecture of the neural network MLP for obtaining V-I curves of PV modules

The calculation procedure to estimate the annual energy produced by the PV generator can be summarized in three steps:

Step 1. - Calculation of cell temperature.

As the input values of the method are the irradiance [G] and cell temperature [T_C], and as the available measured data are irradiance and ambient temperature [T_A], the first thing to do is to calculate the cell temperature by equation (3).

Step 2. - Calculation of the output power of the generator.

For each pair of values of G and T_C, P_{DC} is calculated using the method based on artificial neural network. The result obtained in this step is the mean monthly daily power output of the generator at intervals of ten minutes for each month of the year 2005.

Step 3. - Calculation of annual energy output of the generator.

Since the data available are mean monthly daily values of power, taken at ten minute intervals, the following equation is used to calculate the daily energy from the power values:

$$E = \int_{day} P(t)dt \approx - \sum_{j=1}^{N} P_j \cdot \Delta t \tag{6}$$

Where:
E: energy in kWh,
P(t): power in KW,
P_j: power values taken at intervals of ten minutes,
N=144 (values for each day at intervals of 10 minutes) and
$\Delta t = 24/144 = 1/6$ (hours).

The annual energy is obtained as the sum of mean monthly daily energy, obtained by the expression above, multiplied by the number of days of each month.

4.4 First Results

The mean monthly daily energy (E_{mmd}) obtained by the ANN and the mean monthly daily energy measured for the three PV generators, are shown in Figs. 18, 19 and 20.

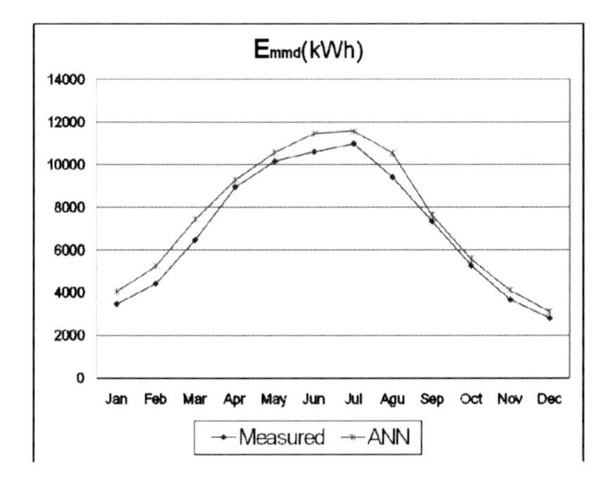

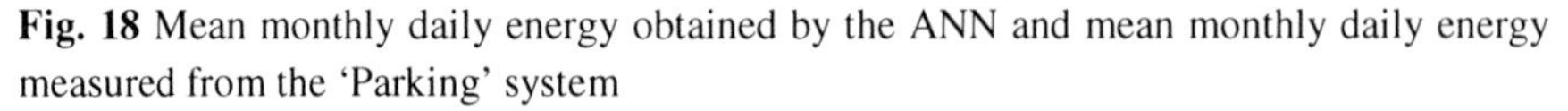

Fig. 18 Mean monthly daily energy obtained by the ANN and mean monthly daily energy measured from the 'Parking' system

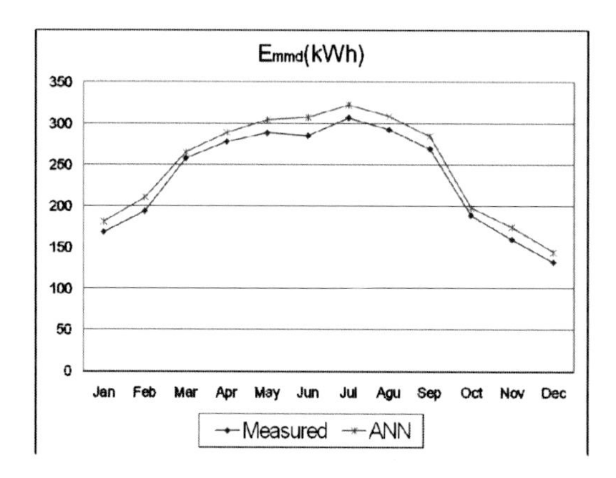

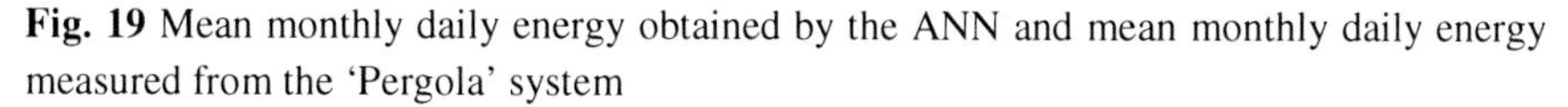

Fig. 19 Mean monthly daily energy obtained by the ANN and mean monthly daily energy measured from the 'Pergola' system

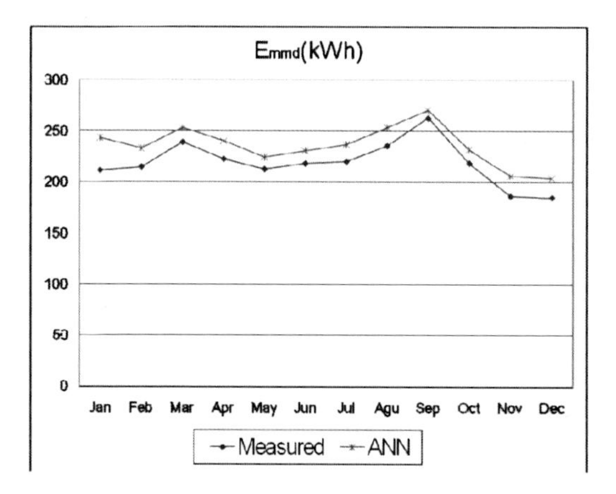

Fig. 20 Mean monthly daily energy obtained by the ANN and mean monthly daily energy measured from the 'Façade' system

The annual energy obtained by the ANN and the annual energy measured for the three PV generators, are shown in Table 5.

Table 5 Annual energy obtained by ANN and the annual energy measured for the three PV generators

	Measured (KWh)	**ANN (KWh)**
'Parking'	83459	90491
'Pergola'	2823	2994
'Façade'	2630	2830

The percentage error between the energy measured and the energy obtained by ANN for the three PV generators are shown in Table 6. It can be observed that the error obtained by ANN is very low.

Table 6 Percentage error between mean monthly daily energy obtains by the ANN and mean monthly daily energy measured for the three PV generators

	'Parking'	**'Pergola'**	**'Façade**
January	16%	8%	15%
February	19%	9%	8%
March	15%	3%	6%
April	4%	4%	8%
May	4%	6%	5%
June	8%	8%	6%
July	5%	5%	8%
August	12%	6%	8%
September	4%	6%	3%
October	6%	5%	6%
November	12%	9%	11%
December	10%	9%	10%
Annual Error	**8%**	**6%**	**8%**

The method for estimating the annual energy produced by a PV generator based on the ANN developed by the R&D Group for Solar Energy and Automatic at the University of Jaen can reduce the rate of error between 6% and 8%. The neural network is trained with the values of PV module measured at real weather conditions, so the neural network has taken into account the second order effects as module temperature, low irradiance losses, nominal power, that other traditional methods used to calculate the energy provided by PV system, ignored.

4.5 Comparative Study Against Other Methods

A comparative study is carried out in order to evaluate the effectiveness of the methods described above. The aim of this study is to compare the results of several methods for estimating annual energy produced by a PV generator. Three of them are classical methods and the forth one is based on artificial neural networks. The methods have been applied in different types of systems and different settings and types of modules.

To perform this study, each method has been used to estimate the annual energy produced by a real photovoltaic generator. The results have been compared with the value of the energy actually obtained during the corresponding year.

4.5.1 Estimation the Power Provided by a Photovoltaic Generator by Osterwald´s Method [Osterwald 1986]

Among the classic methods, Osterwald's method [Osterwald 1986] has been chosen, as this is one of the methods that provide the best results. This method is one of the simplest, and is thoroughly described in [Fuentes et al. 2007]:

$$P_m = P_{m,STC} \cdot \frac{G}{G_{STC}} \left[1 - \gamma \cdot (T_c - 25)\right] \tag{7}$$

Where:
P_m: Cell maximum power (W),
$P_{m,STC}$: Cell maximum power in STC (W),
γ:= Cell maximum power temperature coefficient (°C^{-1}).

Coefficient γ ranges from -0·005 to -0·003°C^{-1} in crystalline silicon. Although this parameter is not provided routinely by the Accredited Independent Laboratory (AIL) certificate of calibration of the module, good results are achieved assuming γ = -0·0035 °C^{-1} [Luque and Hegedus 2003].

4.5.2 Estimation of the Power Provided by a Photovoltaic Generator by Araujo-Green´s Method [Araujo et al. 1982. Green 1982]

This method uses the following eight relations sequentially to obtain the values of maximum power obtained from the operation of the cell:

1.- Cell short circuit current

$$I_{sc} = G(W/m^2) \frac{I_{sc,STC}}{1000W/m^2} \tag{8}$$

2.- Cell open circuit voltage:

$$V_{oc}(V) = V_{oc,STC}(V) - 0{,}0023\left(T_C(^{\circ}C) - 25\right) \tag{9}$$

3.- Thermal voltage:

$$V_t(V) = 0{,}025 \frac{T_C(^{\circ}C) + 273}{300} \tag{10}$$

4.- Standard cell voltage:

$$\upsilon_{OC} = \frac{V_{oc}}{V_t} \tag{11}$$

5.- Form Factor for an ideal cell without considering the series resistance

$$FF_0 = \frac{\upsilon_{OC} - \ln(\upsilon_{OC} + 0.72)}{\upsilon_{oc} + 1} \tag{12}$$

6.- Standard resistance:

$$r_s = 1 - \frac{FF_{stc}}{FF_0} \tag{13}$$

7. - Cell voltage and cell current at maximum power point

$$V_m = V_{oc} \cdot \left[1 - \frac{b}{\upsilon_{oc}} \cdot \ln a - r_s \cdot \left(1 - a^{-b}\right)\right] \tag{14}$$

$$I_m = I_{sc} \cdot \left(1 - a^{-b}\right) \tag{15}$$

where: $a = \upsilon_{oc} + 1 - 2 \cdot \upsilon_{oc} \cdot r_s$ and $b = \dfrac{a}{1+a}$

8. - Cell maximum power (P_m)

$$P_m = V_m \cdot I_m \tag{16}$$

From the calculated values for the generator cells, the following values for the generator operation are assumed:

$$I_{MAXG} = I_m \cdot N_{mp} \cdot N_{cp} \tag{17}$$

$$V_{MAXG} = V_m \cdot N_{ms} \cdot N_{cs} \tag{18}$$

$$P_{MAXG} = P_m \cdot N_{mp} \cdot N_{cp} \cdot N_{ms} \cdot N_{cs} \tag{19}$$

4.5.3 Estimation the Power Provided by a Photovoltaic Generator Using the Diode Model [Green 1982]

To calculate the operation values of a PV generator the next expression, can be used. This expression defines the V-I characteristic of the PV generator.
Where:

$$I_G = N_{mp}\, N_{cp}\, I_{sc} \left[1 - \exp\left(\frac{V_G \;/ (N_{cs}\, N_{ms}) - V_{oc} + I_G\, R_s \;/ (N_{cp}\, N_{mp})}{V_t}\right)\right] \tag{20}$$

I_G: PV generator current (A)
V_G : PV generator voltage (V)
N_{cp}: number of parallel cells in PV module
N_{cs} : number of series cells in PV module
N_{mp} : number of parallel modules in PV generator
N_{ms} : number of series modules in PV generator
I_{sc} : short circuit current of a photovoltaic cell module (A)
V_{oc} : open circuit voltage of a photovoltaic cell module (V)
R_s : series resistance of a photovoltaic cell module (Ω)
V_t : thermal voltage (V)

The objective is to obtain the maximum power of the generator for a given pair of values of irradiance and ambient temperature. For each point of the V-I curve, the product of current and voltage represent the output power for these operating conditions. The maximum power output of the cell is obtained for: $d(IV)/dV = 0$.

4.6 Results of Comparative Study

In this section the results of the comparative study are presented. We have evaluated these comparative results for the three systems, but as an example we are going to present the results for the Parking System.

In Fig. 21 the mean monthly daily energy obtained with methods presented above and mean monthly daily energy measured from the 'Parking' system, are plotted. As shown in the figure, again the result obtained by the neural network is the closest to actual system behavior. In this case, the Osterwald´s method is the method that gives the worst performance.

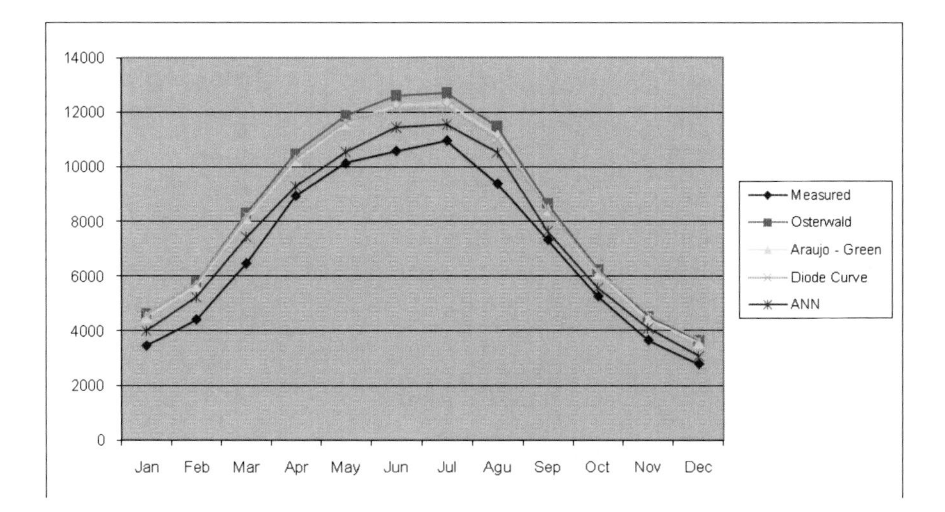

Fig. 21 Values of mean monthly daily energy obtained by different methods and the mean monthly daily energy measured of the 'Parking' system

In table 7, the monthly and annual values of the energy produced by photovoltaic generator obtained from the measured values and those calculated using the above methods, are shown. Again, both monthly and annually, the values provided by the artificial neural network are closest to the measured values.

Table 7 Monthly and annual error values obtained in the estimation of the energy produced by 'Parking' system using the various estimation methods

Energy provided by the "Pergola" System (kWh)													
	Jan	**Feb**	**Mar**	**Apr**	**May**	**Jun**	**Jul**	**Aug**	**Sep**	**Oct**	**Nov**	**Dec**	**Year**
Osterwald	191	221	289	319	335	336	353	338	313	218	184	154	3252
A-G	190	219	285	315	330	330	347	332	308	215	183	153	3206
Diode	202	232	292	320	327	321	335	320	304	223	194	164	3234
ANN	181	211	265	290	305	308	323	309	285	198	175	144	2994
Measured	***169***	***194***	***258***	***278***	***289***	***285***	***307***	***293***	***269***	***189***	***160***	***132***	***2823***

To quantify the difference between the measured values and those obtained with each method, the percentage error between the measured energy and the energy obtained by the different methods for the 'Parking' system, is calculated. The results are presented in Table 8. Both monthly and annually, the values provided by the artificial neural network are closest to the measured values.

Table 8 Monthly and annual percentage error values obtained in the estimation of the energy produced by 'Parking' system using various estimation methods

Parking	Jan	Feb	Mar	Apr	May	Jun	Jul	Aug	Sep	Oct	Nov	Dec	Year
Osterwald	34	31	28	17	17	19	16	22	18	19	24	31	21
Araujo-Green	31	29	26	14	14	16	13	19	15	16	21	28	18
Diode curve	33	31	27	16	15	15	11	17	16	18	28	32	19
ANN	16	19	15	4	4	8	5	12	4	6	12	10	8

The error in the annual prediction of energy generation for the four methods and the three systems considered are shown in Fig. 22. These results show that the method based on artificial neural networks best characterizes the actual behavior of PV systems (with different settings and types of modules). It can be observed that the error obtained by ANN is smaller in all cases.

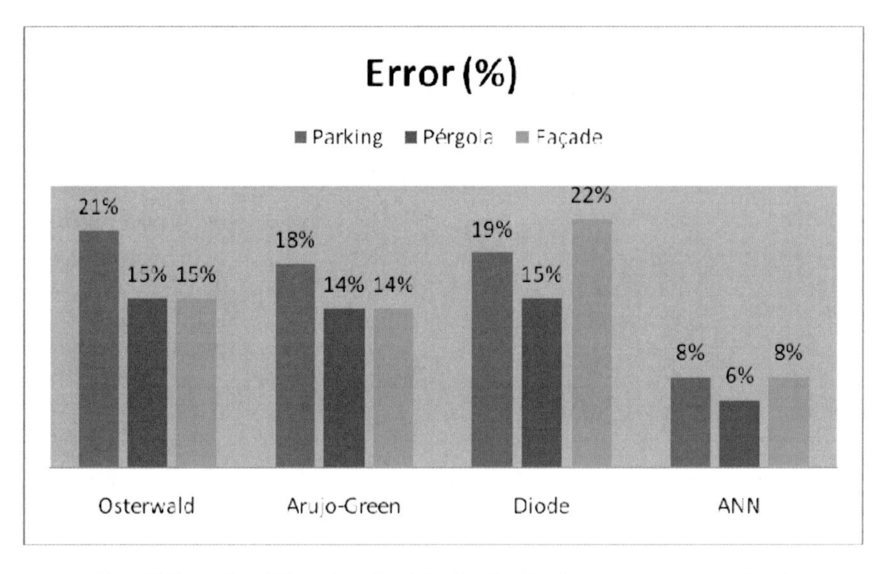

Fig. 22 Error for different methodologies for the three systems considered

This method based in ANN, besides the effect of irradiance and temperature, also takes into account some other second order effects as the different behavior of the module at low temperatures, angular and spectral effects and the difference between rated power and actual power of the module (see Table 9).

Table 9 Losses taken into account with each method

Parameter	Typical percentages of annual energy losses		Losses considered by the method	
	Minimum	Maximal	Classical Methods	ANN
Temperature	5 %	15 %	Yes	Yes
Low irradiance	0,5 %	3 %	No	Yes
Angular and spectral losses	0,5 %	7 %	No	Yes
Tolerance	2 %	5 %	No	Yes
Mismatch	2 %	4 %	No	No
Dirty and dusk	0,5 %	4,5 %	No	No
Ohmic losses	0,5 %	1,5 %	No	No
Shading	0 %	5 %	No	No
Total	**11%**	**45 %**		

5 Conclusions

In this chapter the methodology based on a MLP developed by the IDEA Research Group of the University of Jaén is presented. Initially, a review of the evolution and application of the MLP developed is presented.

The last two research projects carried out in this area and application of this methodology is also presented. For the estimation of the instantaneous PR the two main conclusions are deduced:

- The method based in ANN has been able to predict the instantaneous performance ratio of a PV generator accurately.
- This method can be used for the maintenance of a PV system and to detect its correct operation.

Additionally, for estimating of the energy produced by three PV generators the results presented demonstrate the usefulness of the MLP in this field. As seen, the energy production of a grid-connected PV system depends on various factors. There are a number of reasons that causes a decrease in the expected energy of a PV system. In this study the results of several methods for estimating the annual energy produced by a PV generator (three classical and one based on artificial neural network) in different types of systems with different settings and types of modules have been compared.

The results shown indicate that the method of estimating the annual energy produced by a PV generator based on the ANN developed by the University of Jaén can reduce the error between 6% and 8% compared to other methods currently used for such estimates. This method based on ANN, besides the effect of irradiance and temperature, takes also into account some other second order effects.

The methodology based on ANN can be used also for other applications as the characterization of other parts of the installation, and other PV generators sited in different locations of the world.

References

Aguiar, R., Collares-Pereira, M.: Statistical properties of hourly global radiation. Solar Energy 48(3), 157–167 (1991)

Aguiar, R., Collares-Pereira, M.: TAG: A time-dependent, autoregressive, Gaussian model for generating synthetic hourly radiation. Solar Energy 49(3), 167–174 (1992)

Alonso-Abella, F.C.: A model for energy production estimation of PV grid connected systems based on energetic losses and experimental data. In: On site diagnosis, 19th European Photovoltaic Solar Energy Conference, Paris, pp. 2447–2450 (June 2004)

Almonacid, F., Hontoria, L., Aguilera, J., Nofuentes, G.: Improvement in the Quality Control of PV modules Using Neural Network. In: 21st European Photovoltaic Solar Energy Conference and Exhibition, Dresden, Germany (2006)

Almonacid, L., Hontoria, L., Rus, C., Fuentes, M.: Application of method based in artificial neural networks for characterisation of thin film solar modules. In: 22nd European Photovoltaic Solar Energy Conference and Exhibition, Milan, Italy (2007)

Almonacid, F., Rus, C., Hontoria, L., Fuentes, M., Nofuentes, G.: Characterisation of Si-crystalline PV modules by artificial neural network. Renewable Energy 34, 941–944 (2009)

Almonacid, F., Rus, C., Hontoria, L., Muñoz, F.J.: Characterisation of PV CIS module by artificial neural networks. A comparative study with other methods. Renewable Energy 35, 973–980 (2010)

Araujo, G., Sánchez, E., Marti, M.: Determination of the two-exponential solar cell equation parameters from empirical data. Solar Cells 5, 377–386 (1982)

Balouktsis, A., Tsalides, P.: Stochastic simulation model of hourly total solar radiation. Solar Energy 37(2), 119–126 (1986)

Barra, L., Cataloni, S., Fontana, F., Lavorante, F.: An analytical method to determine the optimal size of a photovoltaic plant. Solar Energy 33(6), 509–514 (1984)

Bartoli, B., Cuomo, V., Fontana, F., Serio, C., Silvestrini, V.: The design of photovoltaic plants: An optimization procedure. Applied Energy 18, 37–47 (1984)

Chapman, R.N. (1987) Sizing handbook for stand alone photovoltaic/storage systems. Sandia National Laboratories (April 1987)

Drif, M., Pérez, P.J., Aguilera, J., Almonacid, G., Gómez, P., de la Casa, J., Aguilar, J.D.: Univer Project-A grid connected photovoltaic system of 200kWp at Jaé n University. Overview and performance analysis. Solar Energy Materials & Solar Cells 91, 670–683 (2007)

Egido, M.A., Lorenzo, E.: The sizing of stand alone PV systems: a review and proposed method. Solar Energy Materials and Solar Cells 26, 51–69 (1992)

Fuentes, M., Nofuentes, G., Aguilera, J., Talavera, D., Castro, M.: Application and validation of algebraic methods to predict the behaviour of crystalline silicon PV modules in Mediterranean climates. Solar Energy 81, 1396–1408 (2007)

Goh, T., Tan, K.: Stochastic modelling and forecasting of solar radiation data. Solar Energy 19(6), 755–757 (1977)

Graham, V.A., Hollands, K.G.T., Unny, T.E.: A time series model for Kt with application to global synthetic weather generation. Solar Energy 40(3), 269–279 (1988)
Graham, V.A., Hollands, K.G.T.: A method to generate synthetic hourly solar radiation globally. Solar Energy 44(6), 333–341 (1990)
Green, M.A.: Solar cells: Operating principles, Technology and System applications. Prentice-Hall, New Jersey (1982)
Hollands, K.G.T., Hughet, R.G.: A probability density function for the clearness index with applications. Solar Energy 30(3), 195–209 (1983)
Hontoria, L., Riesco, J., Zufiria, P.J., Aguilera, J.: Improved generation of hourly solar irradiation artificial series using neural networks. In: 5th International Conference on Engineering Applications of Neural Networks, EANN 1999, pp. 87–92 (1999)
Hontoria, L., Riesco, J., Zufiria, P., Aguilera, J.: Application of neural networks in the solar radiation field. Obtainment of solar radiation maps. In: 16th European Photovoltaic Solar Energy Conference and Exhibition, Glasgow, vol. 3, Paper VD1.35, pp. 2539–2542 (2000)
Hontoria, L., Aguilera, J., Riesco, J., Zufíria, P.: Recurrent neural supervised models for generating solar radiation synthetic series. Journal of Intelligent and Robotic Systems 31, 201–221 (2001)
Hontoria, L., Aguilera, J., Zufiria, P.J.: Generation of hourly irradiation synthetic series using the neural network multilayer perceptron. Solar Energy 72(5), 441–446 (2002)
Hontoria, L., Aguilera, J., Zufiria, P.J.: A tool for obtaining the LOLP curves for sizing off-grid photovoltaic systems based in neural networks. In: Proceedings of the 3rd World Conference on Photovoltaic Solar Energy Conversion on CD-ROM, Osaka, Japan (2003)
Hontoria, L., Aguilera, J., Zufiria, P.J.: A new approach for sizing stand alone photovoltaic systems based in neural networks. Solar Energy 78(2), 313–319 (2005) (Special Issue ISES Solar World Congress 2003)
Hontoria, L., Aguilera, J., Nofuentes, G., Almonacid, F., de la Casa, J.: Contribution to quality control of PV modules: a new standard test conditions (stc) V-I curve conversion method using neural networks. In: Proceedings of World Renewable Energy Congress on CD-ROM, Aberdeen, United Kingdom (2005b)
IEC Standard 61724, Photovoltaic system performance monitoring—Guidelines for measurement, data exchange and analysis (1993)
International Energy Agency Report IEA-PVPS T1-16 (2007)
Knight, K.M., Klein, S.A., Duffie, J.A.: A methodology for the synthesis of hourly weather data. Solar Energy 46(2), 109–120 (1991)
Liu, B.Y.H., Jordan, R.C.: The interrelationship and characteristics distribution of direct, diffuse and total solar radiation. Solar Energy 4(3), 1–19 (1960)
Lorenzo, E., Araujo, G.L., Cuevas, A., Egido, M.A., Miñano, J.C., Zilles, R.: Electricidad solar. Ingeniería de los sistemas fotovoltaicos. Ed. Progensa (1994)
Luque, A., Hegedus, S.: Handbook of PV science and engineering, pp. 915–939. John Wiley & Sons, Chichester (2003)
Mustacchi, C., Cena, V., Rocchi, M.: Stochastic simulation of hourly global radiation sequences. Solar Energy 23(1), 47–51 (1979)
Osterwald, C.R.: Translation of device performance measurements to reference conditions. Solar Cells 18, 269–279 (1986)
Pérez, et al.: Experiences in real-time telemonitoring using internet. In: 20th European Photovoltaic Solar Energy Conference and Exhibition, Paris. France (2004)

Priestley, M.B.: Non-Linear And Non-Stationary Time Series Analysis. Academic Press, London (1988)

Rumelhart, D., MacClelland, J.L.: Learning internal representations by error backpropagation. In: Parallel Distributed Processing. Foundations, vol. 1, ch. 8, The MIT Press, Cambridge (1986)

Rus, C., Almonacid, F., Pérez Higueras, P.J., Hontoria, L., Muñoz, F.J.: Artificial neural networks methodology for maintenance of grid connected photovoltaic systems. In: 24nd European Photovoltaic Solar Energy Conference and Exhibition, Hamburg, Germany (2009)

Sidrach-de-Cardona, M., López, M.L.: A simple model for sizing stand alone photovoltaic systems. Solar Energy Materials and Solar Cells 55(3), 199–214 (1998)

Sidrach-de-Cardona, M., López, M.L.: A general multivariate qualitative model for sizing stand alone photovoltaic systems. Solar Energy Materials and Solar Cells 59(3), 185–197 (1999)

Weigend, A.S., Rumelhart, D.E., Huberman, B.A.: Back-propagation, weight-elimination and time series prediction. In: Proceedings of the 1990 Connectionist models Summer School on CD-ROM, Morgan Kaufmann, San Francisco (1990)

Werbos, P.: Beyond Regression: New Tools for Predicting and Analysis in the Behavioural Sciences, Ph. D. thesis, Harvard University (1974)

Zufiria, P.J., Vázquez, A., Riesco, J., Aguilera, J., Hontoria, L.: A neural network approach for generating solar irradiation artificial series. In: Sánchez-Andrés, M.J.V. (ed.) IWANN 1999. LNCS, vol. 1607, pp. II-874–II-883. Springer, Heidelberg (1999)

Optimization of Fuzzy Logic Controller Design for Maximum Power Point Tracking in Photovoltaic Systems

Lawrence K. Letting[1], Josiah L. Munda[1], and Yskandar Hamam[1,2]

[1] Tshwane University of Technology, Pretoria, South Africa
[2] ESIEE-Paris Paris-Est University, LISV, UVSQ, France
{LettingLK,MundaJL,HamamA}@tut.ac.za

Abstract. This chapter presents the design and optimization of a fuzzy logic controller (FLC) with a minimum rule base for maximum power point tracking in photovoltaic (PV) systems. A strategy for automated design and optimization of the FLC using genetic algorithms is proposed. An optimal Takagi-Sugeno FLC with a rule base of only 9-rules is realized and compared to the conventional design of 49 or 25 rules. Two FLCs, one using Gaussian input membership functions (MFs) and the other using trapezoidal MFs are designed and their performance compared. Expert knowledge for tuning the FLC is extracted from a PV module model under varying solar radiation, temperature, and load conditions. The proposed method is implemented using C language as a dynamic linked library (*.dll* format) and simulated using LabVIEW. Simulation results are used to compare the performance of the optimized FLCs in terms of speed, accuracy, and robustness. It is shown that the optimization algorithm produces an optimal FLC for both Gaussian and trapezoidal MFs.

1 Introduction

Photovoltaic (PV) power generation is a reliable and economical source of electricity in rural areas, especially in developing countries where the population has low incomes and the grid power supply is not fully extended due to viability and financial constraints. The efficiency of PV modules depends on the material used in solar cells and the technology used in arranging the solar cells to form a module. Currently, PV modules have very low efficiencies with only about $12-29\%$ efficiency in their ability to convert sunlight to electrical power [Ocran 2005]. The efficiency can drop further due to other factors such as PV module temperature and load conditions. In order to maximize the power derived from the PV module it is important to operate the module at its optimal power point. To achieve this, a maximum power point tracking (MPPT) controller is required.

Many maximum power point (MPP) tracking strategies have been proposed such as perturb and observe and incremental conductance. Recently artificial

K. Gopalakrishnan et al. (Eds.): Soft Comput. in Green & Renew. Ener. Sys., STUDFUZZ 269, pp. 233–260.
springerlink.com

intelligence based methods using genetic algorithms, neural networks, and fuzzy logic have been introduced in order to improve on the tracking efficiency. Fuzzy logic is appropriate for nonlinear control because it does not use complex mathematical equations. The behaviour of a FLC depends on the shape of membership functions and the rule base. However, there is no formal method to determine accurately the fuzzy parameters to yield optimum operating point. The conventional design of a fuzzy controller and its performance therefore depends on the experience of the designer. This chapter proposes an automated method for choosing the FLC parameters using genetic algorithms. It will be shown that it is possible to design and optimize a minimum rule base of 9 rules and attain good transient and steady state performance.

1.1 An Overview on MPPT Algorithms

MPPT algorithms can be generally categorized into three groups: 1) perturbation and observation methods [Hua 1998, Koutroulis 2001, Enslin 1992]; 2) incremental conductance methods [Bodur 1992, Sullivan 1993]; 3) artificial intelligence based methods [Kohata 2009, Larbes 2009, Chen 2002, Veerachary 2002]. An overview of each method is presented next.

1.1.1 Perturb and Observe

The perturbation and observation (P&O) method, also known as the hill-climbing method, is popular because of its ease of implementation. This method tracks the maximum power point (MPP) by repeatedly increasing or decreasing (perturbing) the module voltage and comparing the output power with that at the previous perturbing cycle. Various problems occur in this method when acquiring the maximum power. It cannot track the MPP during low solar radiation levels and when radiation changes rapidly. It also oscillates around MPP instead of directly tracking it [Ocran 2005, Hohm 2003]. As oscillations always appear in the method, the power loss may be increased. Several improvements of the P&O algorithm have been proposed. One approach involves the use of the short-circuit current or the open-circuit voltage to determine the direction in which to perturb the module voltage. Methods based on this approach can be considered as variations of the standard perturb and observe algorithm since instead of observing the change in PV module power, change in either module short-circuit current or open-circuit voltage is used. The "short circuit current method" [Noguchi 2002] performs MPPT control using the PV module short circuit current as a control input. Although this method does not have oscillations like those appearing in the standard P&O method, the power loss may increase since the short circuit current flows whenever MPPT control is performed. Furthermore, it becomes difficult to perform MPPT control during periods of low solar radiation because short-circuit current decreases with solar radiation. The "open circuit voltage method" [Enslin 1997] utilizes the fact that the operating voltage is almost linearly proportional to open-circuit voltage of the PV module at MPP. It is simple, cost-effective, and avoids power loss associated with the short-circuit current method. A limitation of this method is the fact that the reference voltage does not change between samplings [Ocran 2005].

1.1.2 Incremental Conductance Method

The incremental conductance algorithm is a technique used to reduce the oscillation around the MPP. This method calculates the direction in which to perturb the module's operating point and it can determine when it has actually reached the MPP [Hohm 2003]. It is however, computationally intensive and the speed at which it approaches the MPP depends on a fixed perturbation step. The perturbation step is difficult to choose when dealing with trade-off between steady state performance and fast dynamic response. The control circuit is also complex resulting in a higher system cost [Koutroulis 2001].

1.1.3 Artificial Intelligence Based Methods

Artificial intelligence based methods using genetic algorithms, neural networks, and fuzzy logic have been introduced in order to improve on the tracking efficiency. With the neural network based method, the solar radiation, temperature, module voltage and current are measured and used to identify the maximum power point of the PV module [Kohata 2009, Ocran 2005, Baghat 2004]. Although this method can predict the maximum power point, the data acquisition and memory space requirements are very intensive and greatly affects the performance of the algorithm. Fuzzy logic is appropriate for nonlinear control because it does not use complex mathematical equations. The behaviour of a FLC depends on the shape of membership functions, input and output scale factors, and size of the rule base. However, there is no formal method to determine accurately the fuzzy parameters to yield optimum operating point and a good control system depends on the experience of the designer.

1.2 Principle of Maximum Power Point Tracking

The power produced from a PV module depends on the operating voltage of the load to which it is connected, solar radiation level, and cell temperature. This is illustrated in Fig. 1 and Fig. 2 using BP solar SX 75TU PV module. The electrical characteristics for this module are given in Table 1 [BP solar 2002].

If a variable load resistance R, is connected across the module's terminals, the operating point is determined by the intersection of module I-V curve and the load I-V characteristic. Fig. 3 illustrates the operating characteristic of a PV module. It consists of two regions: Zone I is the current source region, and Zone II is the voltage source region. In Zone I, the internal impedance of the module is high, while in Zone II the internal impedance is low. The maximum power point P_{mp}, is located at the knee of the power curve. An increase in solar radiation at constant temperature causes a decrease in internal impedance as it causes an increase in short-circuit current. An increase in temperature at constant solar radiation causes a decrease in internal impedance since it causes a decrease in open circuit voltage.

The power delivered to the load is maximum when the source internal impedance matches the load impedance. The load characteristic is a straight line with a slope of $1/R$. If R is small, the module operates in the region AB only and behaves like a constant current source at a value close to the short-circuit current, I_{sc}. If R is large, the module operates in the region CD behaving like a constant voltage source, at a value almost equal to the open-circuit voltage, V_{oc}. Maximum power point tracking is therefore based on load line adjustment under varying atmospheric and load conditions by searching for optimal equivalent output impedance. A dc-dc converter is used to perform load-line adjustment by varying the converter duty cycle using a controller. The converter can be buck, boost, or buck-boost depending on the application.

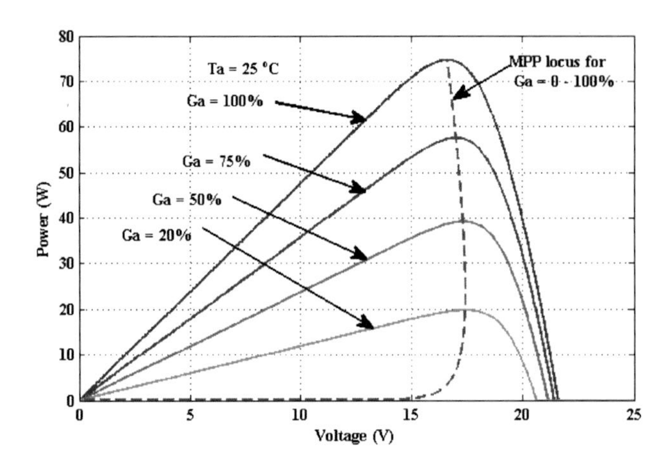

Fig. 1 Effects of ambient solar radiation for constant temperature

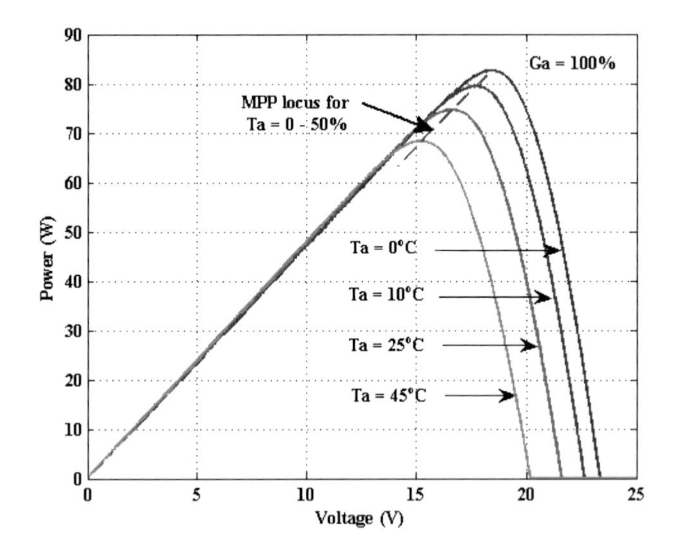

Fig. 2 Effects of ambient temperature for constant solar radiation

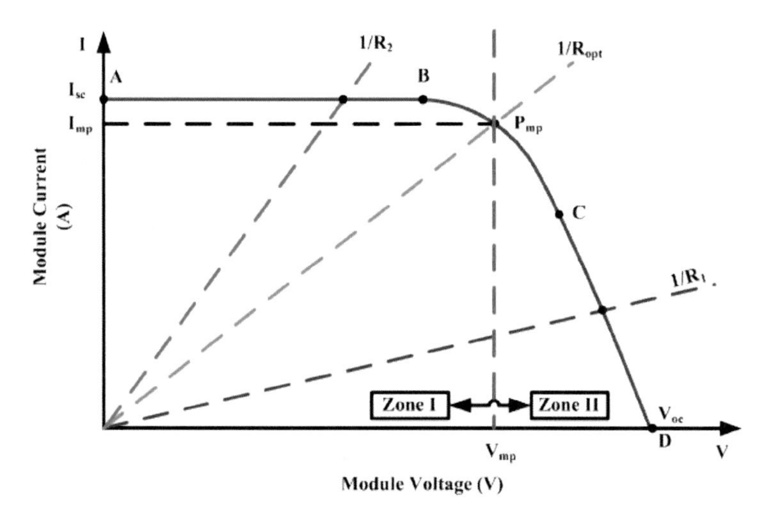

Fig. 3 Tracking the maximum power point by varying load resistance

1.3 Fundamentals of Fuzzy Logic Controllers

Fuzzy Logic is a branch of Artificial Intelligence. It owes its origin to Lofti Zadeh, a professor at the University of California, Berkley, who developed fuzzy set theory in 1965 [Bose 2001]. The basic concept underlying fuzzy logic is that of a linguistic variable, that is, a variable whose values are words rather than numbers (such as *small* and *large*). Fuzzy logic uses fuzzy sets to relate classes of objects with unclearly defined boundaries in which membership is a matter of degree.

Table 1 Characteristics of BP SX 75TU PV module

BP SX 75TU Photovoltaic Module	
Type: Silicon Multicrystalline	
Number of Cells in series	36
Number of Cells in parallel	1
Maximum Power (P_{max})	75 W
Voltage at P_{max} (V_{mp})	17.3 V
Current at P_{max} (I_{mp})	4.35 A
Short-circuit current (I_{sc})	4.75 A
Open-circuit voltage (V_{oc})	21.8 V
Temperature co-efficient of I_{sc}	$(0.065 \pm 0.015)\% \,/\, {}^{o}C$
Temperature co-efficient of voltage	$-(80 \pm 10) mV \,/\, {}^{o}C$
Nominal Cell Operating temperature (NOCT)	$47 \pm 2\,{}^{o}C$

1.3.1 Fuzzy Controller Structure

The general structure of a fuzzy logic controller is presented in Fig. 4. It comprises of four principal components; fuzzification, knowledge base, inference engine, and defuzzification. The fuzzification interface converts input data into suitable linguistic values using a membership function while the knowledge base consists of a database with the necessary linguistic definitions and the control rule set. The inference engine deduces the fuzzy control action using the knowledge of the control rules and the linguistic variable definitions. The last stage is the defuzzification interface which converts an inferred output into a non-fuzzy control action.

1.3.2 Need for Fuzzy Logic in MPPT Control

MPPT controller design is an intriguing subject due to the nonlinearity of dc-dc converters and PV modules. This is because an accurate model of the plant and the controller is necessary while formulating the control algorithm. The nonlinear behaviour of dc-dc converters is caused by the switching device. Depending on the state of the switch (ON/OFF) the plant structure exhibits very different functioning modes, resulting in a severe nonlinearity. PV modules also have nonlinear current-voltage (I-V) characteristics that are dependent on solar radiation, temperature, and degradation due to environmental effects. Therefore, their operating point that corresponds to the maximum output power varies with the environmental and load conditions. Fuzzy logic offers a design approach that avoids precise mathematical modelling of the plant and the controller. However it leads to the new problem of determining optimal FLC parameters.

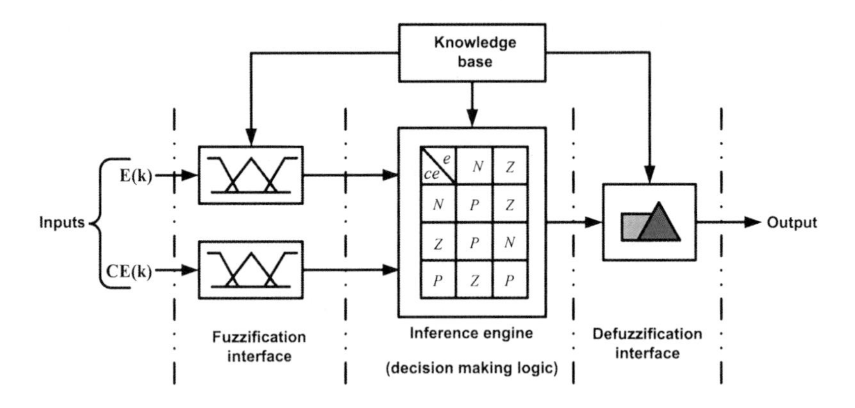

Fig. 4 Structure of a fuzzy logic controller

1.4 Automation of Fuzzy Logic Controller Design

In the recent past the selection of fuzzy membership functions and size of the rule base has been automated using genetic algorithms (GAs) and particle swarm optimization. Larbes et al [2009] presented optimization of a 25-rule Mamdani based

FLC for MPPT using genetic algorithms. They used a combination of trapezoidal and triangular membership functions. Otieno et al [2009] presents a fuzzy controller with 21 rules tuned using adaptive neural fuzzy inference system (ANFIS). Recent publications have also presented tuning of the FLC for MPPT using swarm intelligence. In Khaehintung [2010] , a 25 rule base FLC for MPPT with bifurcation control is tuned using particle swarm optimization.

Automated tuning is crucial because a fuzzy controller consists of a relatively large number of parameters. It is noted in the refereed works that optimization is done by first selecting a fixed rule base size and type of membership functions. The FLC parameters are then optimized using the identified method. This study shows that it is possible to optimize a rule base of 9 rules and meet the desired performance using genetic algorithms.

2 Modelling of the MPPT Controller

This section presents modelling of the MPPT controller for formulating and testing the performance of the formulated fuzzy controllers. Modelling of the PV module, buck-boost converter, and the complete system implementation in LabVIEW is presented. The models are implemented in C language and compiled as dynamic linked library (dll) for compatibility with the LabVIEW user interface.

2.1 Modelling of the PV Module

The PV module was modelled using equations in *Hybrid2* theory manual [Manwell 2006]. *Hybrid2* is a computer simulation model for hybrid power systems developed by the University of Massachusetts. A PV module is composed of individual solar cells connected in series - parallel and mounted on a single panel. The goal is to calculate the power output from a PV module based on an analytical model that defines the current-voltage (I-V) relationship based on the electrical characteristics of the module. The one diode solar cell model of Fig. 5 forms the basic circuit used to establish the I-V curve. The diode current I_T and the current through the shunt resistance I_{sh} are given by Equations (1) and (2) respectively; where, m is the idealizing factor, k is Boltzmann's gas constant, T_c is the absolute cell temperature, q is the electronic charge, V is the voltage imposed across the cell, and I_o is the cell reverse saturation current. The module current I_m, under arbitrary operating conditions is given by Equation (3). $I_{G,m}$ is the module's light generated current, $I_{o,m}$ is the module reverse saturation current, V_m is the module voltage, and $R_{s,m}$ is the module series resistance, and A is the curve fitting parameter.

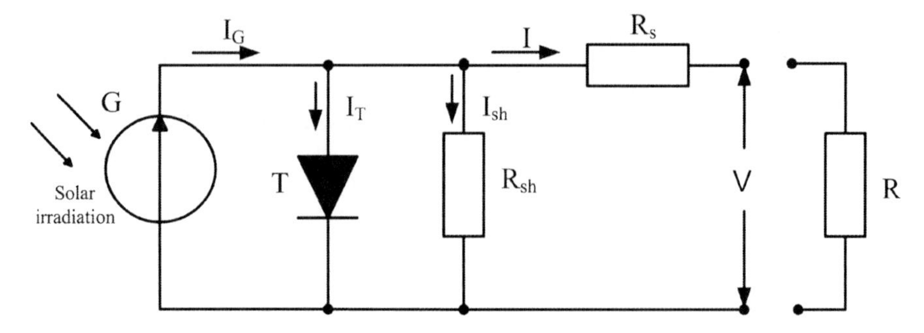

Fig. 5 Equivalent circuit of a solar cell

$$I_T = I_o \left\{ \exp\left[\frac{q}{mkT_c} (V + IR_s) \right] - 1 \right\} \tag{1}$$

$$I_{sh} = \frac{V + IR_s}{R_{sh}} \tag{2}$$

$$I_m = I_{G,m} - I_{o,m} \left\{ \exp\left[\frac{(V_m + I_m R_{s,m})}{A} \right] - 1 \right\} \tag{3}$$

The model was implemented as a C function whose inputs are the ambient solar radiation G_a, ambient temperature T_a, and the load resistance R. The model outputs are: module operating voltage V_m, output current I_m, voltage at maximum power V_{mp}, and the maximum power P_{mp}. The PV module model is obtained from the solar cell model using manufacturer supplied data. The model was validated using manufacturer supplied data for BP solar SX 75TU PV module. The LabVIEW block diagram of the model is shown in Fig. 6.

2.2 Converter Modelling

A buck-boost converter was chosen for the MPPT because of its ability to perform maximum power tracking in both zones I and II of Fig. 3. The circuit of a buck-boost converter is shown in Fig. 7. It consists of four basic components; transistor Q, diode T, inductor L, and capacitor C. The inductor is modelled as an ideal inductor in series with a resistance R_L. The capacitor is modelled as an ideal capacitor in series with a resistance R_C. R_L and R_C are used to model the power losses in the inductor and capacitor respectively. The transistor has an on-state resistance, R_t while the diode has a forward voltage drop V_T. A state space model of the converter was formulated using the principle of state-space averaging [Erickson 2001]. The obtained model was implemented as C language code. The converter transforms the load resistance using the controller generated duty ratio into an equivalent input resistance for the PV module.

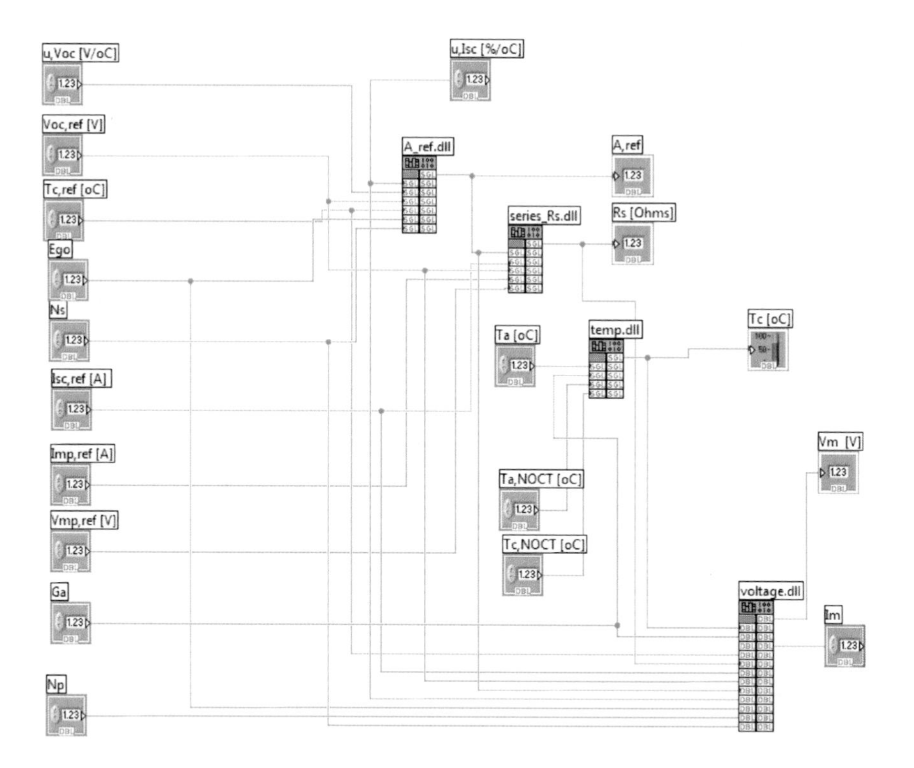

Fig. 6 LabVIEW block diagram of the PV module

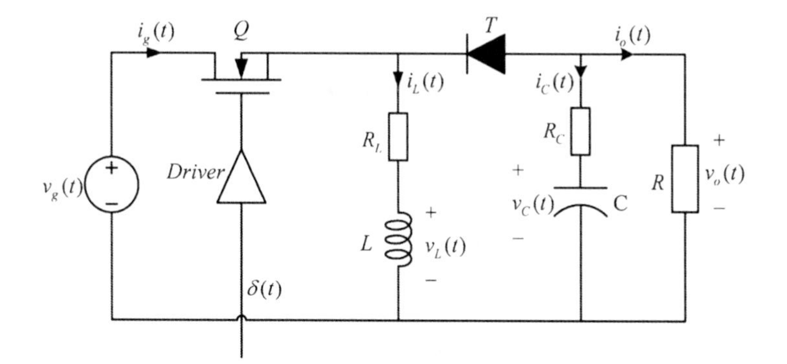

Fig. 7 Circuit of a buck-boost converter

2.3 Complete MPPT Model

The complete LabVIEW block diagram model of the maximum power point tracker was implemented as shown in Fig. 8. The PV module model of Fig. 6 was modified into a single function *PVmodule.dll* that outputs the module current I_m, voltage V_m, and the possible maximum power P_{mp} for given solar radiation, temperature, and load.

3 Fuzzy Logic Controller Design

The structure of the FLC and formulation of the optimization criteria is presented in this section. Two FLCs are designed and optimized. One fuzzy controller (GFLC) uses Gaussian input membership functions while the other controller (TFLC) uses trapezoidal membership functions.

3.1 Input Variables

There are two input variables, error $E(k)$, and change of error $CE(k)$ at the k_{th} sampling instant as defined in Equations (4) and (5). $P_m(k)$ is the instantaneous power of the PV module and $E(k)$ is the gradient of the P-V curve of Fig. 1. The sign of $E(k)$ gives the operating mode. When $E(k) > 0$ the system is moving towards the MPP ; at $E(k) = 0$ the system is operating at the MPP; and for $E(k) < 0$ the system is moving away from the MPP.

$$E(k) = \frac{P_m(k) - P_m(k-1)}{V_m(k) - V_m(k-1)} \tag{4}$$

$$CE(k) = E(k) - E(k-1) \tag{5}$$

3.2 Membership Functions

The computational effort, simulation time, and quality of the results need to be considered when choosing the variables to be optimized. In this study, optimization is considered using Gaussian and trapezoidal membership functions. For each input the number and type of MF is fixed. The only variable is the area covered by each MF. The area of the MF is optimized by varying the defining points shown in Fig. 9 and Fig 10 for the Gaussian and trapezoidal MFs respectively.

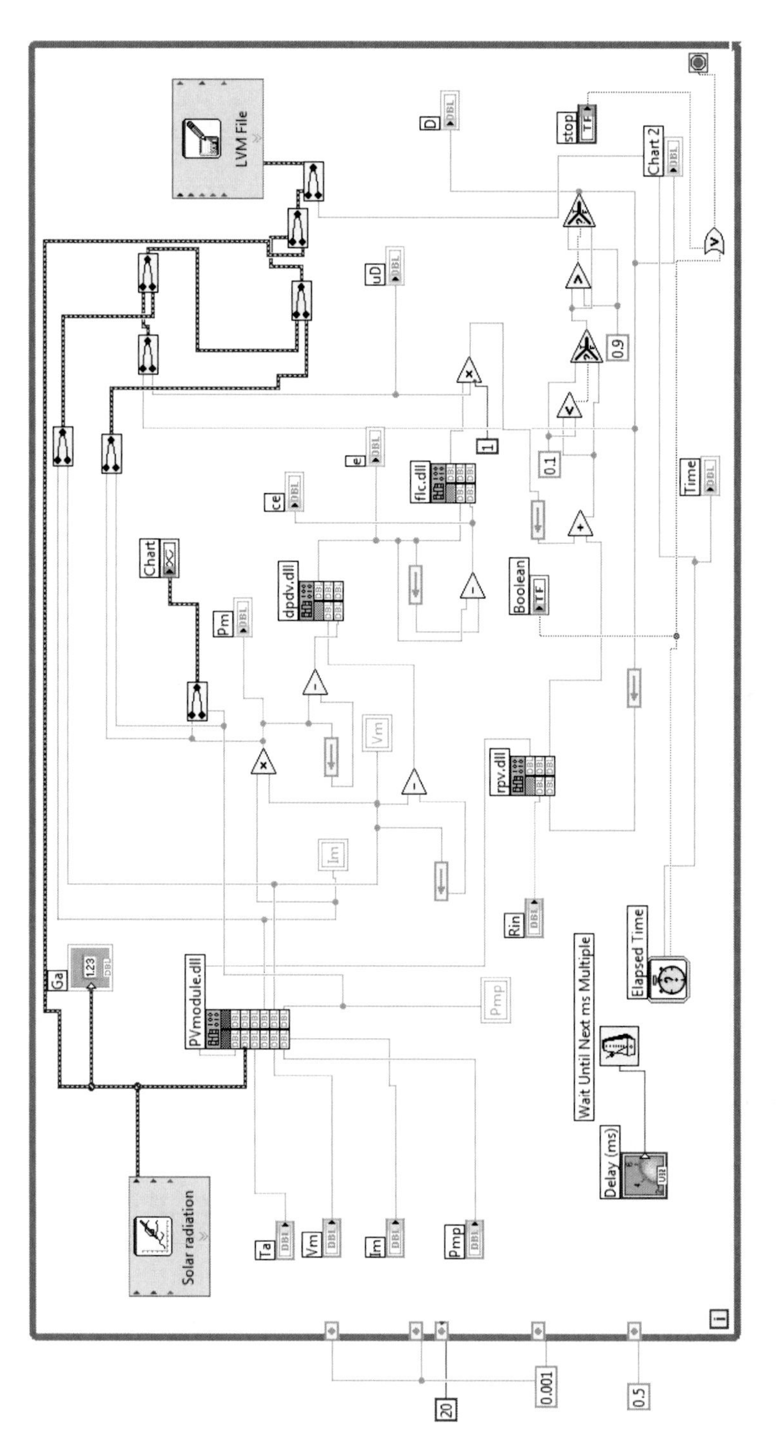

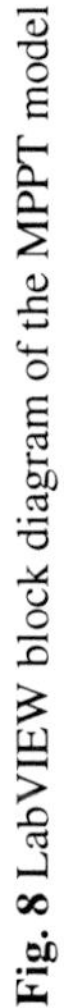
Fig. 8 LabVIEW block diagram of the MPPT model

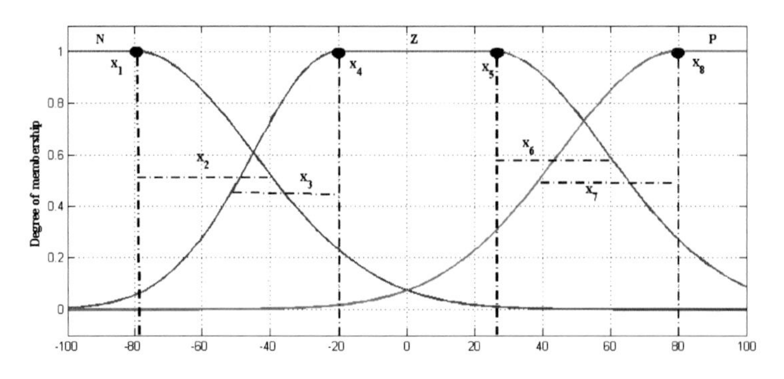

Fig. 9 Encoding input membership using *gauss2mf* function

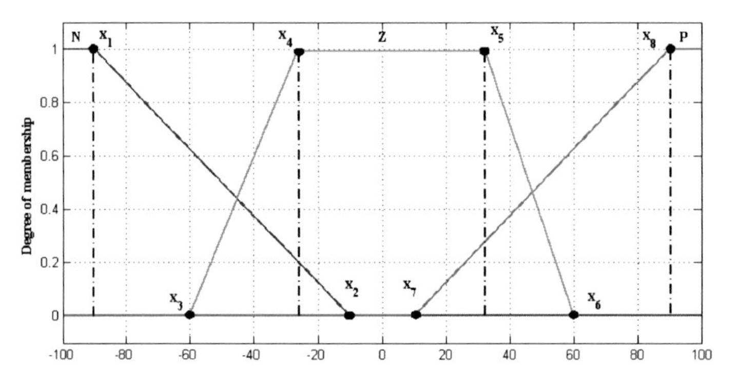

Fig. 10 Encoding of membership using *trapmf* function

3.3 Simulation Software

The fuzzy control algorithm was developed in C language to avoid limitations in the Matlab fuzzy logic toolbox. The toolbox does not allow the area of a membership function to be reduced to zero during online tuning. It was observed that during optimization it is computationally efficient to allow the area of an MF to vary rather than making a whole MF active or inactive. A large rule base is required when rules have to be weighted and this translates to more memory requirements and the system takes long to converge. The Takagi-Sugeno inference system was chosen as it is more compact and has a computationally efficient representation than a Mamdani system [Bose 2002].

3.4 Encoding of Optimization Parameters

The input MFs for $E(k)$ and $CE(k)$ are defined using *gauss2mf* and *trapmf* functions defined in Matlab. The MFs are as shown in Fig. 9 and Fig. 10. A gauss2mf function consists of two Gaussian functions and hence four parameters need to be identified. For example, to encode the *Zero* MF the parameters are: *sigma 1* (x_3), *centre 1* (x_4), *centre 2* (x_5) and *sigma 2* (x_6). A trapmf function is defined using the same number of parameters. The FLC has only nine rules as shown in Table 2 and the output of each rule is the change in converter duty cycle given by Equation (6). The constants *a*, *b*, and *c* are to be determined for each output MF in the rule base of Table 2.

$$\mu D(k) = aE(k) + bCE(k) + c \tag{6}$$

The optimization parameters are encoded as a vector S given by Equation (7) where sub-vectors X_i and Y_i each consist of 8 parameters that correspond to the input MFs for *E(k)* and *CE(k)* respectively. Z_i consists of linear Sugeno output MF parameters of Equation (6). The rule base has 27 parameters to be optimized, and hence a complete fuzzy logic controller has 43 optimization parameters. The encoding of one FLC is therefore given by (7).

$$S_i = [X_i, Y_i, Z_i] \tag{7}$$

where,

$$X_i = [x_1, x_2, \ldots, x_8] \tag{8}$$

$$Y_i = [y_1, y_2, \ldots, y_8] \tag{9}$$

$$Z_i = [z_1, z_2, \ldots, z_{27}] \tag{10}$$

Table 2 FLC rule base

E\CE	*N*	*Z*	*P*
N	*MF1*	*MF2*	*MF3*
Z	*MF4*	*MF5*	*MF6*
P	*MF7*	*MF8*	*MF9*

3.5 Optimization Criterion

The mean-square-error defined in Equation (11) is used as the fitness function. P_m is the attained PV module power, P_{mp} is the maximum power, and *N* is the number of iterations.

$$J = \frac{1}{N}\sum_{k=1}^{N}\left(P_m(k) - P_{mp}(k)\right)^2 \tag{11}$$

The number of iterations *N* consists of an outer loop and inner loop that run for N_1 and N_2 iterations respectively. Solar radiation, temperature, and load resistance are held constant in the inner loop but allowed to vary in the outer loop. The inner loop determines the steady state performance while the outer loop determines the transient performance of the controller during optimization. This is illustrated in the flowchart of Fig. 11.

4 Optimization of the FLC Using Genetic Algorithms

Genetic algorithms (GAs) and their application to optimization problems using the principle of natural evolution was developed by Holland in 1975 [Sivanandam and Deepa 2008]. GAs consist of three basic steps: selection, crossover (mating), and mutation. The parameters that specify the potential solutions of the optimization problem are encoded as population of chromosomes. The boundaries of the optimization parameters define the space of potential solutions also known as the search space. A fitness function is used to evaluate the quality of each potential solution. A comprehensive overview and introduction to GAs is presented in [Sivanandam and Deepa 2008, Haupt and Haupt 2004].

4.1 Generation of the Population

In this design, the population consists of fuzzy logic controllers. Each FLC is modelled as a single chromosome with 43 genes where each gene represents an optimization parameter. The initial population is randomly generated using the parameters of Table 3. At the end of each iteration the cost of each chromosome is evaluated and ranking is done. 50% of the individuals with the least cost are selected to form the next population. The remaining half is reproduced through mating of the selected individuals. The parents for mating are selected using rank-weighting and the offspring is generated using single-point crossover [Haupt 2004]. Finally, random mutations are carried out on the population in order to ensure that the entire cost surface is explored. A mutation rate of 20% is applied to the population except the best chromosome.

Table 3 Initialization of GA Parameters

Parameter	**Value**
Population (N)	30
No. of iterations (I_{max})	50
No. of bits	8
Selection rate	0.5
Mutation rate	0.2

4.2 Simulation Steps

At the start of the simulation, the initial population of chromosomes is generated in binary format. Each chromosome (S_i) is then decoded into a fuzzy inference structure (FIS) and the results are passed to the MPPT controller. The controller evaluates the fitness of the FIS using the fitness function of Equation (11). The genetic algorithm then generates the next population using the crossover and mutation rates of Table 3. The new population is again evaluated for fitness and used to generate the population for the next iteration. The procedure is repeated for the set number of iterations I_{max} as illustrated in the flowchart of Fig. 11.

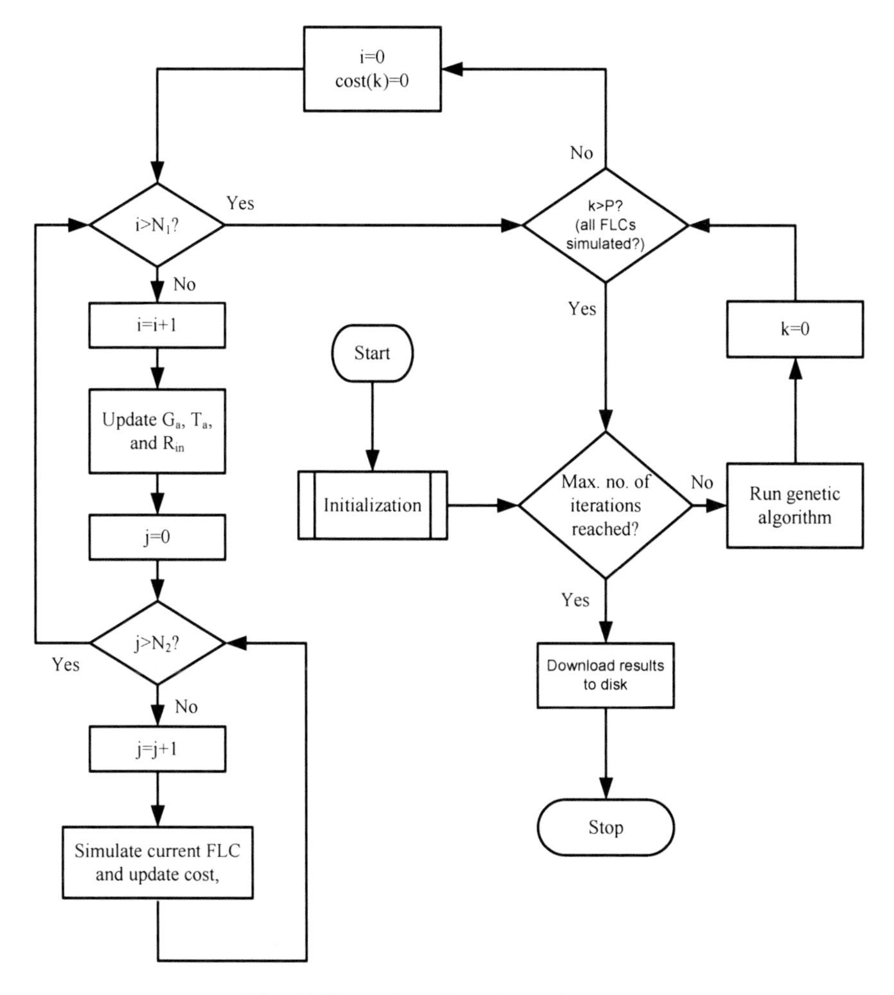

Fig. 11 Illustration of the simulation steps

4.3 Optimal FLC Structure

The system converges to an optimal solution which is obtained from the best chromosome at the end of the simulation. The input MFs of the best solution for the FLC structure using Gaussian functions (GFLC) is presented in Fig. 12 and Fig. 13. The optimal structure of the FLC using trapezoidal MF functions (TFLC) is as shown in Fig. 14 and Fig. 15. The defining points of the MFs shown in Fig. 9 and Fig. 10 were constrained during optimization in order to ensure uniform partitioning. The optimized rule surface for the GFLC and the TFLC are shown in Fig. 16 and Fig. 17 respectively. The contour maps are shown at the bottom of each surface plot.

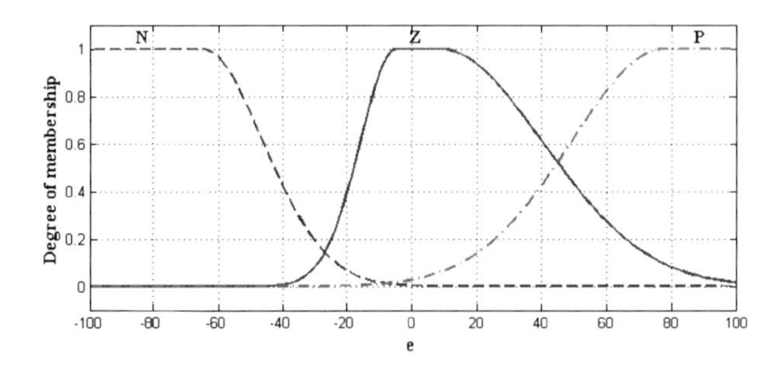

Fig. 12 Gaussian MF for error (E) after optimization

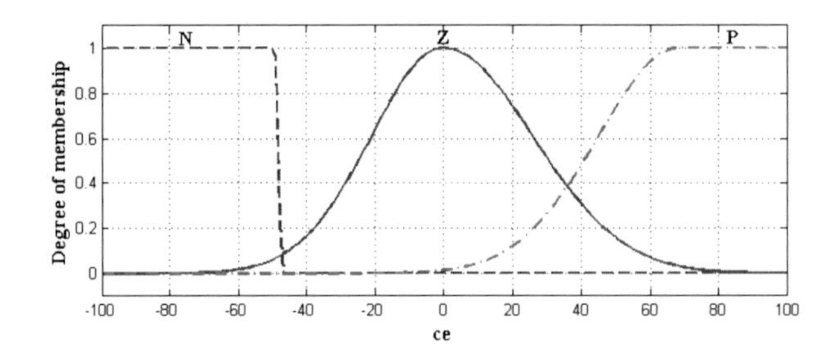

Fig. 13 Gaussian MF for change of error (CE) after optimization

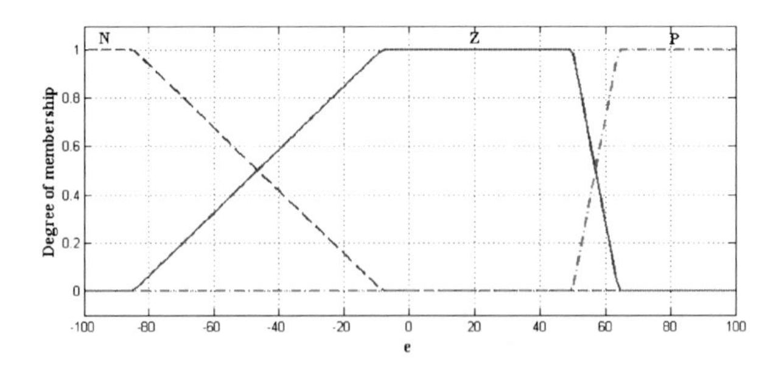

Fig. 14 Trapezoidal MF for error (E) after optimization

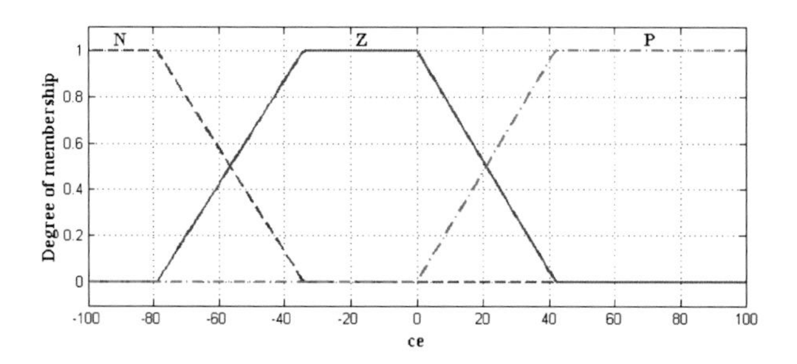

Fig. 15 Trapezoidal MF for change of error (CE) after optimization

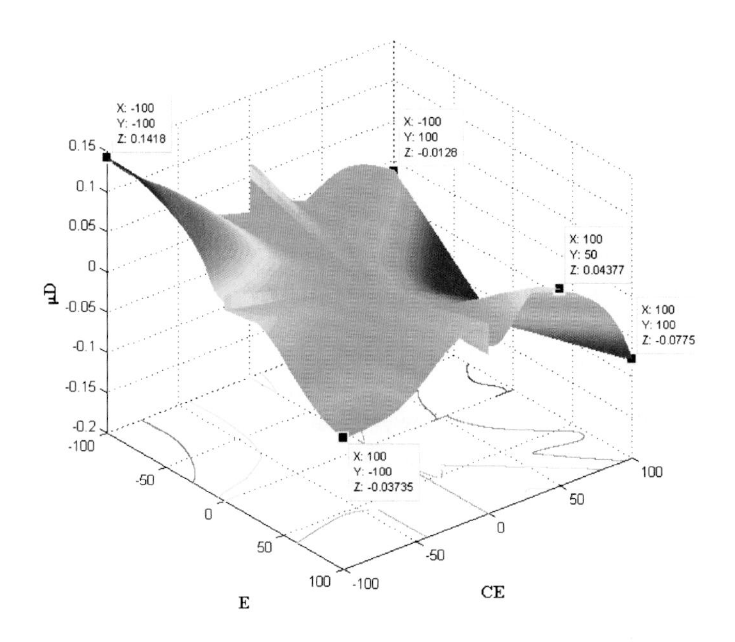

Fig. 16 Rule surface of FLC using *gauss2mf* (GFLC) after optimization

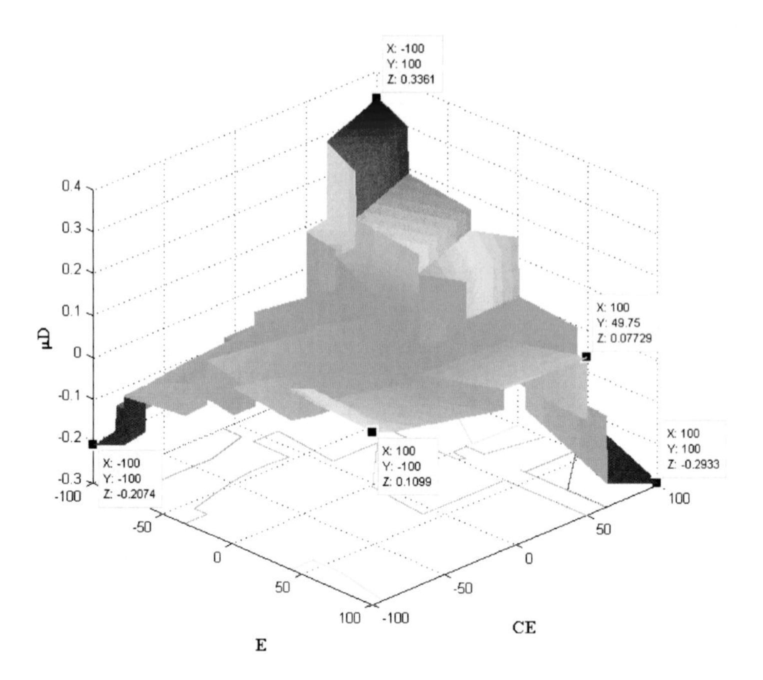

Fig. 17 Rule surface of FLC using *trapmf* (TFLC) after optimization

5 Simulation Results and Discussion

The optimal membership functions of both the GFLC and the TFLC in Fig. 12 to 15 show that the Zero MF for error has a relatively larger area on the positive side compared to the negative side. However, the Zero MF for change of error is skewed to negative side in the TFLC. The zero centered MFs correspond to normal controller operation and determine the accuracy in tracking and maintaining operation at the maximum power point. These MFs are therefore expected to experience a high firing frequency when the controller is running. This is due to the fact that the Zero MFs contribute to the firing of 5 out of 9 rules in the FLC rule base shown in Table 2. A positive error indicates movement towards the MPP while a negative change in error indicates that the previous change in control effort resulted in decreased change in PV module power.

It is observed that the rule surfaces of both the GFLC and the TFLC in Fig. 16 and 17 depict small changes in control effort when the error (E) is positive. This is expected because $E > 0$ means that the controller is operating below the MPP, and $CE > 0$ shows the previous perturbation in control effort resulted in a net increase in module power. The variation of E and CE during maximum power point tracking is illustrated in Fig. 18. Fig. 16 shows that when E is close to 100 and CE is increasing from 0 to 50, the GFLC slowly increases the duty cycle but starts to decrease it smoothly when both E and CE are approaching the maximum value. This is a necessary feature because successive increase in CE indicates fast increase in solar radiation. However for the TFLC surface in Fig. 17, the region corresponding to positive E and CE is flat with a constant duty cycle with sudden decrease in duty cycle when $CE > 50$. It can therefore be predicted that the TFLC will experience overshoots in duty cycle variation during fast change in solar radiation. This is confirmed by the simulation plots of Fig. 19 to 22. The control signal for the GFLC shows a smooth control signal curve while the TFLC presents overshoots.

The rule surface plots also indicate that the change in duty cycle is smaller when the change in error is positive compared with when it is negative. A successive negative increase in CE indicates that the tracking is fast moving away from MPP. It can also indicate a fast decrease in solar radiation. When this scenario occurs, it indicates that the controller is operating in Zone II of Fig. 1 where the gradient is very steep. The performance of the two fuzzy controllers under a fast decrease in solar radiation from $1000W/m^2$ to $200W/m^2$ is presented in Fig. 22. It is observed that the TFLC reaches the MPP slightly faster than the GFLC. However, the control signal of the TFLC is not smooth as observed earlier.

The performance of the optimized FLCs under fast change in temperature was observed to be similar as shown in Fig. 23. The performance under step changes in load is also presented in Fig. 24 and Fig. 25. It is generally observed the TFLC exhibits a faster response than the GFLC, but both have good steady state performance. The drawback of the TFLC is the non-smooth control signal variation.

The results show that an optimized fuzzy logic controller has improved performance and is more robust than the conventional P&O controller. It is also concluded that it is possible to optimize a fuzzy logic controller with a minimum rule base of nine rules while still attaining good transient and steady state performance.

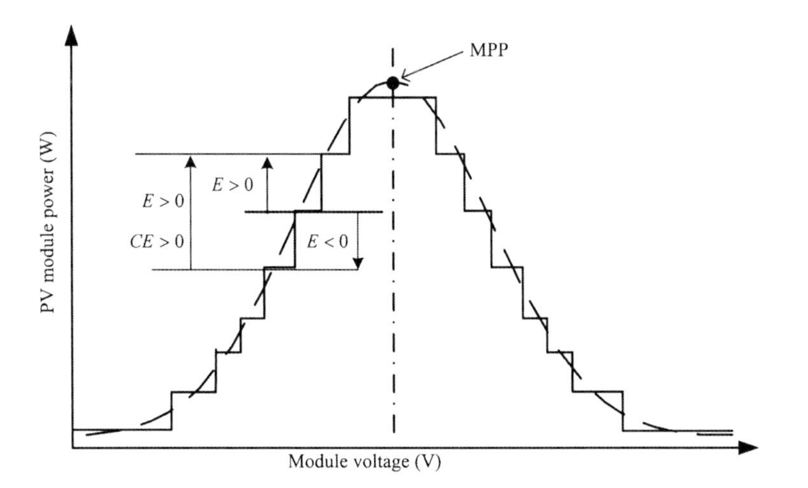

Fig. 18 Illustration of variation of error and change of error

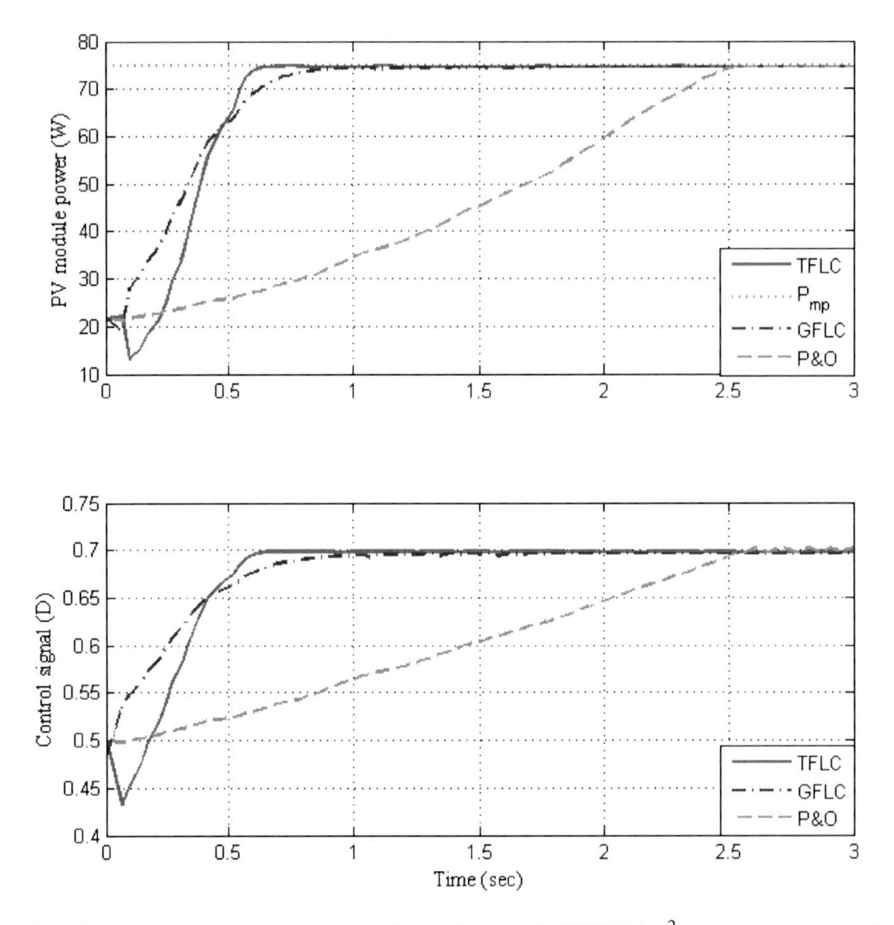

Fig. 19 Response during turn on at solar radiation of $1000W / m^2$ and temperature of $25^o C$

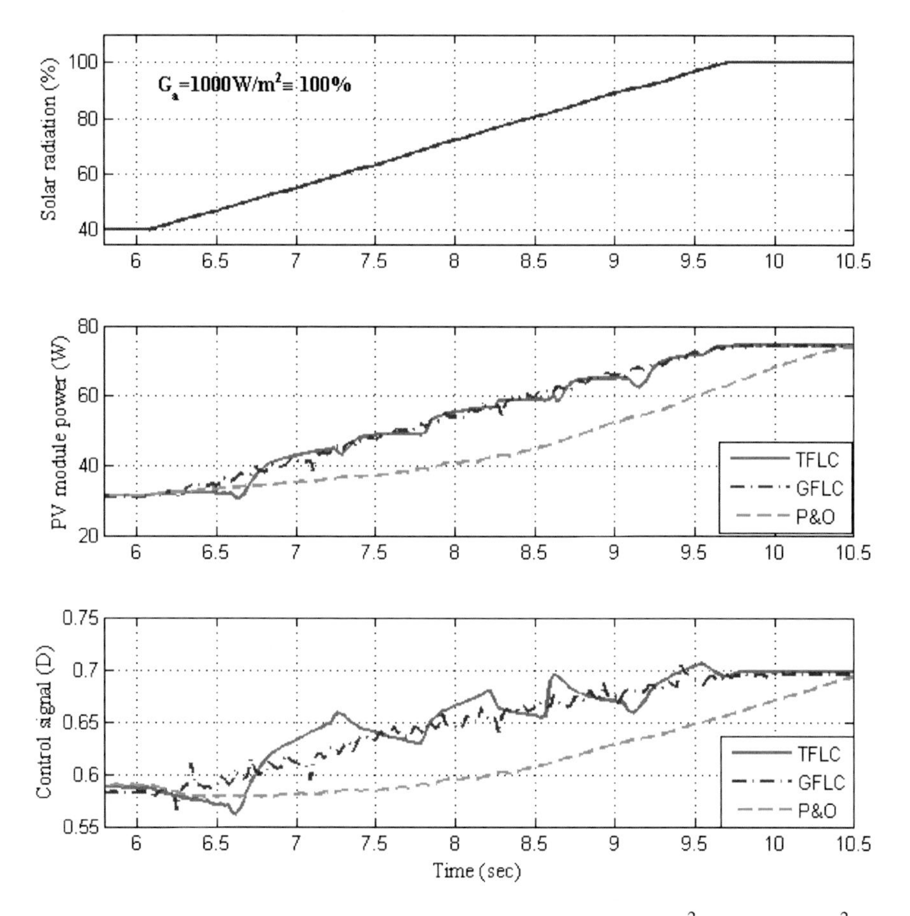

Fig. 20 Response for a fast increase in solar radiation from $400W/m^2$ to $1000W/m^2$ in 3s at a constant temperature of 25^oC

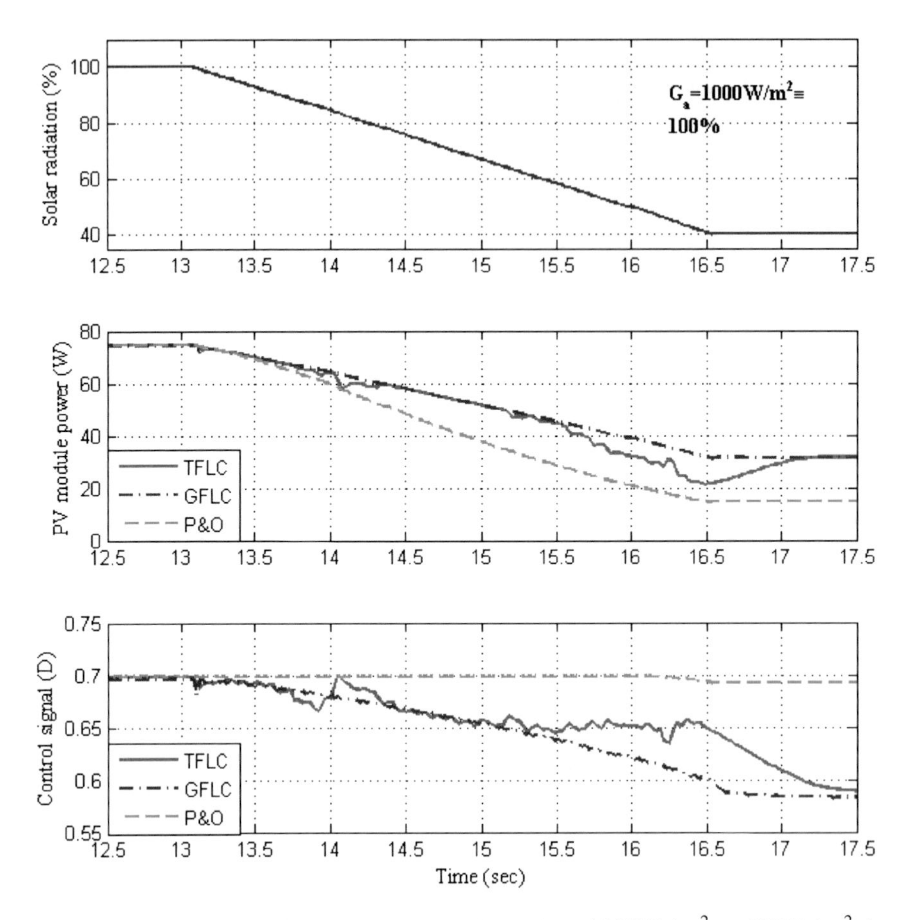

Fig. 21 Response for a fast decrease in solar radiation from $1000W/m^2$ to $400W/m^2$ in 3s at a constant temperature of $25^o C$

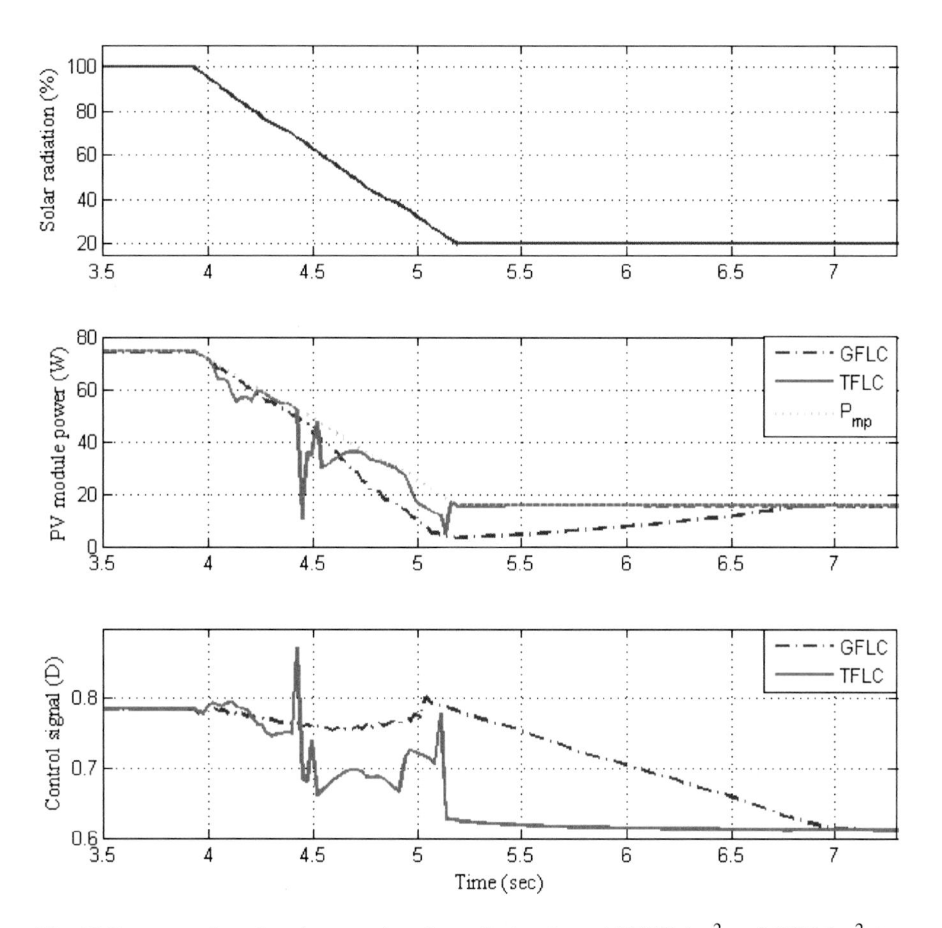

Fig. 22 Response for a fast decrease in solar radiation from $1000W/m^2$ to $200W/m^2$ in 1s at a constant temperature of 25^oC

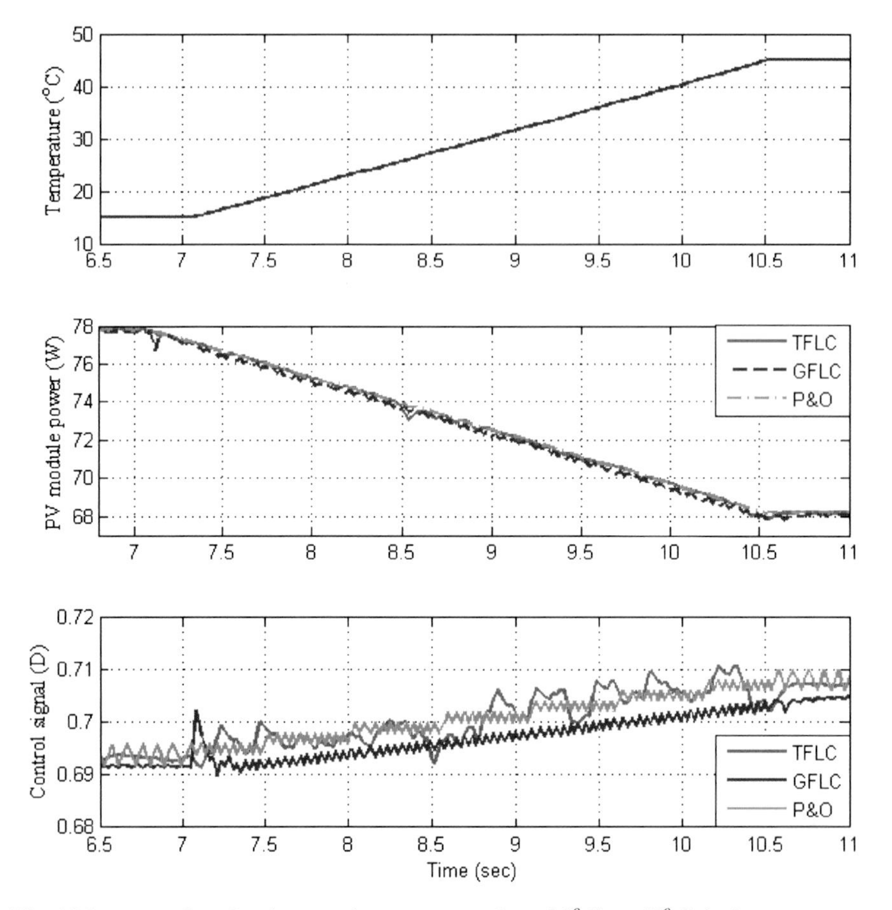

Fig. 23 Response for a fast increase in temperature from $15^{o}C$ to $45^{o}C$ in 3s at a constant solar radiation of $1000W / m^{2}$

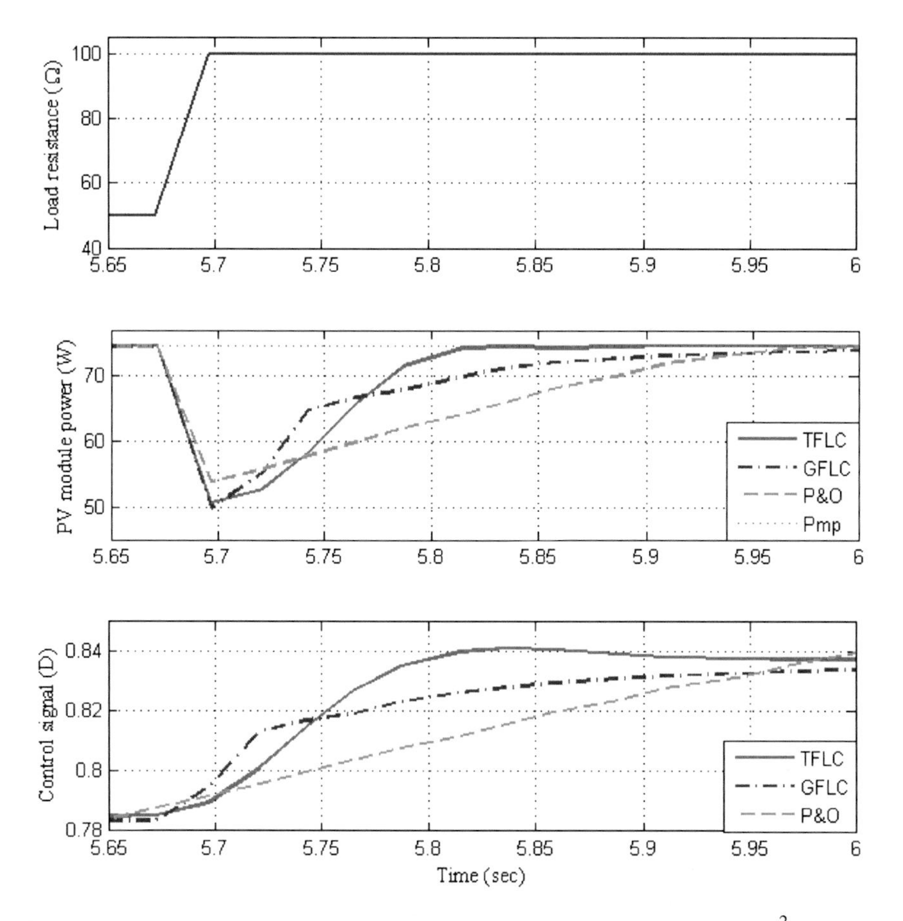

Fig. 24 Response during step increase in load at solar radiation of $1000W / m^2$ and temperature of $25^o C$

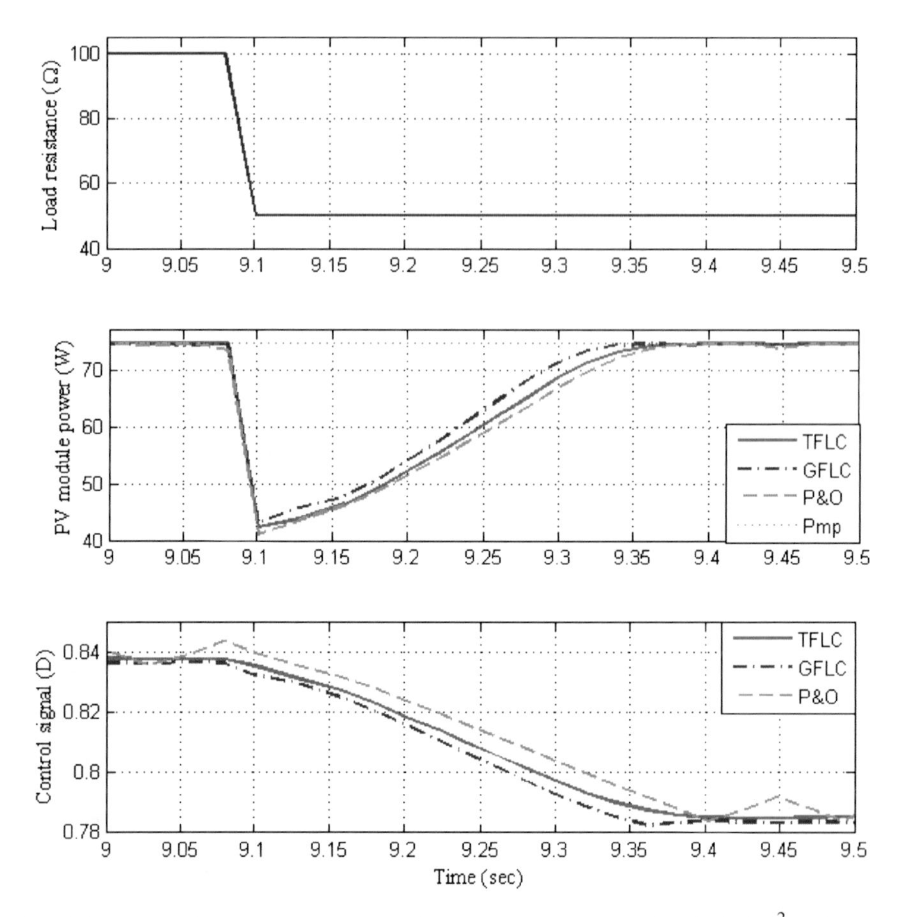

Fig. 25 Response during step decrease in load at solar radiation of $1000W/m^2$ and temperature of 25^oC

6 Conclusion

In this chapter the design and optimization of a Takagi-Sugeno fuzzy logic controller for MPPT in a PV system using genetic algorithms is presented. The PV module model and the FLC were implemented using C language and simulated in LabVIEW using the *.dll* format. Two optimal FLCs each with a rule base of only 9-rules are realized. One FLC uses Gaussian input MFs while the second uses trapezoidal MFs. The expert knowledge for tuning the FLCs is extracted using a fitness function by simulating the PV module under varying solar radiation, temperature, and load conditions. Simulation results are used to compare the performance of the optimized FLCs in terms of speed, accuracy, and robustness. Simulation results have shown that the proposed optimized FLC is robust and provides fast and accurate tracking of the maximum power point compared to the conventional FLC. It is also observed that the proposed optimization strategy produces an optimal FLC using either trapezoidal or Gaussian MFs. The FLC using Gaussian input MFs shows a smoother variation of the control signal compared to the one using trapezoidal input MFs.

References

Baghat, A.B.G., et al.: Maximum power point tracking controller for PV systems using neural networks. Renewable Energy 30(2005), 1257–1268 (2004)

Bose, B.K.: Modern Power Electronics and AC Drives. Prentice-Hall, Englewood Cliffs (2001)

BP Solar Global Marketing, BP Solar SX 75TU PV module data sheet (2002), http://www.solardepot.com/

Hua, C., Shen, C.: Comparative study of peak power tracking techniques for solar storage systems, in Proc. IEEE Appl. In: Power Electron. Conf. and Expo., vol. 2, pp. 676–683 (February 1998)

Sullivan, C.R., Powers, M.J.: A high-efficiency maximum power point trackers for photovoltaic array in a solar-powered race vehicle. In: Proc. IEEE PESC, pp. 574–580 (1993)

Otieno, C.A., Nyakoe, G.N., Wekesa, C.W.: A Neural Fuzzy Based Maximum Power PointTracker for a Photovoltaic System. In: IEEE Africon, pp. 1–6 (September 2009)

Hohm, D.P., Ropp, M.E.: Comparative Study of Maximum Power Point Tracking Algorithms. Progress in Photovoltaics: Research and Applications 11, 47–62 (2003), doi:10.1002/pip.459

Koutroulis, E., Kalaitzakis, K., Voulgaris, N.C.: Development of a Microcontroller Based Photovoltaic Maximum Power Point Tracking Control System. IEEE Transactions on Power Electronics 16(1) (2001)

Manwell, J.F., et al.: Hybrid2 - A hybrid system simulation model theory manual (2006), http://ceere.org/rerl/projects/software/hybrid2/Hy2theorymanual.pdf

Enslin, J.H.R., Snyman, D.B.: Simplified feed-forward control of the maximum power point tracker for photovoltaic applications. In: Proc. Int. Conf. IEEE Power Electron. Motion Control, vol. 1, pp. 548–553 (1992)

Enslin, J.R., Wolf, M.S., Snyman, D.B., Sweigers, W.: Integrated photovoltaic maximum power point tracking converter. IEEE Trans. Ind. Electron 44(6), 769–773 (1997)

Bodur, M., Ermis, M.: Maximum power point tracking for low power photovoltaic solar panels. In: Proc. IEEE Electro Tech. Conf., vol. 2, pp. 758–761 (1992)

Veerachary, M., Senjyu, T., Uezato, K.: Feed-forwardmaximum power point tracking of PV systems using fuzzy controller. IEEE Trans. Aerosp. Electron. Syst. 38(3), 969–981 (2002)

Khaehintung, N., Kunakorn, A., Sirisuk, P.: A Novel Fuzzy Logic Control Technique tuned by Particle Swarm Optimization for Maximum Power Point Tracking for a Photovoltaic System using a Current-mode Boost Converter with Bifurcation Control. International Journal of Control, Automation, and Systems 8(2), 289–300 (2010), http://www.springer.com/12555, doi:10.1007/s12555-010-0215-7.

Erickson, R.W., Maksimovic, D.: Fundamentals of Power Electronics, 2nd edn. Kluwer Academic Publishers, Dordrecht (2001)

Haupt, R.L., Haupt, S.E.: Practical Genetic Algorithms. John Wiley & Sons, Inc., Hoboken (2004)

Sivanandam, S.N., Deepa, S.N.: Introduction to Genetic Algorithms. Springer, Heidelberg (2008)

Noguchi, T., Togashi, S., Nakamoto, R.: Short-current pulse-based maximum power point tracking method for multiple photovoltaic-and converter module system. IEEE Trans. Ind. Electron. 49(1), 217–223 (2002)

Ocran, T.A., et al.: Artificial Neural Network Maximum Power Point Tracker for Solar Electric Vehicle. Tsinghua Science & Technology 10(2), 204–208 (2005)

Kohata, Y., Yamauchi, K., Kurihara, M.: Quick Maximum Power Point Tracking of Photovoltaic Using Online Learning Neural Network. In: Leung, C.S., Lee, M., Chan, J.H. (eds.) ICONIP 2009. LNCS, vol. 5863, pp. 606–613. Springer, Heidelberg (2009)

Application of Artificial Neural Networks for the Prediction of a 20-kWp Grid-Connected Photovoltaic Plant Power Output

Adel Mellit[1], Alessandro Massi Pavan[2], and Soteris A. Kalogirou[3]

[1] Department of Electronics, Faculty of Sciences and Technology, Jijel University, Ouled-aissa, P.O. Box .98, Jijel, 18000, Algeria
mellitadel@yahoo.fr

[2] Department of Materials and Natural Resources, University of Trieste Via A. Valerio, 2 – 34127 Trieste, Italy

[3] Department of Mechanical Engineering and Materials Science and Engineering, Cyprus University of Technology, P.O. Box 50329, Limassol 3603, Cyprus
Soteris.kalogirou@cut.ac.cy

Abstract. Due to various seasonal, hourly and daily changes in climate, it is relatively difficult to find a suitable analytic model for predicting the output power of Grid-Connected Photovoltaic (GCPV) plants. In this chapter, a simplified artificial neural network configuration is used for estimating the power produced by a 20kWp GCPV plant installed at Trieste, Italy. A database of experimentally measured climate (irradiance and air temperature) and electrical data (power delivered to the grid) for nine months is used. Four Multilayer-perceptron (MLP) models have been investigated in order to estimate the energy produced by the GCPV plant in question. The best MLP model has as inputs the solar irradiance and module temperature. The results show that good effectiveness is obtained between the measured and predicted power produced by the 20kWp GCPV plant. The developed model has been compared with different existing regression polynomial models in order to show its effectiveness. Three performance parameters that define the overall system performance with respect to the energy production, solar resource, and overall effect of system losses are the final PV system yield, reference yield and performance ratio.

1 Introduction

The technology for power production from renewable energy sources (RES) is now widely available, reliable and matured. The use of renewable energy systems, such as photovoltaics (PV) is rapidly expanding and has an increasing role in electricity generation, providing pollution-free and secured power [Mellit 2009].

The growth of photovoltaics (PV) for electricity generation is one of the highest in the field of renewable energies and this tendency is expected to continue in the

K. Gopalakrishnan et al. (Eds.): Soft Comput. in Green & Renew. Ener. Sys., STUDFUZZ 269, pp. 261–283.
springerlink.com

next years. As shown in Fig. 1 [IEA 2006], the installation of Grid-Connected photovoltaic (GCPV) plants is growing at an exponential rate. The world annual rate of growth of the cumulative installed capacity is around 40% and is strongly stimulated by the economical incentives given by governments to investors and by the increased attention to environmental problems created by the burning of fossil fuels [Lughi et al. 2008, Pavan et al. 2007, Pavan et al. 2010]. For this reason, the need to understand how these plants work becomes more important. The reason for monitoring GCPV plants arises from the need of assessing productivity (in terms of energy delivered to the grid) and operative conditions [Mellit and Pavan 2010a]. Both climate and electrical factors have an impact on productivity. Photovoltaic modules are sensitive to climate factors, namely irradiance and temperature. However, electrical factors can also have an impact, especially if some quantities deviate from nominal ranges; this may occur in case of certain perturbations, which can be determined both by control issues (DC or AC side) and by grid events [Mellit and Pavan 2010b]. Recorded data from GCPV plants offer a valuable source of performance information to researchers in their effort to improve the performance of these systems.

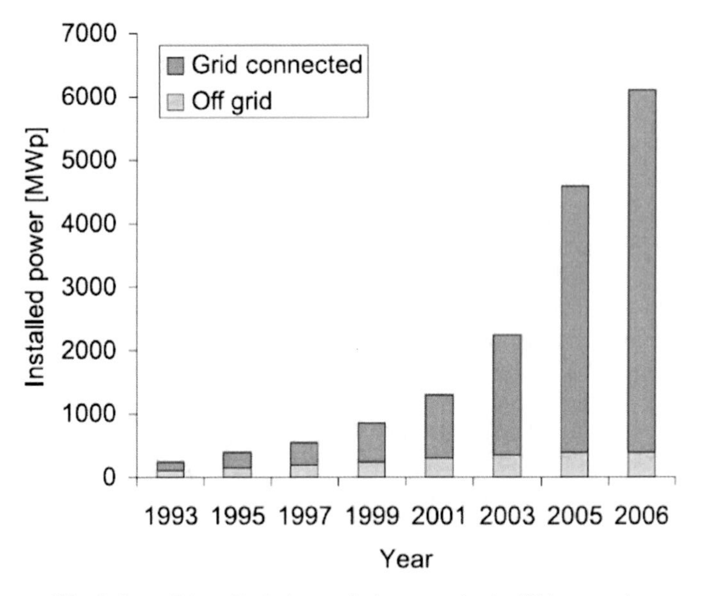

Fig. 1 Overall installed photovoltaic power in the IEA countries

Designers need a reliable tool to predict energy production from photovoltaic panels under all conditions. Several models have been developed for predicting the output power from the PV modules. Existing models can be classified into 4 categories [Mayer et al. 2008]:

- Simple models [Menicucci and Fernandez 1988];
- First and second order physical models [De Soto et al. 2006, Mavromatakis et al. 2010, Zhou et al. 2007];

- Complex physical models [King et al. 1998, Mayer et al. 2008];
- Polynomial regression models [Whitaker et al. 1998, PVUSA 1991, Gianolli-Rossi and Krebs 1988, Meyer and Dyk 2000, Rosell and Ibanez 2006, Mayer et al. 2008];

However, most of these models require the availability of several parameters that are not always available except the simple model. In addition, some models have complicated structures (e.g. Sandia, presented by King et al. [1998]), which do not permit easy manipulation of the system performance. Furthermore, the provided data (open circuit voltage, V_{oc}; open circuit current, I_{sc}; current at maximum power point, I_{mp}; voltage at maximum power point, V_{mp}; and nominal operating cell temperature, T_c) by manufactures are calculated at Standard Rating Conditions (SRC).

Simplicity and practicality are the main advantages of the polynomial regression models when a large amount of experimental data is available to characterize the PV system or module under study.

The key characteristic of an Artificial Neural Network (ANN) is its ability to learn from examples. If a convenient mathematical model that describes a data set is already known, a neural network is unlikely to be required. But, when the rules that underlie the data are only partially known, or not known at all, a neural network may discover interesting relationships as it rambles through the database [Livingstone 2008]. ANN approach is also helpful in order to determine a prediction of the power produced by the PV plant which can be confronted with the logged power trend. ANNs represent a way to solve this kind of problems and can be used for modelling, prediction and optimization of complex systems. ANN techniques have been widely used in energy and renewable energy systems, such as in modelling, simulation, sizing, control and diagnosis of different kinds of the energy systems, including stand-alone, grid-connected, and hybrid PV systems [Mellit and Kalogirou 2008].

The main objective of this chapter is to investigate the suitability of the well-known Multilayer Perceptron (MLP) network for predicting the produced power of a 20 kWp GCPV plant installed at the roof top of the municipality of Trieste building in Italy. In order to do this, four MLP configurations have been investigated and discussed. A comparison between the proposed MLP-models and simple polynomial regression models is presented and analysed.

2 Overview of the Existing Models

2.1 Simple Model

The electrical power *P* produced by the PV system can be calculated as [Mayer et al. 2008]:

$$P = A.f.H.\eta.\eta_{inv} \tag{1}$$

Where A is the net area of the PV array, f is the fraction of array area with active solar cells, H is the irradiance on the plane of the array, η is the module conversion efficiency, and η_{inv} is the inverter (DC to AC) conversion efficiency.

A simplified algebraic equation was proposed by Menicucci and Fernandez [1988] to give the maximum power:

$$P_{mp} = \frac{H}{H_{ref}} P_{mp,ref} \left(1 + \gamma(T - T_{ref})\right) \tag{2}$$

Where H is the incident irradiance, P is the power output, T is the temperature, subscript 'mp' refers to maximum power, subscript 'ref' refers to standard testing conditions ($H_{ref} = 1000$ W/m^2, $T_{ref} = 25$°C) and γ is the maximum power correction factor for temperature.

2.2 First and Second Order Physical Models

Different modifications of the well-known five parameter model have been developed by De Soto et al. [2006]. According to the authors, the predictions from the five-parameter model are shown to agree well with both the King model results and the NIST (National Institute of Standards and Technology) measurements for all four cell types over a range of operating conditions.

Mavromatakis et al. [2010] developed a model for estimating the produced power by a PV array. The later is based upon the nominal power of the array under study, the temperature coefficient of the modules, the solar irradiance at the plane of the array, the air temperature and wind speed.

Zhou et al. [2007] proposed a novel and simple model to predict the PV module performance for engineering applications. Five parameters have been introduced in this model to account for the complex dependence of the PV module performance upon solar-irradiance intensity and PV module temperature. The model's accuracy is demonstrated by comparing the predictions with field measured data. The results demonstrate an acceptable accuracy of the model for modelling PV array outputs under various environmental conditions.

2.3 Complex Physical Models

King et al. [1998] develop an accurate model (Sandia) to predict energy production. Five equations are used to describe the variation of short-circuit current I_{sc}, open-circuit voltage V_{oc}, and maximum power point current I_{mp} and voltage V_{mp}, as a function of irradiance H, cell temperature T_c, absolute air mass AM and solar angle-of-incidence AOI on the PV array [Mayer et al. 2008]. However, it requires parameters that are normally not available from the manufacturer.

2.4 Polynomial Regression Models

A US government and utility sponsored activity called PVUSA has developed a test method [Whitaker et al. 1998, PVUSA 1991] that relates PV system performance to the prevailing environmental conditions (solar irradiance, ambient temperature and wind speed) for a variety of technologies. These dependencies are combined in the following equation [Mayer et al. 2008]:

$$P = H_i(A + B.H_i + C.T_a + D.WS) \tag{3}$$

Where P is the PV array or inverter output, H_i is the plane-of-array solar irradiance, T_a is the ambient temperature, WS is the wind speed and A, B, C and D are the regression coefficients. Systems and climatic conditions are monitored for several weeks and once a sufficient data set is obtained, data are filtered and fitted to obtain the regression coefficients.

Gianolli-Rossi and Krebs [1988] developed a regression model (ENergy RAting: 'ENRA') to compute the power rating:

$$P = A.H + B.H^2 + C.Ln(H) \tag{4}$$

The coefficients A, B and C of the model were obtained from data above 500 W/m^2 only.

Meyer and Dyk [2000] developed a regression model based on daily irradiation and maximum ambient temperature (Energy rating at Maximum Ambient Temperature 'EMA'). The model is given as:

$$P = A.H + B.HT_{\max}^{-2} + C.T_{\max} \tag{5}$$

Where P is the total daily energy produced by the module in Wh/day, H is the total daily irradiation in $Wh/m^2/day$, T_{max} is the maximum ambient temperature in °C, and A, B and C the regression coefficients. This model is able to predict the daily module energy production based on these two parameters only. The data used in this study were collected over a 15-month period at the University of Port Elizabeth (UPE), South Africa.

Rosell and Ibanez [2006] proposed a methodology for estimating the PV electrical production from outdoor testing data. It is based on the adjustment of a well known I–V model curve slightly modified and a new maximum power output expression. The method is validated for a wide range of operating conditions using outdoor and indoor testing data. The following expression is proposed to determine the maximum power output in operating conditions using the parameters A, B, C, D and m:

$$P_{mp} = A.H + B.T + C\left[\ln(H)\right]^m + D.T\left[\ln(H)\right]^m \tag{6}$$

Where A, B, C, D, and m are the coefficients of the model determined by least square fits.

The following generic polynomial regression model has been described in [Mayer et al. 2008]:

$$P = A + B \cdot T_{mod} \cdot H_i + C \cdot H_i + D \cdot H_i^2 \tag{7}$$

Where T_{mod} is the PV module temperature; H_i is the on-plane global irradiance; A, B, C and D are polynomial constants determined by least square fits.

3 Artificial Neural Networks

Artificial neural networks have been used widely in many application areas. Most applications use a Multilayer perceptron (MLP) network with the back-propagation (BP) training algorithm. There are numerous variants of the classical BP algorithm and other training algorithms. MLP networks consist of units arranged in layers with only forward connections to units in subsequent layers [Yu and Jenq-Neng 2001]. The connections have weights associated with them. Each signal traveling along the link is multiplied by a connection weight. The first layer is the input layer, and the input units distribute the inputs to units in subsequent layers. In subsequent layers, each unit sums its inputs, adds a bias or threshold term to the sum and nonlinearly transforms the sum to produce an output. This nonlinear transformation is called the activation function of the unit. The output layer units often have linear activations. In the remainder of this section, linear output layer activations are assumed. The layers sandwiched between the input layer and output layer are called hidden layers and units in hidden layers are called hidden units. The architecture of such a network is shown in Fig. 2.

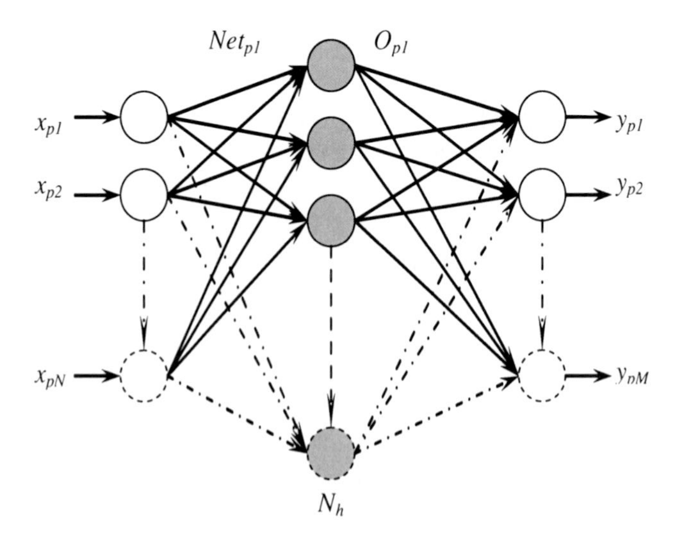

Fig. 2 Feed-forward neural network

The training data set consists of N training patterns $\{(x_p, t_p)\}$, where p is the pattern number. The input vector x_p and desired output vector t_p have dimensions N and M respectively; y_p is the network output vector for the pth pattern. The thresholds are handled by augmenting the input vector with an element $x_p(N + 1)$ and setting it equal to one.

For the jth hidden unit, the net input $net_p(j)$ and the output activation $O_p(j)$ for the pth training pattern are:

$$net_p(j) = \sum_{i=1}^{N+1} w(j,i).x_p(i), \quad \text{for } 1 \le j \le N_h \tag{8}$$

$$O_p(j) = f(net_p(j)) \tag{9}$$

where $w(j, i)$ denotes the weight connecting the ith input unit to the jth hidden unit. For MLP networks, a typical activation function f is the sigmoid, given by:

$$f(net_p(j)) = \frac{1}{1 + exp(-net_p(j))} \tag{10}$$

For trigonometric networks, the activations can be the sine and cosine functions. The kth output for the pth training pattern is y_{pk} and is given by:

$$y_{pk} = \sum_{i=1}^{N+1} w_{io}(k,i).x_p(i) + \sum_{j=1}^{N_h} w_{ho}(k,j).O_p(j), \quad \text{for } 1 \le k \le M \tag{11}$$

Where $w_{io}(k, i)$ denotes the output weight connecting the ith input unit to the kth output unit and $w_{ho}(k, j)$ denotes the output weight connecting the jth hidden unit to the kth output unit. The mapping error for the pth pattern is:

$$E_p = \sum_{k=1}^{Np} \left(t_{pk} - y_{pk}\right)^2 \tag{12}$$

Where t_{pk} denotes the kth element of the pth desired output vector. In order to train a neural network in batch mode, the mapping error for the kth output unit is used defined as:

$$E(k) = \frac{1}{N_v} \sum_{p=1}^{Nv} \left(t_{pk} - y_{pk}\right)^2 \tag{13}$$

The overall performance of an MLP neural network, measured as mean square error (MSE), can be written as:

$$E = \sum_{k=1}^{M} E(k) = \frac{1}{N_p} \sum_{p=1}^{Nv} E_p \tag{14}$$

The key distinguishing characteristic of a MLP with the back-propagation learning algorithm is that it forms a nonlinear mapping from a set of input stimuli to a set of outputs using features extracted from the input patterns. The neural network can be designed and trained to accomplish a wide variety of nonlinear mappings, some of which are very complex. This is because the neural units in the neural network learn to respond to features found in the input. By applying the set of formulations of the backpropagation (BP) algorithm, presented above, the calculation procedure of the learning process summarized in the Appendix is employed [Gupta et al. 2003].

In the procedure listed in the Appendix, several learning factors such as the initial weights, learning rate, number of hidden neural layers and number of neurons in each layer, may be readjusted if the iterative learning process does not converge quickly to the desired point. Although, the BP learning algorithm provides a method for training MLPs to accomplish a specified task, in terms of the internal nonlinear mapping representations, it is not free from problems. Many factors affect the learning performance and must be dealt with in order to have a successful learning process. Mainly, these factors include the initial parameters, learning rate, network size and learning database. A procedure to select these parameters is presented by Kalogirou [2001]. A good choice of these items may greatly speed up the learning process to reach the target, although there is no universal answer for these issues [Gupta et al. 2003]. Advanced methods for learning and adaptation in MLPs are presented in [Haykin 1999, Lakhmi and Martin 1998].

4 Description of the GCPV Plant and Dataset

4.1 Description of the GCPV Plant

With reference to Fig. 3, the GCPV plant considered is the one installed at the rooftop of Trieste local government building in Italy. The 174 photovoltaic modules composing the field are oriented south and tilted at 34°.

Fig. 3 The GCPV plant installed at the roof top of the Trieste local government building

The electrical schematic of the plant, which consists of 12 photovoltaic strings, is shown in Fig. 4. Each string is made of 14 or 15 series connected EC-115 Evergreen Solar photovoltaic modules. The main characteristics of the modules and strings are listed in Tables 1 and 2.

Table 1 EC-115 Evergreen module

Technology	Polycrystalline Si
Peak power	115 W
Open circuit voltage at STC	21.5 V
Maximum power point voltage at STC	17.3 V
Short circuit current at STC	7.26 A
Maximum power point current at STC	6.65 A
Number of cells	72
Nominal Operating Cell temperature	44°C
Voltage-temperature coefficient	-0.53%/°C
Current-temperature coefficient	0.049%/°C
Power-temperature coefficient	-0.49%/°C

Table 2 String made of 14 PV modules

Number of modules	14
Peak power	1610 W
Open circuit voltage at SRC	301.0 V
Maximum power point voltage at SRC	242.2 V
Short circuit current at SRC	7.26 A
Maximum power point current at SRC	6.65 A

The 12 strings are subdivided into three groups of four strings. Each group is connected to the inverter input stage, as depicted in Fig. 4. Each inverter is endowed with 4 Maximum Power Point Tracking (MPPT) systems, so that the maximum power point can be tracked for each string [i.e., in this case a string architecture [Pavan et al. 2007] has been used]. Each inverter has a single phase AC output (the first one connected to the R phase of the grid, the second to the Y, and the third to the B phase). As shown in Fig. 5(a), the inverter used is the Mastervolt QS-6400 whose electrical data are reported in Table 3. The monitoring of the photovoltaic plant is made by using two different data loggers shown in Fig. 5(b); one is dedicated to the climate data (Danfoss ComLynx Monitor), while the second one is used to record the electrical data (Mastervolt QS Data Control Premium) for six groups of two strings called QS1, QS2, QS3, QS4, QS5, and QS6.

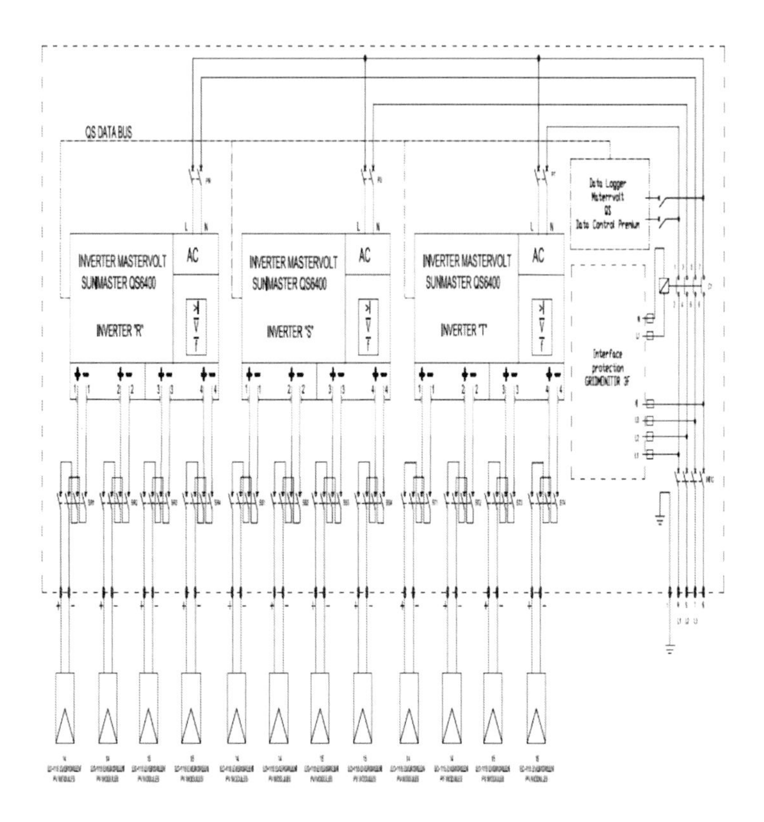

Fig. 4 Electrical schematic of the 20kWp GCPV plant

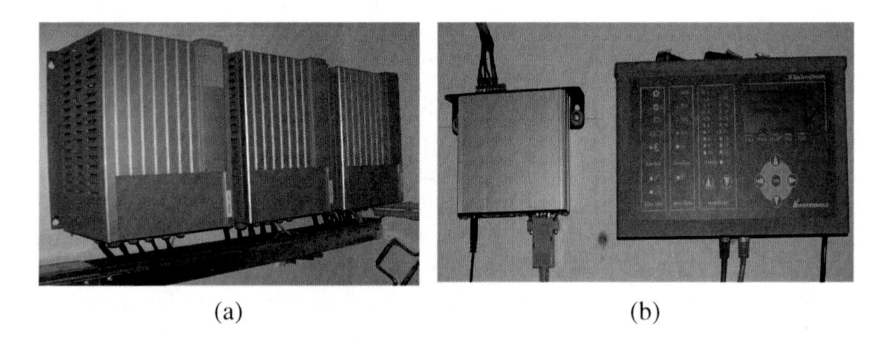

(a) (b)

Fig. 5(a) The inverters used in the plant, **(b).** Data loggers used for the climate data (left side) and for the electrical data (right side)

Table 3 Mastervolt inverter electrical data

Number of MPPT	4
Output voltage	230V, 50Hz
Nominal power	5500W
European efficiency	94%
Maximum efficiency	95%
MPPT voltage range	100-380V
Maximum DC voltage	450V
Nominal DC current	7.5A

4.2 Data Used for the ANN Application

The climate data which are recorder are: the irradiance on the array plane (H), the module temperature measured at the backside of a reference module (T_{PV}) and the ambient temperature at array side (T_a). Figs 6(a–c) show the calibrated reference cell used for measuring *H*, and the temperature sensors (PT 100) for measuring T_{PV} and T_a according with [IEC 1999]. With reference to the electrical data, the quantities which are recorded for each string are: the operating voltage (V_{str}), current (I_{str}), power (P_{str}) and the AC power (P_{grid}). Finally, the Mastervolt QS Data Control Premium records also the grid voltage (V_{grid}) and its frequency (f), and the energy produced both for the DC and the AC side. An example of the recorded data is presented in Fig. 7(a) showing the evolution of the measured climate and electrical data (H, T_a , T_{pv} and P_{grid}) from January 1st to June 30th 2009 with a time scale of 10 min.

A correlation between the produced power AC and solar irradiance is shown in Fig. 7(b) and as can be seen the power produced by the PV system strongly depends on the solar irradiance.

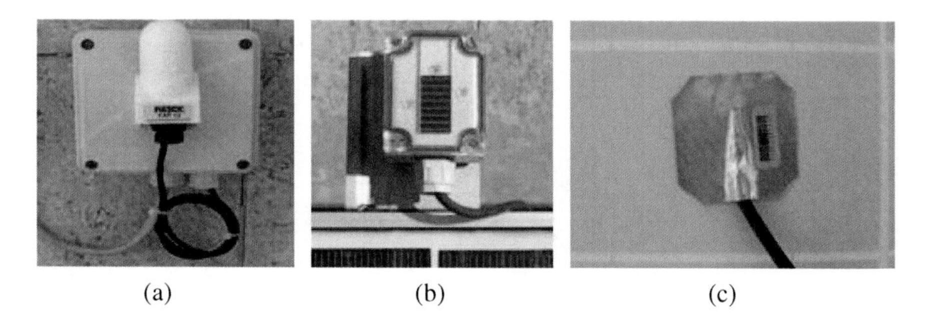

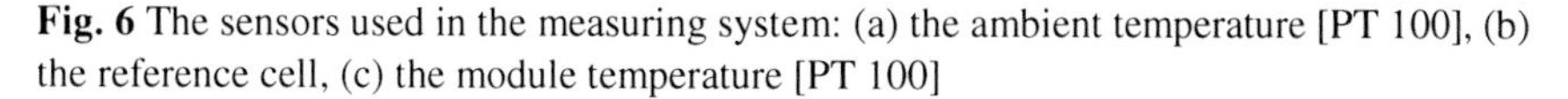

(a) (b) (c)

Fig. 6 The sensors used in the measuring system: (a) the ambient temperature [PT 100], (b) the reference cell, (c) the module temperature [PT 100]

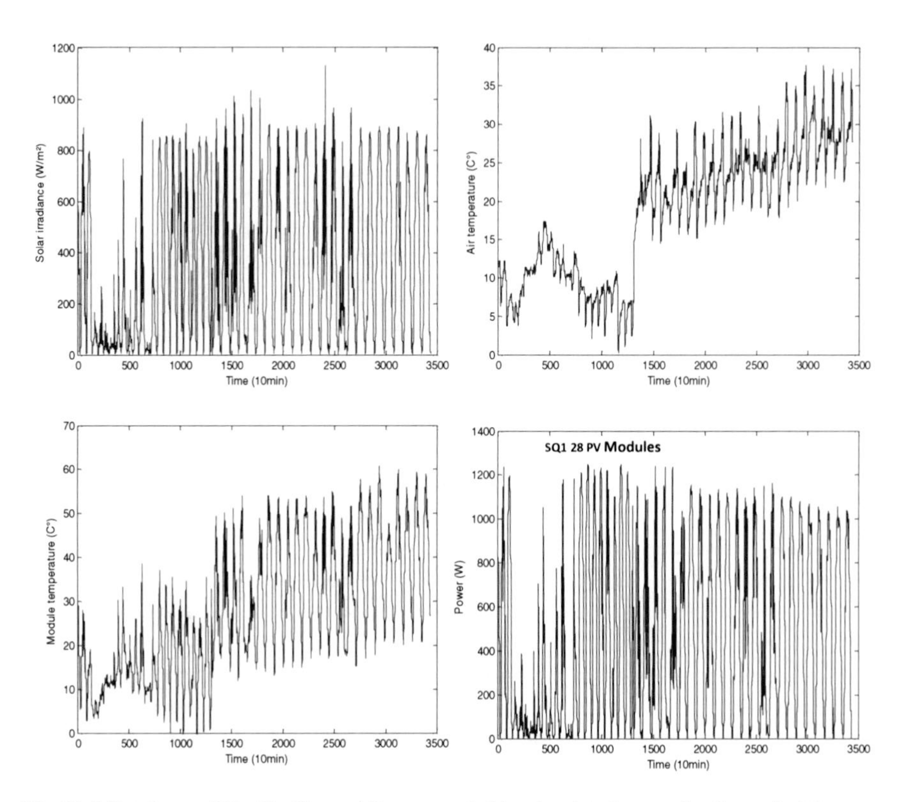

Fig 7(a) Database of H , Ta, T_{pv} and P_{grid} recorded by the data logger for the period January 1st to June 30th 2009

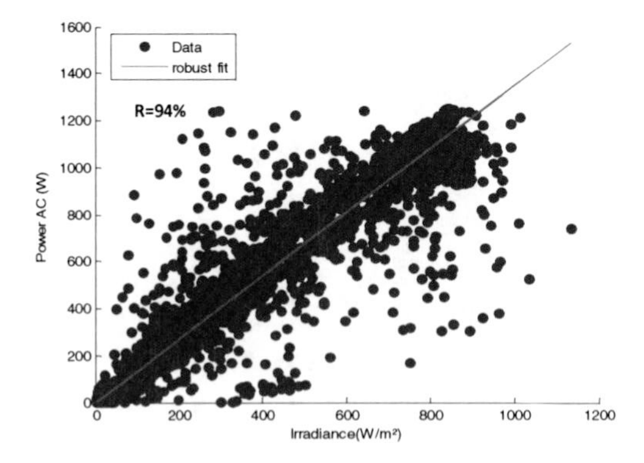

Fig 7(b) Correlation between produced AC power and solar irradiance for the period January 1st to June 30th 2009

5 ANN Method and Discussion

5.1 Model Development

The well-known MLP network is used for predicting the power produced by the GCPV system. Four MLP-structures have been proposed:

- The first MLP structure { $(P_{grid}) = \tilde{f}(H, T_a, T_{PV})$ } has as inputs H, T_a and T_{pv} and provide the power produced (P_{grid}) by the GCPV plant.
- The second MLP structure { $(P_{grid}) = \tilde{f}(H, T_{PV})$ } has as inputs H and T_{PV} and provide the power produced (P_{grid}) by the GCPV plant.
- The third MLP structure { $(P_{grid}) = \tilde{f}(H, T_a)$ } has as inputs H and T_a and provide the power produced (P_{grid}) by the GCPV plant.
- The fourth MLP structure { $(P_{grid}) = \tilde{f}(H)$ } has only one input H and provide the power produced (P_{grid}) by the GCPV plant.

Where $\tilde{f}$ is a non-linear approximation function which can be estimated based on the weights and the bias of the optimal MLP structure.

The numbers of the neurons within the hidden layers are optimized during the learning process of the network, according with a specified criterion such as a Root Mean Square Error (RMSE). Figure 8 shows the MLP structure used for predicting the power produced by the GCPV plant (first MLP structure).

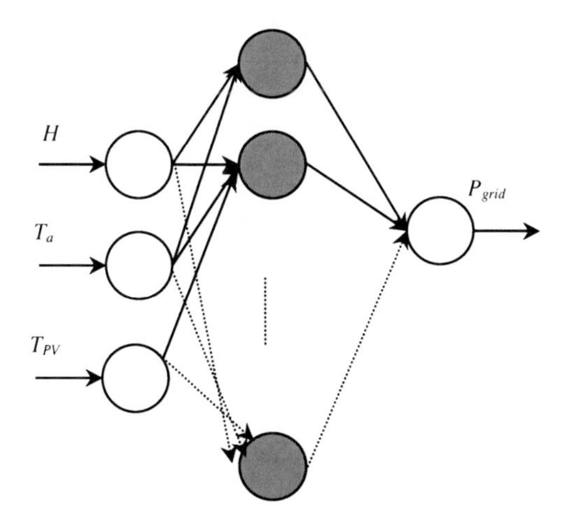

Fig. 8 MLP structure used for predicting the power produced by the CGPV plant

As described in Section 4 and with reference to the way the electrical data are logged, the GCPV plant consists of three groups of two strings with 28 PV modules (called group SQ1, SQ3, and SQ4), and another three groups of two strings with 30 PV modules (called group SQ2, SQ5, and SQ6). In order to find which MLP model can better predict the power produced by the GCPV, the data recorded on group SQ1 are used.

A database of 3437 patterns is available and is divided into two parts: a set of 3079 patterns is used for training the two MLP models, while the other set of 358 patterns is used for testing and validating the MLP models developed.

A soft computing program for predicting the power produced by the GCPV plant based on the Levenberg-Marquard (LM) algorithm has been implemented on MatLab® (Ver. 7.5). The LM training algorithm (*trainlm*) is used to adjust the weights such that the neural network produces the required output for the given input data.

Before applying the training algorithm, the data (input/output) should be normalized to [-1,1] using Eq. (15). Traditional normalization techniques use linear or logarithmic scaling, which requires the designer to supply practical estimates of maximum and minimum values of normalized variables to improve the neural network performance.

$$y = y_{\min} + \frac{x - x_{\min}}{x_{\max} - x_{\min}}\left(y_{\max} - y_{\min}\right) \tag{15}$$

Where $x \in [x_{\min}, x_{\max}]$ and $y \in [y_{\min}, y_{\max}]$; x is the original data value, and y is the corresponding normalized variable. Finally, y_{min} = -1, and y_{max} = 1 have been assumed.

5.2 *Results and Discussion*

The simulation results are shown in Fig. 9. The graphs on the left show the superposition curves between measured and predicted power produced by the GCPV plant using the different MLP-models. As can be seen, a good agreement is obtained between the measured and the predicted data as the correlation coefficient (r) is in the range 97.85-98.60%. In the graphs on the right of Fig. 9, scatter plots of the measured and predicted power using the different MLP-models are shown in order to illustrate the correlation between the measured and predicted values. The plot giving a correlation coefficient of 98.63% shows the strength of the second MLP-model.

A comparison between the measured (actual) and ANN predicted power given by the different MLP models is shown in Table 4.

Table 4 Comparison between the measured and predicted power given by the four MLP-models

MLP-Model	Architecture (input, hidden, output neurons)	Measured (actual) mean power (W)	Predicted mean power (W)	AMRE (%)	r-value (%)
$(P_{grid}) = \tilde{f}(H, T_a, T_{PV})$	{3X9X1}	556.23	528.90	4.90	97.85
$(P_{grid}) = \tilde{f}(H, T_{PV})$	{3X7X1}	556.23	545.36	1.97	98.60
$(P_{grid}) = \tilde{f}(H, T_a)$	{3X7X1}	556.23	566.94	1.92	98.20
$(P_{grid}) = \tilde{f}(H)$	{3X11X1}	556.23	582.58	4.74	98.00

As can be seen from the results presented in Table 4, the second $(P_{grid}) = \tilde{f}(H, T_{PV})$ and the third $(P_{grid}) = \tilde{f}(H, T_a)$ models provide more accurate results than the other models $(P_{grid}) = \tilde{f}(H, T_a, T_{PV})$ and $(P_{grid}) = \tilde{f}(H)$. The absolute mean relative error (AMRE) is lower in the second and the third MLP-models (less than 2%) whereas, for the first and the fourth model the MRE is more than 4%.

It should be noted that the results obtained with the fourth MLP model are also important as this model can be used when only one input parameter, the solar irradiance H, is available.

The developed MLP-models are based on the following formula:

$$\tilde{y} = \sum_{k=1}^{M}\left(\left[\frac{2}{1+\exp(-\left(\sum_{r=1}^{M}\sum_{i=1}^{N}\left(w_1(i,r)x(i)\right)+b_1(r)\right))}-1\right]\right)w_2(k)+b_2 \tag{16}$$

Where w_1, w_2, b_1 and b_2 are the weights and the bias of the networks respectively, while x represents the inputs data which can be the couple solar irradiance and air temperature, or only the solar irradiance. M and N are the number of neurons in the hidden layer and in the input layer respectively, and finally $\tilde{y}$ corresponds to the output, which is the predicted power for the GCPV plant.

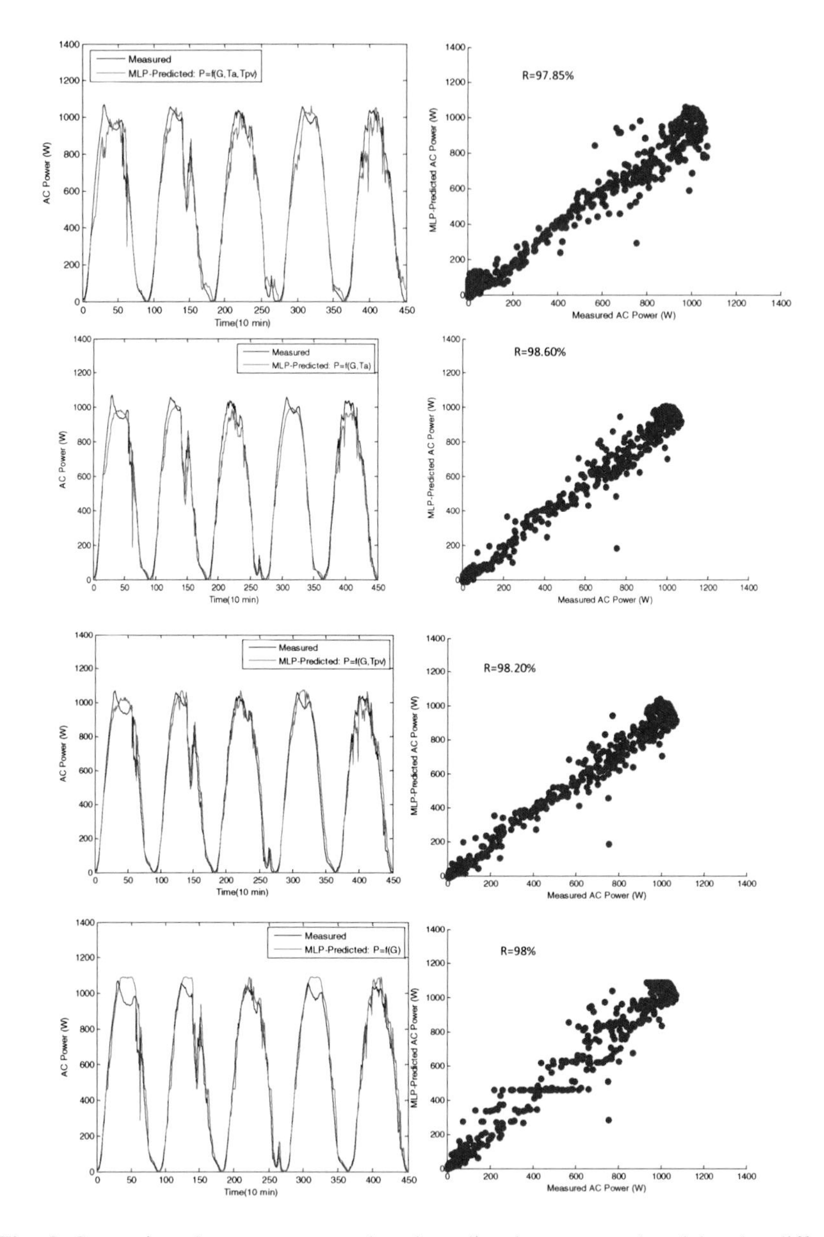

Fig. 9 Comparison between measured and predicted power produced by the different MLP-models

5.3 Comparison between Different Regression Models

In order to verify the effectiveness of the developed MLP-model, a comparison between the power predicted by the second MLP-model $P_{MLP} = \tilde{f}(H, T_a, w_{ij}, b_i)$ with some existing polynomial regression models presented in section 2 is given. These are:

$$P_1 = A.f.H.\eta.\eta_{inv}$$

$$P_2 = A + B \cdot T_{mod} \cdot H_i + C \cdot H_i + D \cdot H_i^2 \text{ [Mayer et al. 2008]}$$

$$P_3 = A.H + B.T + C_3 [Ln(H)]^m + D.T.[Ln(H)]^m \text{ [Rosell and Ibanez 2006]}$$

$$P_4 = A.H + B.HT_{max}^{-2} + C.T_{max} \text{ [Meyer and Dyk 2000]}$$

$$P_5 = \mathrm{A.H} + \mathrm{B.H}^2 + \mathrm{C.Ln(H)} \text{ [Gianolli-Rossi and Krebs 1988]}$$

The different coefficients have been determined by least square fits.

The IEC standard 61724 [1998] defines three performance parameters for assessing the overall operation of a PV system: the reference yield Y_r, the system yield Y_f and the performance ratio PR. An evident limitation for the purposes of this work is that above the parameters are clearly influenced by weather [Marion et al. 2005]:

- Y_r : the ratio between the total in-plane irradiance and the reference irradiance – has a month-to-month and year-to-year weather variability;
- Y_f : the ratio between the produced energy and the nominal power of the PV generator – is influenced by solar radiation;
- PR: the ratio between the system yield and the reference year – is influenced to a lower extent by the weather as its value is normalized with respect to solar radiation, but is still influenced by seasonal variations in temperature and plant availability.

A statistical test between the measured and predicted power produced is summarized in Table 5. Figure 10 shows the plot of the predicted power produced by using different polynomial regression models versus the experimental ones. With reference to Fig. 10, it is worth noticing that the predicted power by the different polynomial regression models is close to the measured ones, since the correlation coefficients are between 97% and 98%.

Table 5 Comparison between the predicted power produced by the second MLP-model and different existing regression models.

Model	Measured (actual) mean power (W)	Predicted mean power (W)	AMRE (%)	PR (%)	r (%)
$(P_{grid}) = \tilde{f}(H, T_{PV})$, {3X7X1}	556.23	545.36	2	89	98.6
$P_1=1.037*14*0.111*0.8*H$	556.23	500.61	10	80	97.5
$P_2 = -11.216 - 0.002T_cH$ $+1.858H - 0.0005H^2$	556.23	545.10	2	89	98.4
$P_3 = 0.0443H + -0.0058T_c$ $+0.0018Ln(H)^{7.003}$ $-0.0001T_cLn(H)^{7.003}$	556.23	461.67	17	75	98
$P_4 = 1.2754\ H + 53.3249\frac{H}{(T_{max})^2}$ $+1.0915T_{max}$	556.23	555.67	0.1	87.7	98
$P_5 = 0.7552H + -0.0009\ H^2$ $+ 0.1962H.Log(H)$	556.23	539.54	3	89	98.4

The calculated PR for the testing period is 89%. As can be seen from Table 10 the performance ratio (PR) for all polynomial regression models varies in the range between 75% and 89%. It should be noted that, the second regression P_2 and the fifth P_5 regression model provide acceptable results compared with the developed MLP-model. In addition, the performance ratio for the second and fifth regression models is the same as that of the MLP-model.

Finally, the database described above is used for estimating the degradation rate by using the second polynomial regression model. Figure 11 depicts the polynomial AC power rating of the GCPV plant against time from January through June 2009. Linear fit to the power rating curve indicate that the GCPV roof plant has degraded at a rate 0.3% over six months.

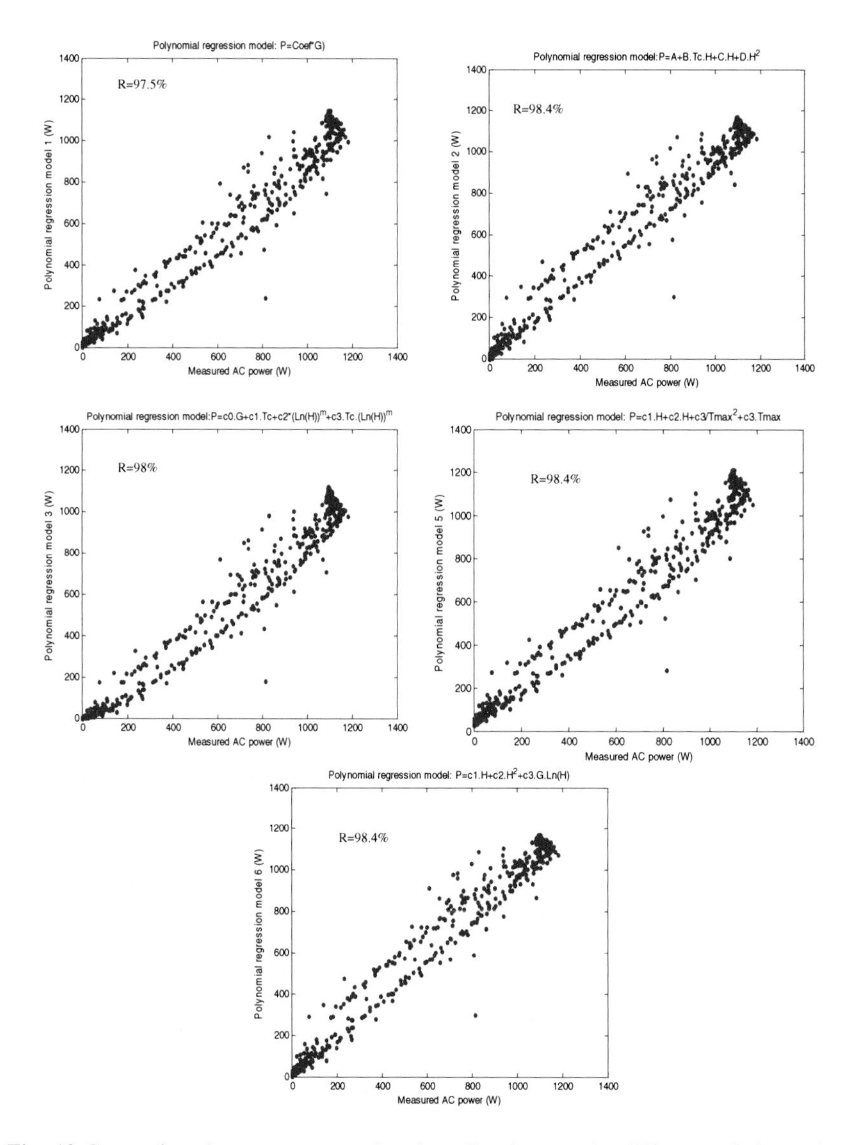

Fig. 10 Comparison between measured and predicted power by different existing polynomial regression models

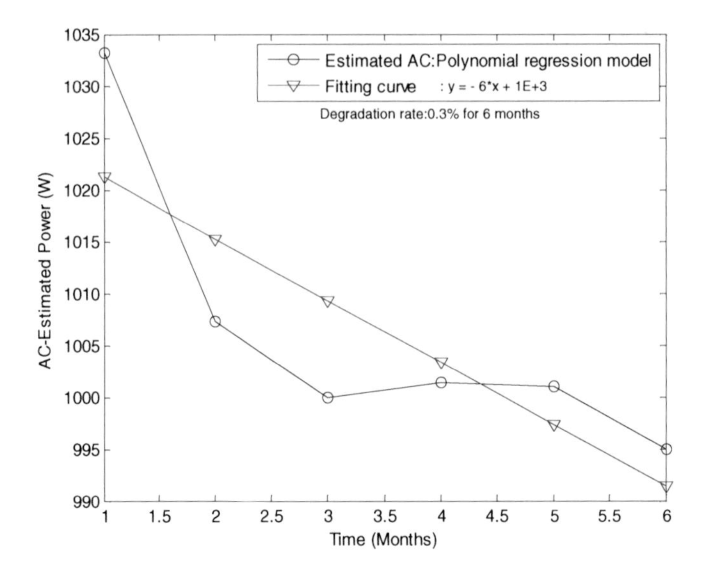

Fig. 11 AC-estimated power against time

6 Conclusion and Future Work

Development of accurate models for predicting the power produced from grid-connected PV systems (GCPVs) is important for performance analysis and energy management.

In this chapter, four MLP configurations have been investigated in order to predict the power produced from a grid-connected PV plant. It has been demonstrated that the MLP-model which has as input the solar irradiance and the module temperature gave accurate results compared to the other examined MLP-configurations.

Comparison between the MLP-model designed and some regression polynomial models shows that the second MLP-model provides accurate results. The second and the fifth regression models also provide accurate results compared with the developed MLP-model.

The degradation rate of the GCPV plant was determined to be 0.3% per six months according to the second polynomial model.

The MLP has been chosen because its implementation is easy in particular if compared with a hybrid ANN (e.g. ANFIS, GA-ANN, etc.) [Mellit and Kalogirou 2008]. Furthermore, all functions used are available in the neural networks toolbox 6.0.2 of MatLab®. Therefore, readers interested in this type of predictions can use the MatLab® ANN-toolbox for developing their own models (e.g. *newff*, *train*, *sim*, etc.). Moreover, it is worth noticing that other hybrid ANN-architectures are difficult to use by those who are not familiar with these techniques.

As ANNs can help to understand how a GCPV works with respect to climate conditions, the next step is to develop a software tool for GCPV plants performances analysis. This will be developed with simple user interface for requiring, for example, a spreadsheet file containing climate and electrical data and provide to the user the correlation coefficient, the mean bias error or the performance ratio, which can be viewed as quality parameters of the PV system operation.

References

De Soto, W., Klein, S.A., Beckman, W.A.: Improvement and validation of a model for photovoltaic array performance. Solar Energy 80, 78–88 (2006)

Gianolli-Rossi, E., Krebs, K.: Energy rating of PV modules by outdoor response analysis. In: 8th E.C. PV Solar Energy Conference, Florence, Italy (1988)

Gupta, M.M., Liang, J., Noriyasu, H.: Static and dynamic neural networks. In: Zadeh, L.A. (ed.) Fundamentals to Advanced Theory. IEEE press, John Wiley and Sons Ltd. (2003)

Haykin, S.: Neural networks: A comprehensive foundation, 2nd edn. Macmillan, Basingstoke (1999)

IEA (2006) Trends in photovoltaic applications. Survey report of selected IEA countries between 1992 and 2006. International Energy Agency

IEC - CEI EN Std. 61724, Photovoltaic system performance monitoring - Guidelines for measurement, data exchange and analysis, International Electro-technical Commission (1999)

IEC 61724, Photovoltaic System Performance Monitoring - Guidelines for Measurement, Data Exchange and Analysis (1998)

Kalogirou, S.A.: Artificial Neural Networks in Renewable Energy Systems: A Review. Renewable & Sustainable Energy Reviews 5(4), 373–401 (2001)

King, D.L., Kratochvil, J.A., Boyson, W.E., Bower, W.I.: Field Experience with a New Performance Characterization Procedure for Photovoltaic Arrays. Presented at the 2nd World Conference and Exhibition on Photovoltaic Solar Energy Conversion, Vienna, Austria, July 6-10 (1998)

Lakhmi, C.J., Martin, N.M.: Fusion of neural networks, fuzzy systems and genetic algorithms: Industrial Applications. CRC Press, LLC, Boca Raton (1998)

Livingstone, D.J.: Artificial neural networks: Methods and Applications. Humana Press (2008)

Lughi, V., Massi Pavan, A., Quaia, S., Sulligoi, G.: Economical analysis and innovative solutions for grid connected PV plants. In: Proceeding of Speedam, International Conference on Power Electronics, Electrical Drives. Automation and Motion, Ischia (Italy), June 11-13, pp. 211–216 (2008)

Marion, A., Adelstein, J., Boyle, K., Hayden, H., Hammond, B., Fletcher, T., Canada Narang, B., Shugar, D., Wenger, D., Kimber, H., Mitchell, A., Rich, L., Townsend, G.: Performance parameters for grid-connected PV systems. In: 31st IEEE Photovoltaics Specialist Conference and Exhibition, Lake Buena Vista, Florida (2005)

Mavromatakis, F., Makrides, G., Georghiou, G., Pothrakis, A., Franghiadakis, Y., Drakakis, E., Koudoumas, E.: Modeling the photovoltaic potential of a site. Renewable Energy 35, 1387–1390 (2010)

Mayer, D., Wald, L., Poissant, Y., Pelland, S.: Performance prediction of grid-connected photovoltaic systems using remote sensing. Report IEA-PVPS T2-07, p. 18, http://www.trpvplatform.org

Mellit, A.: Recurrent neural network-based forecasting of the daily electricity generation of a Photovoltaic power system. In: Ecological Vehicle and Renewable Energy (EVER), Monaco, March 26-29, pp. 265–270 (2009)

Mellit, A., Kalogirou, S.A.: Artificial intelligence techniques for photovoltaic applications: A review. Progress in Energy and Combustion Science 34, 547–632 (2008)

Mellit, A., Massi Pavan, A.: A 24-hours forecast of solar irradiance using artificial neural network: Application for performance prediction of a grid-connected PV plant at Trieste, Italy. Solar Energy 84, 807–821 (2010a)

Mellit, A., Massi Pavan, A.: Performance prediction of 20kWp grid-connected photovoltaic plant at Trieste (Italy) using artificial neural network. Energy Conversion and Management 51, 2431–2441 (2010b)

Menicucci, D., Fernandez, J.P.: User's manual for PVFORM: a photovoltaic system simulation program for stand-alone and grid interactive applications; Sandia National Laboratories, SAND85-0376, Albuquerque, USA (1988)

Meyer, E.L., van Dyk, E.E.: Development of energy model based on total daily irradiation and maximum ambient temperature. Renewable Energy 21, 37–47 (2000)

PVUSA Technical Speci®cation, 19271-EMT-3 (August 1991)

Pavan, M.A., Castellan, S., Quaia, S., Roitti, S., Sulligoi, G.: Power Electronic Conditioning Systems for Industrial Photovoltaic Systems: Centralized or String Inverters? In: Proc. of ICCEP Int. Conf. on Clean Electrical Power, Capri (Italy), May 21-23, pp. 208–214 (2007)

Pavan, M.A., Lughi, V., Roitti, S., Mellit, A.: Impiego di reti neurali per la previsione dell'irradianza solare e della potenza prodotta da un impianto fotovoltaico, 10° Convegno Nazionale AIMAT, Capo Vaticano (VV), September 5-8 (2010)

Rosell, J.I., Ibanez, M.: Modelling power output in photovoltaic modules for outdoor operating conditions. Energy Conversion and Management 47, 2424–2430 (2006)

Whitaker, C., Townsend, T.U., Newmiller, J.D., King, D.L., Boyson, W.E., Kratochvil Collier, J.A., Osborn, D.E.: Application and validation of a new PV performance characterization method. In: Proceedings of the 26th IEEE PV Specialist Conference, Anaheim, USA (1997)

Yu, H.H., Jenq-Neng, H.: Handbook of neural network signal processing. CRC Press, Boca Raton (2001)

Zhou, W., Yang, H., Fang, Z.: A novel model for photovoltaic array performance prediction. Applied Energy 84, 1187–1198 (2007)

Appendix: Back Propagation (BP) Algorithm

Given a finite length input pattern $x_1(k), x_2(k)....x_n(k) \in \Re$ ($1\leq k\leq K$):

Step 1: Select the total number of layers M, the number n_i (i=1,2,...N-1) of the neurons in each hidden layer, and an error tolerance parameter $\varepsilon > 0$

Step 2: Randomly select the initial value of the weight vectors $w_{aj}^{(i)}$ for I=1,2,...,M and j=1,2,...,n_i

Step 3: Initialization $w_{aj}^{(i)} \leftarrow w_{aj}^{(i)}(0)$, E←0, and k←1

Step 4: Calculate the neural outputs;

$$\begin{cases} s_j^{(i)} = \left(w_{aj}^{(i)}\right)^T x_a^{(i-1)} \\ x_j^{(i)} = \sigma\left(s_j^{(i)}\right) \end{cases} \quad \text{for I=1,2,...,M and j=1,2,...,}n_i$$

Step 5: Calculate the output error $e_j = d_j - x_j^{(M)}$ for j=1,2,…,m

Step 6: Calculate the output delta's $\delta_j^{(M)} = e_j \sigma'\left(s_j^{(M)}\right)$

Step 7: Recursively calculate the propagation errors of the hidden neurons;

$e_j^{(i)} = \sum_{l=1}^{n_i+1} \delta_l^{(i+1)} w_{lj}^{(i+1)}$ from the layer M-1, M-2,… to layer 1.

Step 8: recursively calculate the hidden neurons delta values;

$\delta_l^{(i)} = e_l^{(i)} \sigma'\left(s_l^{(i)}\right)$

Step 9; update weight vector: $w_{aj}^{(i)} = w_{aj}^{(i)} + \eta \delta_j^{(i)} x_a^{(i-1)}$

Step 10: Calculate the error function $E = E + \frac{1}{k} \sum_j^m e_j^2$

Step 11: if k=K then go to step 12; otherwise k←k+1 and go to step 4.

Step 12: if E≤ε then go to step 13; otherwise go to step 3.

Step 13: Learning is completed. Output the weights.

Artificial Neural Networks for the Diagnosis and Prediction of Desert Dust Transport Episodes

Silas Michaelides[1], Filippos Tymvios[1], Dimitris Paronis[2], and Adrianos Retalis[3]

[1] Meteorological Service, Nicosia, Cyprus
{smichaelides,ftymvios}@ms.moa.gov.cy
[2] Institute for Space Applications & Remote Sensing,
National Observatory of Athens, Greece
paronis@noa.gr
[3] Institute for Environmental Research & Sustainable Development,
National Observatory of Athens, Greece
adrianr@meteo.noa.gr

Abstract. Artificial Neural Networks (ANN) are widely used as diagnostic and predictive tools in atmospheric sciences. This Chapter presents how such practical applications of ANN can be employed in the study of various aspects of a quite complex atmospheric phenomenon as the atmospheric pollution by particulate matter, due to dust transport episodes. It is also discussed how ANN can be utilized in assembling a useful predictive tool for such events. The diagnosis and prediction of dust episodes is very important for human welfare: indeed, some severe health issues are related to the presence of particulate matter in the atmosphere. Also, several human operations are affected by widespread dust presence: indeed, transportation and the increasing use of renewable energy systems utilizing solar radiation are profoundly affected.

1 Introduction

The occurrence of high level concentrations of dust originating from deserts is quite common even at locations quite distant from the source region. The mechanisms for lifting the particulates within the source region, the conditions leading to their suspension in the atmospheric air, their transportation to great distances and the eventual deposition (either dry or wet) on the ground comprise a highly complex phenomenon, enticing mankind as it affects several activities (e.g., safety of air-transportation, efficiency of solar energy systems, etc), as well as human health (e.g., population prone to respiratory disorders, eye inflammations, etc). Increased levels of fine dust particles in the air are also linked to other health hazards such as heart disease and lung cancer. In this respect, Lave and Seskin [1973] have made a pioneering work in illustrating the association between mortality rates and air pollution.

K. Gopalakrishnan et al. (Eds.): Soft Comput. in Green & Renew. Ener. Sys., STUDFUZZ 269, pp. 285–304.
springerlink.com

An area which currently receives increasing attention regarding the presence of particulate matter in the atmosphere is that of renewable energy systems that utilize solar radiation (e.g., photovoltaic systems for the generation of electricity). In this respect, the presence of suspended atmospheric dust reduces the intensity of solar radiation reaching the system; also the degradation of systems' performance either by dry or wet deposition of dust and the subsequent need for cleaning is considered as a quite serious problem.

Areas adjacent to extensive deserts but even further away from them are occasionally affected by dust transportation and deposition. The occasional transport of particles from the Sahara desert is particularly important for countries in the southeast Mediterranean region, where this Chapter will be focused (see Michaelides et al. 1999).

This phenomenon has been the subject of numerous studies and an extensive literature on this issue exists. In the following, a brief outline of the four general types of studies that were carried out for the investigation of the long-range transportation of desert dust is presented, together with selected literature which can form a starting point for the interested reader.

In the early studies, the large scale atmospheric mechanisms leading to long-range transportation of dust were identified [e.g., Prospero et al. 1970, Tullet 1978, File 1986]. Desert dust has also received considerable attention because it provides a quite strong aerosol signature in satellite retrievals and several space platforms and sensors have been used [e.g., Fraser 1976, Herman et al. 1997, King et al. 1999, Kaufman et al. 1997, Dulac et al. 1992, Tanré et al. 1997, Chu et al. 2002, Retalis and Michaelides 2009]. Remote sensing (other than satellite) of dust suspended in the atmosphere has also been explored [e.g., Torres et al. 2002, Balis et al. 2004]. Forecasting dust transportation using dynamic atmospheric modeling has also been an area of research that has also operational application [e.g., Nickovic et al. 1996, 2001, Lachanas et al. 1998, Gregoryan and Sofiev 1997].

In this Chapter, a novel approach will be presented, namely the application of Artificial Neural Networks (ANN) for diagnosing and predicting atmospheric pollutant levels over the island of Cyprus, in the eastern Mediterranean, due to the transportation of dust from the adjacent deserts. The rather isolated island of Cyprus (located at a considerable distance from the dust source regions) is ideal for such a study. In this endeavor, employment of synoptic circulation types, satellite data and surface measurements will be made. For the implementation of neural methodologies, Matlab's Neural Network Toolbox was employed [Beale et al. 2010]; Matlab was also used for the development of the regression models, as explained below.

A brief outline is given in Section 2 of the atmospheric conditions leading to dust transportation, thus introducing the reader in the meteorological conditions associated with the phenomenon. This is followed by Section 3 where a presentation is made of the surface measurements of PM10 (particles that are less than 10 μm in aerodynamic diameter), as integrated in this study. The methodology for using ANN in the classification of synoptic patterns and the identification of those of them favoring dust transportation is discussed in Section 4. The exploitation of satellite technology in estimating dust load in the atmosphere is presented in

Section 5; more specifically, the Atmospheric Optical Depth determined by the MODIS (Moderate Resolution Imaging Spectrometer) sensor onboard the Aqua - Terra satellites is considered. Section 6 discusses the application of multiple regression in combination with the synoptic classification for the prediction of dust episodes. A neural network prediction methodology is put forward in Section 7. An integrated approach for the prediction of dust episodes that makes use of either the multiple regression or the neural approaches is considered in Section 8.

2 Weather Conditions Leading to Dust Transportation

It is considered useful to start by giving a brief overview of the atmospheric circulation conditions leading to dust episodes; this sets the scene for the discussion that follows but also provides a justification for the approach adopted with regard to the data selection.

It has long been revealed that the type of synoptic-scale atmospheric circulation which favors wet or dry dust deposits over the eastern Mediterranean is a southerly to south-westerly flow throughout the entire troposphere, extending from the northern Sahara desert well into this area. Generally, the phenomenon starts with the development of a North African low pressure system which generates a dust storm. This low pressure is initiated by an upper-level trough which occurs on the polar front jet, when it overlies a heat low. Alternatively, it is initiated by the presence of a low level frontal system southeast of the Atlas Mountains. For more details on the synoptic and dynamical aspects for the formation of these low pressure systems, the reader is referred to Prezerakos [1990] and Prezerakos et al. [1990]. The above atmospheric circulations are more frequent in late winter and spring [see Kubilay et al. 2000]. Indeed, this is the time of the year when dust events are most frequent over the eastern Mediterranean [Dayan et al. 1991].

The rising dust generated by the dust storm forms a cloud stirring up to a few tens or hundreds of meters; under favorable atmospheric conditions, the lighter grains can be lifted at greater heights, of the order of a few kilometers [see Prezerakos et al. 2010]. When the dust cloud is subsequently embedded in a south-westerly tropospheric flow, it can drift over large distances. Hence, hazy weather conditions are often reported at great distances from the source area, sometimes lasting for several days. Under dry conditions, the drifting dust cloud gradually sediments due to gravity and falls as a dust deposit on the Earth's surface. Under conditions of increased humidity, dust particles mix with rain-droplets and fall on the ground as colored precipitation.

3 Surface Measurements of Dust Deposition

For the needs of this research, PM10 measurements from the Background Representative Station at Ayia Marina Xyliatou in Cyprus were considered. This monitoring station is located between the villages of Ayia Marina and Xyliatos (35 02' 17'' N, 33 03' 28'' E). This station is operated by the Cyprus' Ministry of Labour and Social Insurance and it is located in an area which has relatively low local pollution sources, thus it is considered as an EMEP (i.e., within the protocol of the European Monitoring and Evaluation Program) Background Representative Station; for this reason, a large proportion of the PM10 measured can be ascribed to external sources (e.g., dust transportation). The measurements cover the three year period 2003-2005.

A dust transport episode is considered as a day when the average PM10 measurement exceeds the threshold of 50mg/m^3. In the three-year period mentioned above, 85 such dust deposition events were recorded (out of a total of 1096 days). Figure 1 displays the monthly distribution of these episodes during the three year period 2003-2005. It is evident from this figure that there is a seasonal preference for dust events to occur. Indeed, experience supports the finding that Spring and Autumn are the two seasonal periods favoring dust episodes, whereas Summer appear to be suppressing these events. Dust episodes are rather rare in Winter.

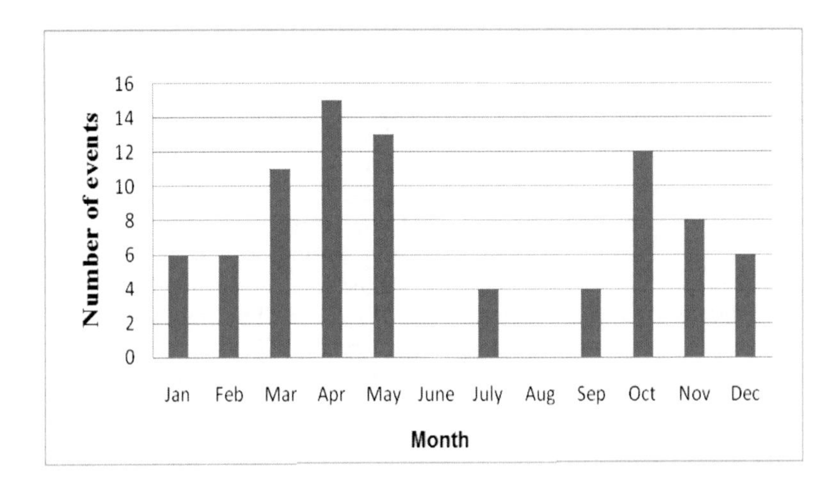

Fig. 1 Number of dust events per month in the three year period 2003-2005

4 Classification of synoptic patterns with Artificial Neural Networks

The systematic use of synoptic weather charts dates back to the beginnings of modern meteorological practices. Synoptic weather stations, scattered all over the world, supply meteorological services with observations of specific parameters at regular and fixed times. Upper-air observing stations report, generally, geodynamic height, speed and direction of wind, as well as temperature and humidity. Conventionally, for the analysis of the prevailing synoptic situation, charts of the geopotential height at selected levels are used. These charts depict the geopotential height at which a given pressure value is found and are usually called isobaric charts (hPa is the pressure unit used); quite commonly, the 500hPa level is analyzed. This level possesses several characteristics that make it distinct from others:

it is well above the friction layer and therefore is not much affected by factors related with the Earth's surface, at least in the short term; it represents the middle troposphere, that is the layer in which most of the weather phenomena take place; it is the level at which roughly half of the mass of the entire atmosphere is found below and half is found above; at this level, on average, divergence of the atmospheric air is negligible, in mid-latitudes.

Meteorologists can identify on such isobaric charts discrete geometric patterns that characterize a synoptic situation of the atmosphere. There is a strong association between large scale atmospheric circulation patterns and regional meteorological phenomena that are observed at the Earth's surface. As a consequence, synoptic upper air charts at certain levels provide a valuable tool for the operational weather forecaster to predict qualitatively occurrences of certain weather phenomena over particular areas [see Tymvios et al. 2010]. One such typical example is the close association between the atmospheric circulation and the onset and maintenance of desert dust transport episodes, which is elaborated in this Chapter.

There are several techniques for weather type classification, developed for different regions and for different purposes. Many of them are based on automatic, objective and consistent methodologies. However, none of the proposed methods in the literature is accepted as universal and applicable for all problems. Each method has its strong and weak points. The method to be selected and its parameters are usually defined by the application itself.

An initial effort for categorization of synoptic situations was made by Lamb [1950], while in literature there exists an abundance of methods of classification [see Hewitson and Crane 1996]. In order to take advantage of these semi-empirical methods and to simplify the statistical processing, stochastic downscaling methods are often applied to the actual weather patterns in order to generate clusters of synoptic cases with similar characteristics. Weather type classifications are simple, discrete characterizations of the current atmospheric conditions and they are commonly used in atmospheric sciences. For a review of various classifications, including their applications, the reader is referred to Key and Crane [1986], El-Kadi and Smithoson [1992], Hewitson and Crane [1996] and Cannon and Whitfield [2002].

Recently, a wide ongoing European effort in evaluating different classification methods within the framework of COST Action 733, which is entitled "Harmonization of weather type classifications in Europe" [http://www.cost733.org/] was completed [Philipp et al. 2010]. The main objective of this Action is to "achieve a general numerical method for assessing, comparing and classifying typical weather situations in European regions, scalable to any European sub-region with time scales between 12 h and 3 days and spatial scales of ca. 200 to 2000 km. The technique described below was partly developed and expanded within this framework.

The effort in this Section is to present a relatively new methodology for the classification of synoptic situations with the use of ANN. More specifically, Kohonen's Self-Organised Features maps (Kohonen 1990, 1997) were used for the classification of distribution of isobaric height on charts of the 500hPa. As a result of this classification, synoptic prototypes can be formulated which will be related to dust events.

While ANN proved to be valuable tools for forecasting purposes, Kohonen's Self Organizing Maps (SOM) technique is a relatively new method for climate research [Main 1997]. Maheras et al. [2000] used a combination of an ANN classifier and semi-empirical methods for the classification of weather types in Greece, while Cavazos [1999] utilized a mixed architecture of a SOM network in line with a feed-forward ANN (FF-ANN) to study extreme precipitation events in Mexico and Texas. The same methodology was successfully applied for wintertime precipitation in the Balkans [Cavazos 2000]. For a research on forecasting the risk of extreme rainfall events, a neural classification was recently implemented [Tymvios et al. 2010]. A neural classification was also used in an attempt to identify possible climatic trends in atmospheric patterns [Tymvios et al. 2010].

The method exploited in the present work for the classification of synoptic patterns is the Kohonen' Self Organizing Maps architecture which is a neural networks method with unsupervised learning (Kohonen, 1990). A detailed description of the method and procedures used is provided below [see also Michaelides et al. 2007].

4.1 Kohonen's Self Organizing Maps

ANN are constructions of artificial neurons (algorithms that mimic the properties of biological neurons); they are commonly used to solve artificial intelligence problems, to simulate and predict the evolution of complex physical systems, to discover hidden structures inside data groups and they are ideal for the classification of individuals into groups of similar properties. All of these are achieved according to the network's architecture and parameter tuning. Details of the classification method are discussed in Tymvios et al. [2010] and Michaelides et al. [2010], hence, a short description is sufficient here.

As mentioned above, the neural network architecture used in this research is the Kohonen' Self Organizing Maps (SOM) [Kohonen 1990]. These networks provide a way of representing multidimensional data in much lower dimensional spaces, usually one or two dimensions. An advantage of the SOM networks over other neural network classification techniques is that the Kohonen technique creates a network that stores information in such a way that any topological relationships within the training set are maintained; for example, even if the Kohonen network associates weather patterns with dust events inaccurately, the error obtained will not be of great amplitude, since the result will be a class with similar characteristics. The process of reducing the dimensionality of height vectors is essential in order to investigate productively the relationship among weather patterns and heavy rainfall. For a recent review of the advantages in using SOM as a tool in synoptic climatology, the reader is referred to Sheridan and Lee [2011].

The geographical area studied is bounded by latitude circles 20°N and 60°N and meridians 20°W and 40°E; it covers Europe, North Africa and the Middle East. The data that were exploited comprise the 500hPa field at 1200 UTC (Universal Time Coordinated) for each day, from 1 January 1980 to 31 December 2005

(26 years), retrieved from the online data base of NCEP (National Centers for Environmental Prediction, USA). The grid distance is 2.5°x2.5°, thus the area in study is covered by 7x25=425 grid points.

The Kohonen SOM algorithm in its unsupervised mode was chosen for building the neural network models, because neither the number of output classes (synoptic classes) nor the desired output are known *a priori*. This is a typical example where unsupervised learning is more appropriate, since the domain expert (in this case a professional meteorologist) will be given the chance to see the results and decide which model gives the best results. The expert's guidance can help to decide the number of output classes that better represent the system (Pattichis et al. 1995).

In unsupervised learning, there are no target values, as in the case of other methods of ANN. Given a training set, $X^{(k)}$, k=1, 2, ... , p, the objective is to discover significant features or regularities in the training data (input data). In our case, the input data were 9497 vectors (i.e., the number of consecutive days in the 26 years 1980-2005). The neural network attempts to map the input feature vectors onto an array of neurons (usually one or two-dimensional), thereby compresses information while preserving the most important topological and metric relationships of the primary data items on the display. By doing so, the input feature vectors can be clustered into c-clusters, where c is less or equal to the number of neurons used [Charalambous et al. 2001]. Input vectors are presented sequentially in time without specifying the desired output [see Schnorrenberg et al. 1996, Michaelides et al. 2001]. The two-dimensional rectangular grid architecture of Kohonen's SOM was adopted in the present research.

The input vector X is connected with each unit of the network through weights w_j, where j = 1, 2,..., M; M equals to 425 grid points, for the area in study. The training procedure utilizes competitive learning. When a training example is fed to the network, its Euclidean distance to all weight vectors is computed. The neuron whose weight vector is closest to the input vector X (in terms of Euclidean distance) is the winner. This neuron is represented with I and is the winner neuron to input X if

$\|w_I - X\| = \min_i \|w_i - X\|, i = 1, \ldots, M.$

Note that $\|w_I - X\| = [(w_{i1} - x_1)^2 + (w_{i2} - x_2)^2 + \ldots + (w_{iN} - x_N)^2]^{½}$ is the Euclidean distance between weight vector w_i and input vector X [see Charalambous et al. 2001]. N refers to the number of output classes.

The weight vectors of the winner neuron, as well as its neighborhood neurons, are updated in such a way that they become closer to the input pattern. Learning follows the following rule:

$$w_i^{(new)} = \left[\begin{array}{ll} w_i^{(old)} + \alpha\left(X - w_i^{(old)}\right), & i \in N(I,R) \\ w_i^{(old)}, & i \notin N(I,R) \end{array} \right. \qquad (1)$$

where, the neighborhood set N(I,R) of neuron I with radius R consists of neurons 1, I±1,..., I±R, assuming these neurons exist, with maximum value being around the winner I, in order for a larger number of neighborhood units to share the experience of learning with the winner unit, and it becomes zero as the distance between the neighborhood units and I increases. The coefficient α in the above relationship is called the learning factor and decreases to zero as the learning progresses. For simplicity, R is considered to have the shape of a geometric area, such as a rectangle or hexagon.

The radius of the neighborhood around the winner unit is relatively large to start with, in order to include all neurons. As the learning process continues, the neighborhood is consecutively shrunk down to the point where only the winner unit is updated [Patterson, 1995].

As more input vectors are represented to the network, the size of the neighborhood decreases until it includes only the winning unit or the winning unit and some of its neighbors. Initially, the values of the weights are selected at random. The method which was used for the SOM to work, can be described as follows:

1. The initial value of the weights is set to small random numbers, as well as the learning rate and the neighborhood. Steps 2 to 4 are repeated until the weights of the network are stabilized.
2. One vector X is chosen from the dataset as an input to the network.
3. The table for unit I with weight vector closest to X is determined by calculating $||w_I - X|| = \min_i ||w_i - X||$.
4. The weight vector in (t + 1) iteration is updated according to:
$w_i(t + 1) = w_i(t) + a(t)(X - w_i(t))$, for units that belong in set N(I,R)
$w_i(t + 1) = w_i(t)$, for units that do not belong in set N(I,R).
5. The neighborhood and the learning rate of the parameters are decreased [Charalambous et al. 2001].

When all vectors in the training set were presented once at the input, the procedure is repeated many times with the vectors presented in order each time. This part of the algorithm at the end organizes the weights of the one-dimensional map, such that topologically close nodes become sensitive to input that is physically similar. Nodes are ordered in a natural manner, reflecting the different classes of the training set [Michaelides et al. 2001].

The number of outputs is not *a priori* determined, but an "optimum" can be adopted by experimentation, in relation to the specific application under study. For several applications, it appears that the optimum number of outputs is around 30, as this exhibits the level of discretization required for the synoptic scale phenomena examined [see Tymvios et al. 2010]. Although several experiments were carried out with varying output nodes, for the association of weather type patterns and dust transportation over the eastern Mediterranean 35 classes were considered.

4.2 Synoptic Pattern Classification

One of the major inherent problems in an endeavor to classify synoptic patterns is the *a priori* determination of the number of different classes that one can expect from such a classification, as mentioned above. In other words, the number of distinctive synoptic patterns over any particular geographical region is by no means fixed. Traditionally, such a synoptic classification was performed by a qualitative inspection of synoptic maps. Professional meteorologists examined a series of plotted synoptic maps and picked out geometric similarities [e.g., Prezerakos et al. 1991]. Nevertheless, it seems that in an attempt to perform a classification with an unknown number of classes, experimentation with various possibilities is a practical procedure, which can lead to some useful considerations. For this reason, it was decided to run a number of experiments and build classification models with different numbers of output nodes (i.e., classes). For the present analysis 35 output nodes are presented, as mentioned above.

The ability of ANN to group synoptic patterns into seasonally dependent clusters was noted by Michaelides et al. [2007]. This seasonal discretization of classes should be an essential attribute of a classification technique.

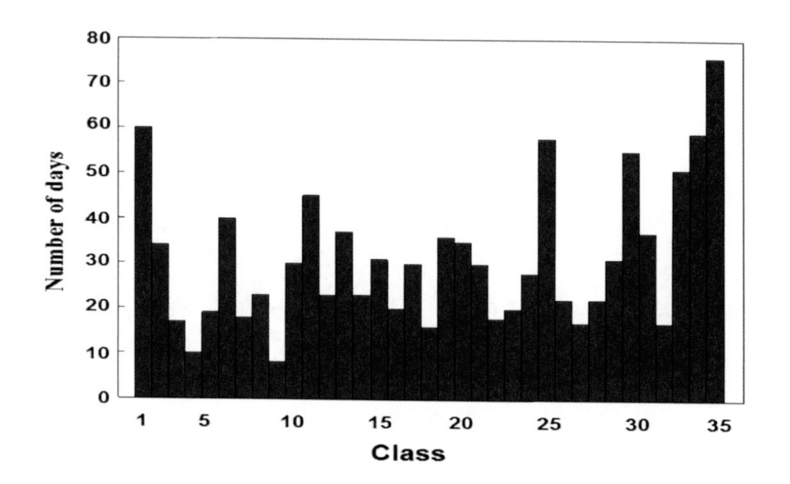

Fig. 2 Number of days per class, for the 35 classes in the synoptic classification, during 2003-2005

Figure 2 shows the frequency of appearance of the synoptic patterns in the 35 classification for the three-year period 2003-2005. Apparently, class 35 is most frequently encountered, followed by classes 1 and 25. Figure 3 is a graphical representation of the assignment of a class to each day in the three year period. It is clear that there is a seasonal "quasi-cyclic" behavior. Certain classes occur, almost exclusively, during summer or winter; during the (Mediterranean) transitional periods of spring and autumn, both the summertime and wintertime patterns can occur. This graphical representation is proposed as a practical visual tool to identify the level of seasonal discretization pursued in adopting a synoptic classification technique.

4.3 Synoptic Classes and Dust Deposition Events

Figure 4 shows how the 85 dust transport events are distributed among the 35 classes in the synoptic classification adopted above. There appears to be a certain preference of classes associated with these events: most prone to dust events is class 1, followed by class 31; representative synoptic situations for these two classes are shown in Fig. 4. Figure 5(a) refers to 1200 UTC 1 February 2003, and 5(b) at 1200 UTC 10 May 2004: in the former, a central Mediterranean upper trough extends well into the north African desert; in the later, the trough axis extends southwards from the Iberian peninsula. In both cases, typical patterns are identified favoring dust raising and its transfer eastwards with the resulting south-westerly airflow over the eastern Mediterranean.

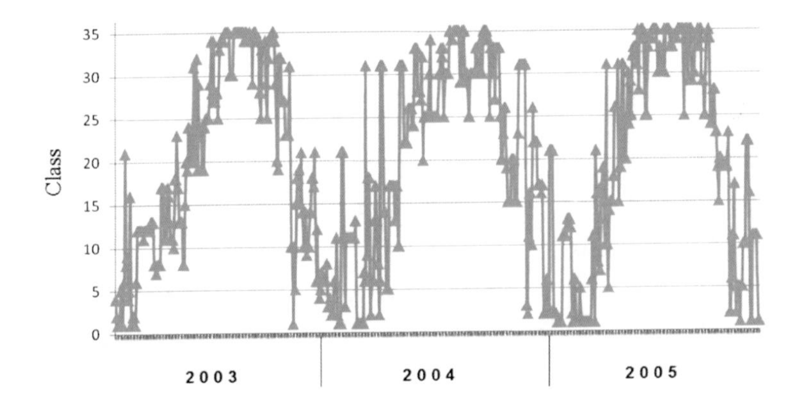

Fig. 3 Daily distribution of 35 classes in the three-year period 2003-2005

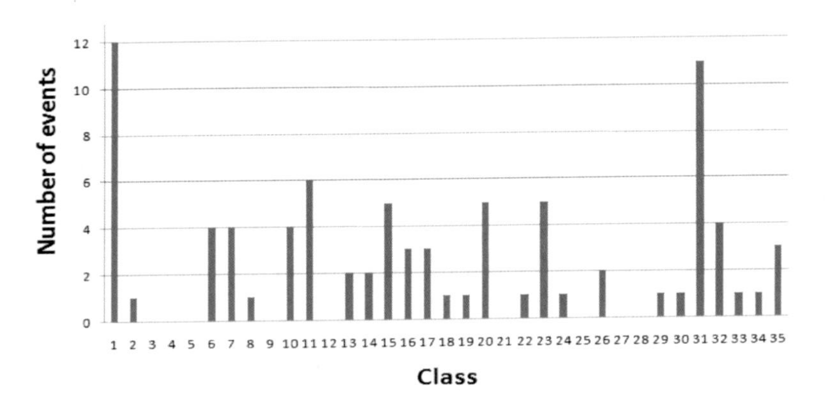

Fig. 4 Distribution of dust deposition events per class, for the 35 classes in the synoptic classification, during 2003-2005

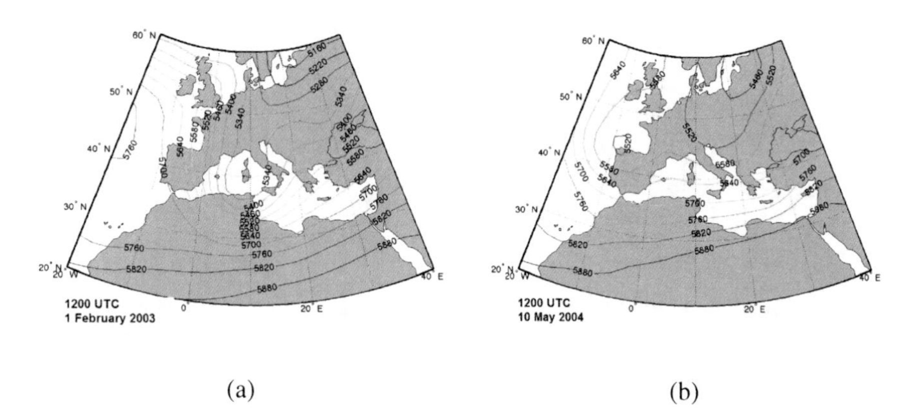

Fig. 5 Representative synoptic situations corresponding to: (a) class 1; (b) class 31. Isolines are drawn for every 60 geopotential meters

5 Dust Load from Satellite Measurements

Satellite remote sensing has been widely used for monitoring and assessment of air pollution. Several sensors have been employed for the retrieval of air pollution products and especially for the estimation of aerosol content. Researchers have been mainly devoted to extract the aerosol content in the total atmospheric column, since there often it is difficult to distinguish the sources of emissions (anthropogenic, natural).

Satellite products are characterized by both their extended spatial coverage and the possibility for real-time air-pollution monitoring against PM ground-based measurements. Several methods to estimate surface PM concentration levels, from aerosol related products have been published in the literature. The majority of these are focused on the derivation of statistical/empirical models for the estimation of PM2.5 (and secondarily of PM10) from satellite derived AOD (Atmospheric Optical Depth, a measure of aerosol loading in the total atmospheric column).

These methods have been applied to different sites of the world and are based on aerosol products from various sensors such as MODIS [Gupta and Christopher 2008, Li et al. 2009], MISR [Liu et al. 2007], SeaWiFS [Vidot et al. 2007], GOES-12 [Liu et al. 2009]. The PM-AOD correlation coefficients reported in the literature vary from low (<0.5) to high values (0.96). This correlation has been found to depend on various factors such as the temporal averaging periods (hourly versus 24-hr), season, aerosol type, satellite AOD retrieval accuracy, meteorological conditions, boundary layer height, station type/location [Hoff and Christopher 2009]. Among the existing satellite aerosol products, MODIS instruments onboard Terra (EOS a.m.) and Aqua (EOS p.m.) satellites have the

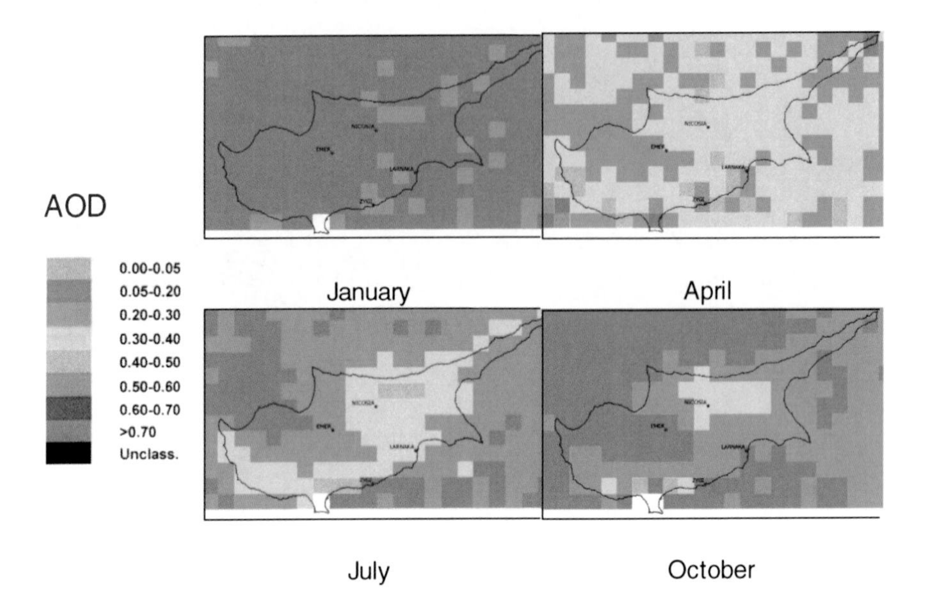

Fig. 6 Spatial distribution of the monthly averaged values of AOD for 2003-2005

advantage to provide images twice a day on a global scale. Thus, the majority of the research efforts are referred to MODIS aerosol products, which already span a period of a decade.

In the study presented here, Terra and Aqua MODIS level-2 (Collection C005, V5.2) daily aerosol products at a spatial resolution of 10×10km were acquired from NASA's Level 1 and Atmosphere Archive and Distribution System (LAADS). The Collection 005 dark-target aerosol product is based on a true inversion that uses three pieces of information: apparent reflectance at 470 nm, 660 nm and 2130 nm to derive AOD and fraction of AOD attributed to non-dust aerosol at 470nm, 670nm and 550nm (interpolated) and the surface reflectance at 2130 nm [Levy et al. 2007, Remer et al. 2009]. For the present work, the AOD values used refer to the dark-target AOD values (at 550nm) extracted from the Scientific Data Set entitled 'Optical_Depth_Land_And_Ocean' and covering the period January 2003-December 2005.

There is a strong seasonal variability of AOD that reflects the seasonality in dust load over dust prone areas, like the eastern Mediterranean. Figure 6 shows the spatial distribution of the AOD values for January, April, July and October, averaged in the 2003-2005 period. As expected, spring exhibits larger values compared to other seasons. Also, Fig. 7 shows some statistical characteristics of AOD, spatially averaged over the area of Fig. 6. The seasonal dependence is also apparent, with April being the most prominent.

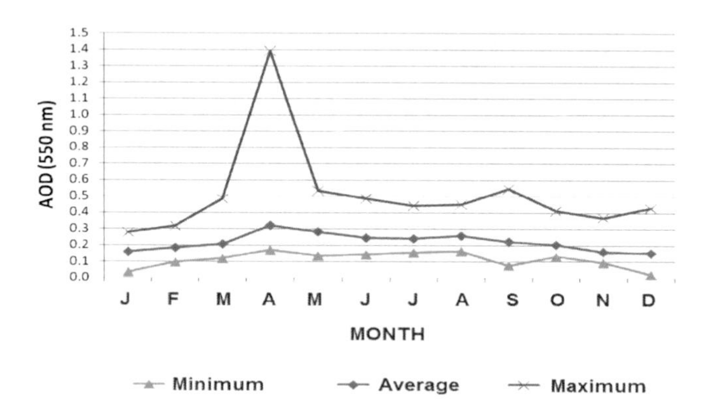

Fig. 7 Monthly average, minimum and maximum values of the spatially averaged AOD during 2003-2005

6 Multiple Regression Models

Surface and satellite measurements were considered in establishing linear regression relationships between PM10 concentrations and satellite estimates of AOD; in this respect, measurements at different times preceding the time for which the prediction of PM10 concentration is sought (all calculations hereafter refer to the EMEP station) were taken into account. It became evident that PM10 measurements and satellite AOD estimates spanning a few days before the actual event yield better results. This is not surprising because the phenomenon that is under consideration is the transfer of dust from remote areas over an island, which has little local widespread sources. Therefore, it is reasonable to assume that the build-up of the concentration of dust particles at ground level is a gradually intensifying process, whereby increasing amounts of dust are noted as the dust plumes move away from the desert source regions.

One of the aims of the research that was carried out was to demonstrate that the neural methodology described above for the classification of synoptic types can be combined with a statistical approach in order to provide estimates of the expected concentration of particulate matter PM10 at ground level. For this purpose, multiple regression models were built, making use of PM10 surface measurements and satellite AOD from MODIS and several multiple regression models were tested; in all of these, the dependent variable was the concentration of PM10 dust particles at ground level. Regarding the independent variables, several combinations were tested. Here, the following combination is presented (see Table 1). The dependent variable is the predicted value of PM10 level on DAY0 (i.e., today); the set of independent variables consist of previous measurements of particles on the ground and satellite estimates: two PM10 measurements for DAY0-1 (i.e., yesterday) and DAY0-2 (i.e., the day before yesterday) and satellite AOD for DAY0-1 and DAY0-2, retrieved from Aqua-Terra satellites. In the data set for the three year period that was used in this study (years 2003, 2004 and 2005, 1096 days), there were 127 days with missing or unreliable data; those were excluded from further processing. As the two satellites are not available at all times, an average of the two measurements from these satellites was used when both satellites were available, whereas one value is used when one of the satellites is available. Finally, in order to take into consideration the classification of the synoptic patterns, different regression models were built for each of the 35 classes.

Table 1 Independent (Input) and Dependent (Output) variables for the regression (neural) prediction models

Independent (Input) variables	Dependent (Output) variable
PM10 for DAY0-2	PM10 for DAY0
PM10 for DAY0-1	
AOD for DAY0-2	
AOD for DAY0-1	

7 A Neural Network Approach for the Prediction of Dust Deposition

In order to implement a neural network methodology for forecasting the deposition of dust at ground level, a modular multi-layer perceptron architecture was adopted. This is graphically shown in Fig. 8, using Matlab's notation [Beale et al. 2010]. Consistently with the regression model described above, the input and output variables for this a network are the same as the independent and dependent variables, respectively, and are also shown in Table 1. Hence, the required output is the predicted value of PM10 concentration on DAY0 (i.e., today); the input consists of two PM10 measurements and two satellite AOD estimates for DAY0-1 (i.e., yesterday) and DAY0-2 (i.e., the day before yesterday).

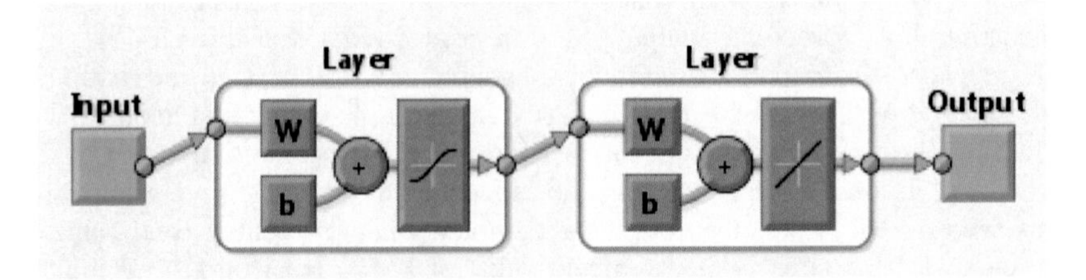

Fig. 8 Modular network architecture: four inputs, one output, two layers with 10 neurons in the first and 1 neuron in the second. The function used in the first layer is *tansig* and in the second layer is *pureline*

As an example of this application, results for the EMEP Background Representative station presented above are given in the following. As mentioned above, out of the 1096 days that were employed in this research, there were some days with missing or unreliable data; those were excluded from further processing. For the training of the network, 646 data sets were used which were randomly fed to the system and the Levenberg-Marquardt methodology was adopted [see Fletcher 1971]. For the verification of the system that was built, the remaining 323 data sets were used.

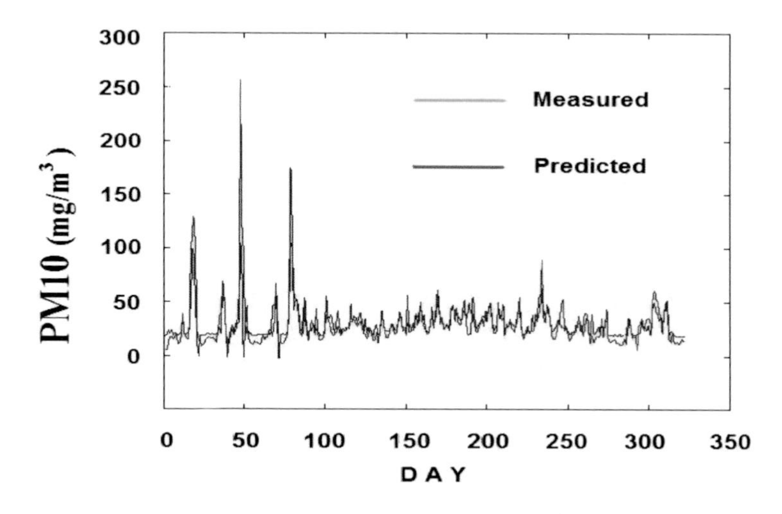

Fig. 9 Measured versus predicted values of PM10 concentration for the verification data set (323 days)

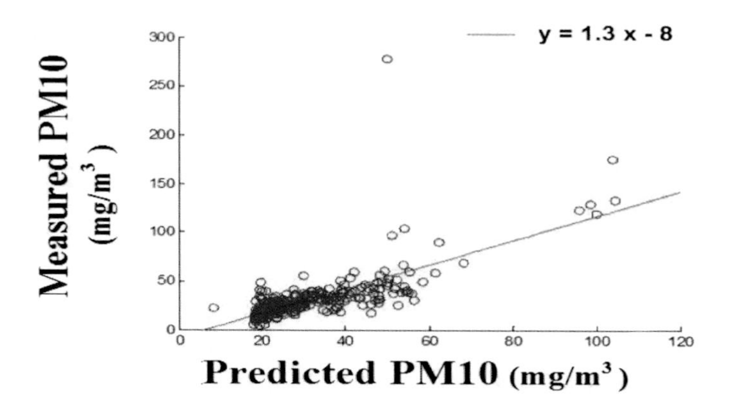

Fig. 10 Measured versus predicted daily average values of PM10 concentration for the verification data set. A linear fit for this relationship is also shown

Figure 9 is a comparative presentation of the measured and predicted values of PM10 for the 323 days in the verification set. The same measured and predicted values are plotted in Fig. 10 together with a linear relationship between the two. Overall, the neural network model seems to underestimate the PM10 concentrations, especially when extreme concentrations are recorded; however, apart from these outliers, the performance of the neural network predictions is acceptable.

8 Dust Transport Event Prediction Tool

All of the above described approaches have been embedded in a user friendly prediction tool, the interface of which is shown in Fig. 11. Input consists of PM10 measurements and AOD from satellite estimates for the two previous days and the synoptic class. The user has an option to perform the concentration prediction by using either the neural network or the multiple regression methodology. The output is the predicted PM10 concentration for today; also the probabilities of exceeding the thresholds of 50, 80 and 100 mg/m^3 are calculated.

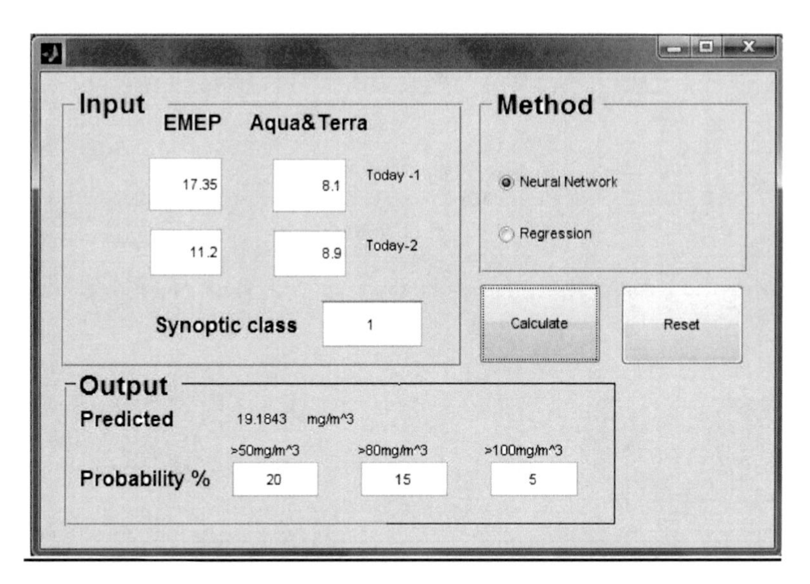

Fig. 11 The interface of the prediction tool

9 Concluding Remarks

The main aim of the research that was presented above was to demonstrate that the use of Artificial Neural Networks can be applied efficiently in the prediction of a highly complex atmospheric phenomenon, namely the dust transportation and its deposition at a distance from the source regions. The data that were used were basically the type of the prevailing synoptic conditions, measurements of dust deposition at ground level (PM10) and Atmospheric Optical Depth (AOD) determined from satellite information (MODIS).

A classification of synoptic types as they are portrayed by the 500 hPa isobaric analyses was performed, in order to treat the well known association between dust transportation and prevailing weather conditions in the middle troposphere. As dust transportation eventually leads to deposition at some distance from the source, the measurements of dust particles provide such an indication, especially where this can be ascribed to a large extent to external sources; hence, such measurements can be used to track dust transportation episodes, although the respective data must be treated with care. Satellite remote sensing of the atmospheric dust load can provide a valuable source of information, especially as regards the spatial and temporal evolution of the transportation of dust.

Although the results of this pilot study are site specific, they indicate that dust transportation can be investigated by using ANN, both for diagnostic and prognostic purposes. As the ANN methodologies are generally highly data demanding, the techniques developed in this research can be updated and, hopefully, improved as more data become available.

An efficient methodology for predicting dust events can have several applications. In the Section 1, several human activities were outlined that can be profited from accurate dust transport predictions. A rather more recent area of application of such predictions is in the planning and running of renewable energy systems based on solar energy exploitation (e.g., photovoltaic systems etc).

Acknowledgements

The methodology that was presented above formed part of the AERAS research project that was partly funded by the Cyprus Research Promotion Foundation. The PM10 data were kindly provided by the Ministry of Labour and Social Security of Cyprus. The data for the synoptic analyses were provided by the National Centers for Environmental prediction, U.S.A. The MODIS data were retrieved from NASA's Atmosphere Archive and Distribution System.

References

Balis, D.S., Amiridis, V., Nickovic, S., Papayannis, A., Zerefos, C.: Optical properties of Saharan dust layers as detected by a Raman lidar at Thessaloniki, Greece. Geophysical Research Letters 31, L13104 (2004), doi:10.1029/2004GL019881

Beale, M.H., Hagan, M.T., Demuth, H.B.: Neural Network Toolbox™ User's Guide. The MathWorks, Inc. (2010), `http://www.mathworks.com/help/pdf_doc/nnet/nnet.pdf`

Cannon, J.A., Whitfield, P.H.: Synoptic map-pattern classification using recursive partitioning and principal component analysis. Monthly Weather Review 130, 1187–1206 (2002)

Cavazos, T.: Large-scale circulation anomalies conducive to extreme precipitation events and derivation of daily rainfall in Northeastern Mexico and Southeastern Texas. Journal of Climate 12, 1506–1523 (1999)

Cavazos, T.: Using self-organizing maps to investigate extreme climate events: An application to wintertime precipitation in the Balkans. Journal of Climate 13, 1718–1732 (2000)

Charalambous, C., Charitou, A., Kaourou, F.: Comparative analysis of neural network models: Application in bankruptcy prediction. Annals of Operations Research 99, 403–425 (2001)

Chu, D.A., Kaufman, Y.J., Ichoku, C., Remer, L., Tanré, D., Holben, B.N.: Validation of MODIS aerosol optical depth retrieval over land. Geophysical Research Letters 29(12) (2002), doi:10.1029/2001GL013205.

Dayan, U., Heffter, J., Miller, J., Gutman, G.: Dust intrusion events into the Mediterranean basin. Journal of Applied Meteorology 30, 1185–1199 (1991)

Dulac, F., Tanré, D., Bergametti, G., Buat-Menard, P., Desbois, M., Sutton, D.: Assessment of the african airborne dust mass over the western Mediterranean sea using Meteosat data. Journal of Geophysical Research 97, 2489–2506 (1992)

El-Kadi, A.K.A., Smithoson, P.A.: Atmospheric classifications and synoptic climatology. Progress in Physical Geography 16, 432–455 (1992)

File, R.F.: Dust deposit in England on 9 November 1984. Weather 41, 191–195 (1986)

Fletcher, R.: A Modified Marquardt Subroutine for Nonlinear Least Squares. Rpt. AERE-R 6799, Harwell (1971)

Fraser, R.S.: Satellite measurement of mass of Sahara dust in the atmosphere. Applied Optics 15, 2471–2479 (1976)

Grigoryan, S.A., Sofiev, M.A.: Numerical modeling of dust elevation during storm episodes and its long-range atmospheric transport. In: Proceedings of the 1st LAS/WMO International Symposium on Sand and Dust Storms, WMO Programme on Weather Prediction Research Report Series Project no. 10, WMO Technical Document no. 864, Damascus, Syria, pp. 88–98 (November 1997)

Gupta, P., Christopher, S.A.: Seven year particulate matter air quality assessment from surface and satellite measurements. Atmospheric Chemistry and Physics 8, 3311–3324 (2008)

Herman, J.R., Bhartia, P.K., Torres, O., Hsu, N.C., Seftor, C.J., Celarier, E.: Global distribution of UV-absorbing aerosols from Nimbus-7/TOMS data. Journal of Geophysical Research 102, 16911–16922 (1997)

Hewitson, B.C., Crane, R.G.: Climate downscaling: Techniques and application. Climate Research 7, 85–95 (1996)

Hoff, R.M., Christopher, S.A.: Critical Review - Remote sensing of particulate pollution from space: have we reached the promised land? Journal of the Air & Waste Management Association 59, 645–675 (2009)

Kaufman, Y.J., Tanré, D., Remer, L.A., Vermote, E.F., Chu, D.A., Holben, B.N.: Operational remote sensing of tropospheric aerosol over the land from EOS-MODIS. Journal of Geophysical Research 102, 17051–17061 (1997)

Key, J., Crane, R.J.: A Comparison of Synoptic classification schemes based on "objective" procedures. Journal of Climatology 6, 375–388 (1986)

King, M.D., Kaufman, Y.J., Tanré, D., Nakajima, T.: Remote sensing of tropospheric aerosols from space: Past, present and future. Bulletin of the American Meteorological Society 80, 2229–2259 (1999)

Kohonen, T.: The Self-Organizing Map. Proceedings IEEE 78, 1464–1480 (1990)

Kohonen, T.: Self-Organizing Maps, 2nd edn. Series in Information Sciences, vol. 30. Springer, Heidelberg (1997)

Kubilay, N., Nickovic, S., Moulin, C., Dulac, F.: An illustration of the transport and deposition of mineral dust onto the eastern Mediterranean. Atmospheric Environment 34, 1293–1303 (2000)

Lachanas, A., Evripidou, P., Michaelides, S.: Regional weather prediction on small network of workstations. In: Proceedings 24th EUROMICRO Conference, Vasteras, Sweden, pp. 1053–1060 (1998)

Lamb, H.H.: Types and spells of weather around the year in the British Isles: Annual trends, seasonal structure of the year, singularities. Quarterly Journal of the Royral Meteorological Society 76, 393–429 (1950)

Lave, L.B., Seskin, E.P.: An Analysis of the Association Between U.S. Mortality and Air Pollution. Journal of the American Statistical Association 68, 284–290 (1973)

Levy, R.C., Remer, L.A., Mattoo, S., Vermote, E.F., Kaufman, Y.J.: Second-generation operational algorithm: Retrieval of aerosol properties over land from inversion of Moderate Resolution Imaging Spectroradiometer spectral reflectance. Journal of Geophysical Research 112, D13211 (2007), doi:10.1029/2006JD007811

Li, H., Faruque, F., Williams, W., Al-Hamdan, M., Luvall, J., Crosson, W., Rickman, D., Limaye, A.: Optimal temporal scale for the correlation of AOD and ground measurements of PM2.5 in a real-time air quality estimation system. Atmospheric Environment 43, 4303–4310 (2009)

Liu, Y., Koutrakis, P., Kahn, R.: Estimating fine particulate matter component concentrations and size distribution using satellite-retrieved fractional aerosol optical depth: Part 1 – method development. Journal of the Air & Waste Management Association 57, 1351–1359 (2007)

Liu, Y., Paciorek, C.J., Koutrakis, P.: Estimating regional spatial and temporal variability of PM2.5 concentrations using satellite data, meteorology, and land use information. Environmental Health Perspectives 117, 886–892 (2009)

Maheras, P., Patrikas, I., Karakostas, T., Anagnostopoulou, C.: Automatic classification of circulation types in Greece: methodology, description, frequency, variability and trend analysis. Theoretical and Applied Climatology 67, 205–223 (2000)

Main, J.P.L.: Seasonality of circulation in Southern Africa using the Kohonen self organizing map M.S. thesis, Department of Environmental and Geographical Sciences, University of Cape Town, South Africa, p. 84 (1997)

Michaelides, S., Evripidou, P., Kallos, G.: Monitoring and predicting Saharan desert dust events in the eastern Mediterranean. Weather 54, 359–365 (1999)

Michaelides, S.C., Pattichis, C.S., Kleovoulou, G.: Classification of rainfall variability by using artificial neural networks. International Journal of Climatology 21, 1401–1414 (2001)

Michaelides, S.C., Liassidou, F., Schizas, C.N.: Synoptic classification and establishment of analogues with artificial neural networks. Journal of Pure and Applied Geophysics 164, 1347–1364 (2007)

Michaelides, S., Tymvios, F., Charalambous, D.: Investigation of trends in synoptic patterns over Europe with artificial neural networks. Advances in Geosciences 23, 107–112 (2010)

Nickovic, S., Dobricic, S.: A model for long-range transport of desert dust. Monthly Weather Review 124, 2537–2544 (1996)

Nickovic, S., Kallos, G., Papadopoulos, A., Kakaliagou, O.: A model for prediction of desert dust cycle in the atmosphere. Journal of Geophysical Research 106, 18113–18130 (2001)

Patterson, D.W.: Artificial Neural Networks, Theory and Applications. Prentice-Hall, Englewood Cliffs (1995)

Pattichis, C.S., Schizas, C.N., Middleton, T.M.: Neural Network Models in EMG Diagnosis. IEEE Transactions on Biomedical Engineering 42(5), 486–495 (1995)

Philipp, A., Bartholy, J., Beck, C., Erpicum, M., Esteban, P., Fettweis, X., Huth, R., James, P., Sylvie, J., Kreienkamp, F., Krennert, T., Lykoudis, S., Michalides, S.C., Pianko-Kluczynska, K., Post, P., Alvarez, D.R., Schiemann, R., Spekat, A., Tymvios, F.S.: Cost733cat – A database of weather and circulation type classifications. Physics and Chemistry of the Earth 35, 360–373 (2010)

Prezerakos, N.G.: Synoptic flow patterns leading to the generation of north-west African depressions. International Journal of Climatology 10, 33–48 (1990)

Prezerakos, N.G., Michaelides, S.C., Vlassi, A.S.: Atmospheric synoptic conditions associated with the initiation of north-west African depressions. International Journal of Climatology 10, 711–729 (1990)

Prezerakos, N., Michaelides, S.C., Vlassi, A.S.: Atmospheric synoptic conditionsassociated with the initiation of north-west African depressions. Internatinal Journal of Climatology 10, 711–729 (1991)

Prezerakos, N.G., Paliatsos, A.G., Koukouletsos, K.V.: Diagnosis of the Relationship between Dust Storms over the Sahara Desert and Dust Deposit or Coloured Rain in the South Balkans. In: Advances in Meteorology 2010 (2010), Article ID 760546, 14 pages, doi:10.1155/2010/760546

Prospero, J.M., Bonatti, E., Schubert, C., Carlson, T.N.: Dust in the Caribbean atmosphere traced to an African dust storm. Earth and Planetary Science Letters 9, 287–293 (1970)

Remer, L., Tanré, D., Kaufman, Y.J., Levy, R.C., Mattoo, S.: Algorithm for Remote Sensing of Tropospheric Aerosol from MODIS for Collection 005: Revision 2 (2009), http://modis-atmos.gsfc.nasa.gov/_docs/ATBD_MOD04_C005_rev2.pdf

Retalis, A., Michaelides, S.C.: Synergetic use of TERRA/MODIS imagery and meteorological data for studying aerosol dust events in Cyprus. International Journal of Environment and Pollution 36, 139–150 (2009)

Schnorrenberg, F., Pattichis, C.S., Kyriacou, K., Vassiliou, M., Schizas, C.N.: Computer-aided classification of breast cancer nuclei. Technology and Health Care 4, 147–161 (1996)

Sheridan, S.C., Lee, C.C.: The self-organizing map in synoptic climatological research. Progress in Physical Geography 35, 109–119 (2011)

Tanré, D., Kaufman, Y.J., Herman, M., Mattoo, S.: Remote sensing of aerosol over oceans from EOS-MODIS. Journal of Geophysical Research 102, 16971–16988 (1997)

Torres, O., Bhartia, K., Herman, J.R., Sinyuk, A., Ginoux, P., Holbren, B.: A long-term record of aerosol optical depth from TOMS observations and comparison to AERONET. Journal of Atmospheric Sciences 59, 398–413 (2002)

Tullet, M.T.: A dust fall on 6 March 1977. Weather 33, 48–52 (1978)

Tymvios, F., Savvidou, K., Michaelides, S.C.: Association of geopotential height patterns with heavy rainfall events in Cyprus. Advances in Geosciences 23, 73–78 (2010)

Vidot, J., Santer, R., Ramon, D.: Atmospheric particulate matter (PM) estimation from SeaWiFS imagery. Remote Sensing of the Environment 111, 1–10 (2007)

Author Index